Antenna Radar Cross Section: Theory and Design

Ying Liu · Yongtao Jia · Shuxi Gong

Antenna Radar Cross Section: Theory and Design

Springer

Ying Liu
The National Key Laboratory of Radar Detection and Sensing
Xidian University
Xi'an, Shaanxi, China

Yongtao Jia
The National Key Laboratory of Radar Detection and Sensing
Xidian University
Xi'an, Shaanxi, China

Shuxi Gong
The National Key Laboratory of Radar Detection and Sensing
Xidian University
Xi'an, Shaanxi, China

ISBN 978-981-96-1225-3 ISBN 978-981-96-1226-0 (eBook)
https://doi.org/10.1007/978-981-96-1226-0

This Springer imprint is published by the registered company Springer Nature Singapore Pte Ltd.
The registered company address is: 152 Beach Road, #21-01/04 Gateway East, Singapore 189721, Singapore

Foreword

Stealth technology plays a highly important role in modern electronic warfare, not only from the point of view of reducing the penetration capability of aggressive missiles targeting a fighter aircraft, for instance, but also by improving the performance of weapons platforms. For low-observable platforms, not only the platform itself, e.g., a stealth fighter aircraft or a drone, but the antennas on the platform also typically contribute, quite significantly, to its total radar cross section (RCS). As a consequence, design of stealth antennas often becomes a priority issue in the process of developing a low-observable platform. Recent developments in thin radar absorbing materials enable us to significantly reduce the RCS of the platform when coating is combined with the shaping of the platform. However, the same approach cannot be applied to the problem of RCS reduction of an antenna without compromising its performance in terms of gain and efficiency, which is undesirable. Although much has been written on the subject of RCS reduction of radar targets, the same cannot be said about designing low-observable antennas. This monograph, which is an attempt to fill this gap, focuses on the topics of prediction and reduction of radar cross section of antennas, and it does this by summarizing the activities of the research group headed by Prof. Ying Liu of Xidian University. The book presents a theoretical analysis of the scattering mechanism of an antenna and discusses techniques for computing its RCS. Next, it examines various methods for reducing the RCS of antennas, including shape modification and the use of metasurfaces. Additionally, the book also covers practical aspects of RCS reduction of antennas by using innovative approaches. The research results have been verified, validated and adequately substantiated by researchers active in the field. In my opinion, this book will fill an important void in the field of RCS reduction of antennas and will serve as a valuable resource for researchers active in the field of stealth technology.

Orlando, FL, USA Prof. Raj Mittra

About This Book

This book is divided into nine chapters. Chapter 1 introduces the background and state of the art of antenna RCS reduction and discusses the development directions of it. Chapter 2 gives fundamental theoretical formula of antenna scattering, from which the theoretical innovations and reference value of antenna RCS reduction are obtained. Chapter 3 analyzes the RCS of single antenna. In this chapter, we propose an analysis model to calculate the antenna mode RCS and structural mode RCS separately, which solves the problem of obtaining the phase difference between the two components. Focusing on the scattering of array antennas, Chap. 4 starts from the model of array antennas, analyzes the mutual coupling of each element, introduces the scattering analysis of an array antenna and an active phased antenna, and provides RCS control methods of a shipborne active phased array antenna. Chapter 5 studies the spatial-domain MoM based on the mixed potential integral equation and the application of the AIM in calculating the electrical characteristics of electrically large objects. Chapter 6 proposed several designs for low-RCS antennas based on the analysis of the distribution of the surface currents and operating properties of the operation modes. Chapter 7 introduces the characteristics of frequency selective surface structures and its application in designing of low-RCS antennas. Chapter 8 introduces a method for designing wideband PRRSs. Chapter 9 introduces the methods for reducing the RCS of an antenna with a PGM.

The whole book is planned by Prof. Ying Liu and Prof. Shuxi Gong. Prof. Yongtao Jia revised the full text and translated it into English.

Contents

About the Authors

Ying Liu received the M.S. and Ph.D. degrees in electromagnetics from Xidian University in 2001 and 2004, respectively. From 2006 to 2007, she was a Post-Doctoral Researcher with Hanyang University, Seoul, South Korea.

Currently, Dr. Liu is a Full Professor and the Lead of the National Key Laboratory of Radar Detection and Sensing, Xidian University. Her research interests include prediction and control of antenna radar cross Section (RCS), and antenna theory and technology.

Yongtao Jia received the B.S. and Ph.D. degrees in electromagnetics from Xidian University, Xi'an, China, in 2012 and 2017, respectively.

Currently, Dr. Jia is a Professor at Xidian University. His research interests include antenna radar cross section (RCS) reduction, metasurface, and wideband phased arrays.

Shuxi Gong received the B.S. and M.S. degrees from Xidian University, Xi'an, China, in 1982 and 1984, respectively, and the Ph.D. degree from Xi'an Jiaotong University, Xi'an, in 1988.

Currently, Dr. Gong's research interests include antenna theory and technology, and electromagnetic scattering theory and technology.

Abbreviations

AIM	Adaptive integration method
APAR	Active phased array radar
AR	Axial ratio
CDSLS-FSS	Composite double square loop slots frequency selective surface
CG	Conjugate gradient
CP	Circularly polarized
CPCMs	Chessboard polarization conversion metasurfaces
CPFP	Circularly polarized Fabry-Perot
CSRR	Complementary split resonant ring
DSLS-FSS	Double square loop slots frequency selective surface
EMCs	Electronic countermeasures
ESMs	Electronic support measures
EW	Electronic warfare
FEM	Finite element method
FFT	Fast Fourier Transform
FMB	Figure of merit for bistatic reduction
FMM	Fast multiple method
FSS	Frequency selective surface
HPA	High-power amplifier
HPBW	Half power beam width
LHCP	Left-hand circularly polarized
LNA	Low-noise amplifier
LP	Linearly polarized
MoM	Method of moments
PCM	Polarization conversion metasurfaces
PDFSS	Polarization-dependent FSS
PFFT	Precorrected fast Fourier transform
PGM	Phase gradient metasurface
PRRSs	Polarization rotation reflective surfaces
RAM	Radar-absorbing material
RCS	Radar cross-section

RCSR	RCS reduction
RWG	Rao-Wilton-Glisson
SIW	Substrate-integrated waveguide

Chapter 1
Introduction

1.1 Antenna RCS Reduction Research Background

Electronic warfare (EW)—also known as electronic countermeasures (EMCs)—is a type of warfare based on modern technology involving electronic reconnaissance, counter-reconnaissance, electronic jamming, anti-interference, electronic stealth, and anti-stealth techniques. Stealth technology plays an important role in modern EW. In view of its major impact on warfare, stealth technology will be increasingly used in future high-tech EW applications. Consequently, more countries around the world are developing it.

Radar cross section (RCS) reduction [1] is a method of controlling and reducing the radar characteristics of military targets, which reduces the combat effectiveness of enemy electronic detection systems and weapon platforms, thus enhancing the penetration and survival capabilities of military equipment. In a narrow sense, RCS reduction (RCSR) can be considered to be antiradar stealth technology. Moreover, the stealth technology aimed at improving RCSR comprises four basic methods—that is, stealth design shape, radar-absorbing materials (RAMs), passive cancellation, and active cancellation methods, the first two methods being the most commonly used. Changing the design shape and using RAMs can reduce the RCS of military platforms. However, for low observable technology, the antenna can contribute much of the total RCS. Consequently, consideration of the antenna RCS has become increasingly important.

Antennas are one of the major contributors to the total RCS of the equipment on an aircraft, so technicians must take steps to reduce the antenna RCS. Moreover, the radar antenna system—comprising a radome, radar antennas, and a radar module—usually produces a strong RCS in the nose cone of an aircraft. Using a typical tactical missile as an example, the RCS of its terminal guidance radar—that is, the Cassegrain antenna—can be as high as 10 dBsm under horizontally polarized incident waves in the X-band. Similarly, the RCS of a planar slotted array antenna used in Doppler radar can reach up to 30 dBsm. Consequently, the RCSR of antenna systems is a

Y. Liu et al., *Antenna Radar Cross Section: Theory and Design*,
https://doi.org/10.1007/978-981-96-1226-0_1

key technology in target stealth applications. However, owing to the limitations of the antenna system's working characteristics, RCSR must ensure the normal reception and transmission of electromagnetic (EM) waves. Thus, conventional stealth methods—such as shape stealth technology and RAMs—cannot be simply applied to antenna RCSR, making the stealth of antenna systems a key problem in stealth technology applications for military aircraft. Currently, there is no ideal way to substantially reduce the antenna RCS without influencing the radiation performance of the radar antenna. Consequently, trade-offs between these factors have to be made in practical engineering applications.

1.2 State-of-the-Art Antenna RCS Reduction

Antenna RCS research can be traced back to the general theory of antenna scattering based on conjugate matching conditions, published by R. B. Green of Ohio University in 1963 [1]. R. E. Collin and R. C. Hansen proposed the antenna scattering theory based on short-circuit and matched-load conditions in 1969 and 1989 [2, 3], respectively. Additionally, the analysis of specific antenna scattering can be traced back to Newman in the United States, who first studied the scattering characteristics of microstrip patches (without considering the feeding structure) in 1987 [4]. Thereafter, the RCSR and control of antennas—especially for microstrip antennas—experienced a period of enthusiastic development [5–16]. The specialized topic of antenna RCS was first discussed in the 1992 IEEE International Symposium on Antennas and Propagation, which included 11 papers, all of which focused on the calculation and reduction of the RCS for microstrip antennas. Thus, it is evident that the scholars of the time demonstrated a keen interest in using such antennas as low-RCS antennas for aircraft.

An antenna is a special kind of scatterer, its scattering usually comprising two components. One component is structural mode scattering, which is independent of the antenna load. Its scattering mechanism is the same as that of ordinary scattering—that is, it is only related to the shape, size, and material of the antenna. The other component is antenna mode scattering, which is related to the antenna load. It is caused by the reflection and reradiation of the received power owing to a mismatch in the antenna's terminating load. This is a scattering component unique to the antenna when used as a loaded scatterer.

In previous studies, the RCSR and calculation methods for antennas have been focused primarily on microstrip patches that lacked feeding structures. However, the scattered field of the antenna is neither structural mode scattering nor antenna mode scattering. Moreover, calculation of the phase difference between the structural mode scattering field and the antenna mode scattering field has long been a difficult problem to solve, making it difficult to obtain exact solutions for the structural and antenna mode components.

The authors' group has been studying the theory and design of antenna RCSR since 2000. When dealing with the RCS of a microstrip antenna, we have considered the feeding structure rather than simply calculating the microstrip patch; therefore, the calculated results are closer to the actual scattering performance of the microstrip antenna. In research on antenna scattering theory, we have derived the basic expression for the scattering field of an antenna based on arbitrary loading conditions through the antenna scattering matrix [17, 18]. Moreover, the antenna scattering field expressions [1–3] proposed by previous scholars can be derived from ours. Consequently, an analysis model of antenna scattering was proposed. The structural mode and antenna mode scattering can be obtained separately using the proposed model, solving the problem of obtaining the phase difference between them.

Microstrip antennas conform easily to the surface of aircraft and have the advantages of their small size, low profile, low cost, and light weight. Scholars around the world have proposed various methods to reduce the RCS of microstrip antennas, which can be divided into three categories—that is, resistance loading, varactor diode loading, and special substrate/superstrate applications.

1. **Resistance loading**: The most common method for reducing the antenna RCS by means of resistive loading is to use resistive loading at the feeding point of the microstrip antenna [5] and to load a lumped impedance at the edge of the microstrip patch [9, 10]. However, such methods only obtain large reductions of the antenna RCS at the resonant frequency of the antenna. To realize antenna RCSR over a wider band, a method of loading a resistive sheet around the patch was proposed [10]. The primary aim of this approach is to reduce the antenna RCS over a broader band and to have less antenna gain deterioration, features that have been validated experimentally.
2. **Varactor diode loading**: The resonant frequency of a microstrip antenna can be tuned by active loading. For example, connecting the metallic patch and ground of a microstrip antenna using varactor diodes can enable it to obtain a wider impedance bandwidth. Additionally, the minimum scattering response of the antenna can be achieved at several frequencies by properly adjusting the bias of the diodes. The varactor diodes are usually loaded at the position with the highest electric field intensity. When the antenna is in working order, its peak RCS can be shifted in the frequency domain by adjusting the bias of the diodes. When the antenna is not operational, turning the bias off can reduce the antenna efficiency considerably, as well as the scattering properties. This method has been reported to realize an RCSR of more than 10 dB.
3. **Using a special substrate/superstrate**: It is well known that the antenna RCS is usually reduced at the cost of deterioration of the antenna gain/radiation efficiency. Therefore, antenna RCSR can be obtained by using a lossy substrate/superstrate as they can reduce the radiation efficiency and input impedance of the antenna. Within the acceptable reduction ranges of radiation efficiency and input impedance, a proper lossy superstrate is usually used to realize antenna RCSR. Because ferrite has an obvious nonreciprocal property with a biased magnetic field, it has different effects on the radiation and scattering characteristics of a microstrip antenna when used as a superstrate. When considering the scattering characteristics of the antenna, ferrite can reduce the electric field intensity of an

incident wave impinging on the patch of the microstrip antenna considerably. As a result, the induced current is reduced, which leads to RCSR of the antenna. By contrast, when considering the radiation characteristics of an antenna, ferrite can increase the electric field intensity radiated by the antenna, so improving the antenna gain [12]. An antenna using ferrite as the superstrate requires a device to provide a biased magnetic field under the antenna ground. Although this device can increase the weight and complexity of the antenna, it affords greater freedom of RCS control compared with reciprocal materials. Additionally, the peak value of the antenna RCS can be shifted by changing the biased magnetic field. Consequently, adjusting the biased magnetic field to control the working characteristics of ferrite may be an effective method for antenna RCSR [13]. Moreover, if electronic support measures (ESMs)—such as the radar signal receiver—are used to detect and obtain the frequency of incident radar signals, tuning the biased magnetic field based on the detected signal can cause the antenna RCS to sag at this frequency, achieving the aim of radar stealth.

The biggest problem in reducing the antenna RCS is balancing its radiation and scattering performance. In previous studies, the antenna RCS was reduced at the expense of the deterioration of the radiation performance (such as gain reduction and loss of bandwidth). However, low-RCS antennas with good radiation performance can be obtained by using one or several methods, including fractal [20–33], slotting [34, 35], aperture coupling [36–39], or ultrawideband antenna [40–42] technologies.

Our research on the fundamental theories of antenna RCS and low-RCS antenna design have resulted in breakthroughs and improved efficiencies. Moreover, related publications have increasingly attracted the research interest of scholars of antenna RCSR around the world. Consequently, the antenna RCS model proposed by the authors has been widely used and cited in antenna RCSR research.

In addition to in-depth RCS studies of individual antennas, research on the RCS of antenna arrays has also attracted considerable attention. The scattering of an antenna array is associated with the contributions of all of the antenna elements and feeding networks in the array. Compared with the RCS of a single antenna, the complexity of the antenna array RCS is caused primarily by the feeding network, as each discontinuous impedance point in the feeding network causes reflections of EM energy to be transmitted within it. The multiple transmission and reflection processes of EM waves can be complex, so the end result is that some EM energy is reradiated by the antenna array, and some is absorbed by the terminal load. The reradiation process can contribute enormously to the total RCS. Consequently, the design of low-RCS antenna arrays can be a difficult problem to solve, as the feeding network must be considered.

The calculation methods for the radiation and scattering of antenna arrays without a feeding network—such as the microstrip patch antenna array and microstrip dipole antenna array—have been investigated [8, 16]. Using the method of moments (MoM) and fast calculation methods, our research group has analyzed microstrip antennas and antenna arrays with both microstrip and coaxial feeding networks [48, 49]. We have also investigated the fundamental scattering theory of antenna arrays and determined that the RCS of an array has the same two components as an individual antenna—that is, a structural mode scattering component and an antenna mode scattering component [18].

However, these studies have not been able to meet the requirements of practical applications. Antennas in practice are often phased array antennas, which have complex feeding networks. Consequently, finding an effective way to design low-RCS antenna arrays remains a crucial problem to solve.

1.3 Summary

This chapter introduced the research background and current state of antenna RCS applications. Research on the antenna RCS remains a complex topic, which scholars around the world should pay more attention to.

References

1. Green, R.B., The General Theory of Antenna Scattering, Report 1223-17, Antenna Laboratory, Ohio State University, November 30, 1963.
2. Collin, R.E., The Receiving Antenna, Antenna Theory, Part1, R.E. Collin and F.J. Zucker(eds.), McGraw-Hill, New York, pp. 123–133, 1969.
3. Hansen, R.C., Relationships Between Antennas as Scatters and as Radiators, Pro. IEEE, Vol.77, pp. 659–662, May 1989.
4. Edward H. Newman, David Forrai, Scattering from a Microstrip Patch, IEEE Trans. on AP, Vol. 35, No. 3, pp. 245–251, 1987.
5. D.M. Pozar, Radiation and Scattering from a Microstrip Patch on a Uniaxial Substrate, IEEE Trans. on AP, Vol. 35, No. 6, pp. 613–621, 1987.
6. D.R. Jackson, The RCS of a Rectangular Microstrip Patch in a Substrate-Superstrate Geometry, IEEE Trans. on AP, Vol. 38, No. 1, pp. 2–8, 1990.
7. James T. Aberle, David M. Poazr, Craig R. Birtcher, Evaluation of Input Impedance and Radar Cross Section of Probe-Fed Microstrip Patch Elements Using an Accurate Feed Model, IEEE Trans. on AP, Vol. 39, No. 12, pp. 1691–1696, 1991.
8. Jin-Ming Jin, John L. Volakis, A Hybrid Finite Element Method for Scattering and Radiation by Microstrip Patch Antennas and Arrays Residing in a Cavity, IEEE Trans. on AP, Vol. 39, No. 11, pp. 1598–1604, 1991.
9. J.L. Volakis, *et al.*, Broadband RCS Reduction of Rectangular Patch by Using Distributed Loading, Electronic Letters, Vol. 28, No. 25, pp. 2322–2325, 1992.
10. J.L. Volakis, *et al.*, Radar Cross Section Analysis and Control of Microstrip Patch Antennas, 1992 PIEEE AP-S International Symposium, pp. 2225–2228, 1992.
11. J.T. Aberle, *et al.*, Scattering and Radiation Properties of Varactor-Tuned Microstrip Antennas, 1992 PIEEE AP-S International Symposium, pp. 2229–2232, 1992.
12. H.Y. Yang, *et al.*, Multifunctional Antennas with Low RCS, 1992 PIEEE AP-S International Symposium, pp. 2240–2243, 1992.
13. David M. Pozar, RCS Reduction for a Microstrip Antenna Using a Normally Biased Ferrite Substrate, IEEE Microwave and Guided Wave Letters, Vol. 2, No. 5, pp. 196–198, 1992.

14. Adrian S. King, Wallace J. Bow, Scattering from a Finite Array of Microstrip Patches, IEEE Trans. on AP, Vol. 40, No. 7, pp. 770–774, 1992.
15. Feng Ling, Jian-Ming Jin, Scattering and Radiation Analysis of Microstrip Antennas Using Discrete Complex Image Method and Reciprocity Theorem, Microwave and Optical Technology Letters, Vol. 16, No. 4, pp. 212–216, 1997.
16. Vikram Jandhyala, Eric Michielssen, Balasubramaniam Shanker, Weng C. Chew, A Fast Algorithm for the Analysis of Radiation and Scattering from Microstrip Arrays on Finite Substrates, Microwave and Optical Technology Letters, Vol.23, No.5, pp.306-310, 1999.
17. Ying Liu, Shu-Xi Gong, De-Min Fu, Theory Study of Antenna Scattering, Acta Electronica Sinica, Vol. 33, No. 9, pp. 1611–1613, 2005.
18. Ying Liu, Shu-Xi Gong, De-Min Fu, Scattering Analysis of Antenna Array, Asia-Pacific Microwave Conference 2003 (APMC'03), pp. 1252–1255, Korea, 2003.
19. Ying Liu, Demin Fu, Shuxi Gong, A novel model for analyzing the RCS of microstrip antenna, Journal of Electromagnetic Waves and Application, Vol. 17, No. 9, pp. 1301–1310, 2003.
20. X. Yang, *et al.*, Fractal Antenna Elements and Arrays, Applied Microwave and Wireless Magazine, Vol. 11, No. 5, pp. 34–46, 1999.
21. C. Puente, *et al.*, Fractal Multiband Antenna Based on the Sierpinski Gasket, IEE Electronics Letters, Vol. 32, No. 1 pp. 1–2, 1996.
22. C. Puente, *et al.*, On the Behavior of the Sierpinski Fractal Antenna, IEEE Trans. on AP, Vol. 46, No. 4, pp. 517–524, 1998.
23. R. Breden, *et al.*, Printed Fractal Antennas, IEE National Conference on Antennas and Propagation, pp. 1–4, April 1999.
24. Liang Xu, Multiband Characteristics of Two Fractal Antennas, Microwave and Optical Technology Letters, Vol. 23, No. 4, pp. 242–245, 1999.
25. G. J. Walker, *et al.*, Fractal Volume Antennas, Electronics Letters, Vol. 34, No. 16, pp. 1536–1537, 1998.
26. C. T. P. Song, *et al.*, Fractal Stacked Monopole with very Wide Bandwidth, Electronics Letters, Vol. 35, No. 12, pp. 945–946, 1999.
27. C. Puente, *et al.*, Small but Long Koch Fractal Monopole, Electronics Letters, Vol. 34, No. 1, pp. 9–10, 1998.
28. M. Sindou, *et al.*, Multiband and Wideband Properties of Printed Fractal Branched Antennas, Electronic Letters, Vol. 35, No. 3, pp. 181–182, 1999.
29. D.H. Werner, *et al.*, Radiation Characteristics of Thin-Wire Ternary Fractal Trees, Electronics Letters, Vol. 35, No. 8, pp. 609–610, 1999.
30. D.H. Werner, *et al.*, Fractal Constructions of Linear and Planar Arrays, IEEE Antennas and Propagation Society, No. 3, pp. 1968–1971, 1997.
31. R. V. Hara Prasad, *et al.*, Microstrip Fractal Patch Antenna for Multi-band Communication, Electronics Letters, Vol. 36, No. 14, pp. 1179–1180, 2000.
32. Ying Liu, Shuxi Gong, Hongbo Zhang, A Novel Fractal Slot Microstrip Antenna with Low RCS, 2006 IEEE AP-S International Symposium on Antennas and Propagation and USNC/URSI National Radio Science Meeting, Albuquerque, New Mexico, USA, pp.2603–2606, 2006.
33. S. Dey, R. Mittr, Compact Microstrip Patch Antenna, Microwave and Optical Technology Letters, Vol. 13, No. 1, pp. 12–14, 1996.
34. G. Clasen, R.J. Langley, Meshed Patch Antennas, IEEE Trans. On AP, Vol.52, No.6, pp.1412-1416, 2004.
35. Xiulian He, Shuxi Gong, Qizhong Liu, Meshed Microstrip Patch Antennas with Low RCS, Microwave and Optical Technology Letters, Vol. 46, No. 2, pp. 117–120, 2005.
36. Hall R C, Wu D I, Radar Cross Section of Multilayer Patch and Aperture Coupled Patch Antennas, Antennas and Propagation Society International Symposium, pp. 2025–2028,1995.
37. Peter L. Sullivan and Daviel H. Schaubert, Analysis of an Aperture Coupled Microstrip Antenna, IEEE Transaction on Antenna and Propagation, Vol. 34, No. 8, pp. 977–984, 1986.
38. Biox R R, Losada V, Medina F, Determination of the Radar Cross Section of Stacked Circular Microstrip Patches, Antennas and Propagation Society International Symposium, IEEE, Vol. 1, pp. 16–21, 2000.

39. Yang Xian Hua, Lotfollach Shafai, Characteristics of Aperture Coupled Microstrip Microstrip Antenna with Various Radiating Patches and Coupling Aperture, IEEE Transaction on Antenna and Propagation, Vol. 37, No. 4, pp. 72–78, 1985.
40. J.X. Liang, C.C. Chiau, X.D. Chen, and C.G. Parini, Study of a Printed Circular Disc Monopole Antenna for UWB Systems, IEEE Trans. Antennas Propagation, Vol. 53, No. 11, pp. 3500–3504, 2005.
41. S. Sadat, M. Fardis, Fatemeh Geran Kharakhili, G. Dadashzadeh, A Compact Microstrip Square-ring Slot Antenna for UWB Applications, Progress In Electromagnetics Research, PIER 67, pp. 173–179, 2007.
42. Sanming Hu Honghui Chen, Choi Look Law, Zhongxiang Shen, Lei Zhu, Wenxun Zhang, Wenbin Dou, Backscattering Cross Section of Ultrawideband Antennas, IEEE Antennas and Wireless Propagation Letters, Vol. 6, pp. 70–73, 2007.
43. Guanfeng Cui, Ying Liu, Shuxi Gong, A Novel Fractal Patch Antenna with Low RCS, Journal of Electromagnetic Waves and Application, Vol. 21, No. 15, pp. 2403–2411, 2007.
44. Hongwei Yuan, Shuxi Gong, Xing Wang, and Wentao Wang, Scattering Analysis of a Printed Dipole Antenna Using PBG Structures, Progress In Electromagnetics Research B, Vol. 1, pp. 189–195, 2008.
45. Tao Wu, Yan Li, Shuxi Gong, Ying Liu, A Novel Low RCS Microstrip Antenna Using Aperture Coupled Microstrip Dipoles, Journal of Electromagnetic Waves and Application, Vol. 22, pp. 953–963, 2008.
46. Yan Li, Ying Liu, Shuxi Gong, Microstrip Antenna Using Ground-cut Slots and Miniaturization Techniques with Low RCS, Progress in Electromagnetic Research Letters, pp. 211–220, 2008.
47. Ying Liu, Shuxi Gong, A Novel UWB Clover-disc Monopole Antenna with RCS Reduction, Journal of Electromagnetic Waves and Application, Vol. 22, pp. 1115–1121, 2008.
48. He Xiulian, Gong Shuxi, Liu Qizhong, On the Scattering of Arbitrary Shape Microstrip Patch, Journal of Electronics (China), Vol. 21, No. 5, pp. 432–436, 2004.
49. Xiulian He, Shuxi Gong, Qizhong Liu, Fast Computation of Spatial Green's Functions of Multilayered Microstrip Antennas, Microwave and Optical Technology Letters, Vol. 45, No. 1, pp. 85–88, 2005.

Chapter 2
Fundamental Theory of Antenna Scattering

2.1 The Radar Cross-Section Concept

When a target is illuminated by an incident wave, the power is scattered in all directions. The vector sum of the scattered and the incident fields forms the total field. The distribution pattern of the scattered power is called the scattering pattern, and it is dependent on the shape, size and structure of the target as well as the frequency and polarization of the incident wave. Objects causing EM scattering are often called scatterers.

Monostatic scattering occurs when the transmitter and receiver are at the same location, as is the case with most radar systems. Conversely, the situation in which the transmitter and receiver are at different locations is called bistatic scattering, the angle between the transmitter and receiver being called the bistatic angle (γ). Forward scattering and monostatic scattering occur when the bistatic angles are 180° and 0°, respectively. In most monostatic scattering measurements, there is usually a small distance between the transmitter and receiver; consequently, it is, strictly speaking, a type of bistatic scattering. However, because the bistatic angle is sufficiently small that the measured result is close to that of monostatic scattering, it is sometimes called quasimonostatic scattering.

The physical quantity used to quantitatively evaluate the scattering intensity of a target can be defined as the effective scattering area of the target to the incident wave, which usually referred to as the RCS [1]. Moreover, the RCS can be considered to be an imaginary target area. A receiving antenna is often regarded as an aperture with an effective receiving area, which can intercept and capture the power passing through it. The received power at the terminal of the receiving antenna is equal to the incident power density multiplied by the illuminated effective receiving area. Similarly, the power reflected or scattered by a target can also be expressed as an effective area multiplied by the incident power density. This effective area is the RCS, which is represented by the symbol σ.

Y. Liu et al., *Antenna Radar Cross Section: Theory and Design*,
https://doi.org/10.1007/978-981-96-1226-0_2

The target is usually far from the transmitting and receiving antennas, the distance being far greater than the size of the target itself. Therefore, the illuminating wave at the target can be regarded as a plane wave, and the target can be considered to be the approximate equivalent of a point scatterer. The RCS can be obtained from the following derivation. The incident power density can be expressed as follows:

$$\omega_i = \frac{1}{2} E^i \cdot H^i = \frac{1}{2Z_0} \left| E^i \right|^2. \tag{2.1}$$

where E^i and H^i denote the electric and magnetic field intensities of the incident wave, respectively, H^{i*} denotes the conjugate of H^i, and $Z_0 = \sqrt{\mu_0/\varepsilon_0} = 377\ \Omega$ denotes the wave impedance in free space.

Consequently, the power captured by the target with an RCS of σ can be expressed as follows:

$$P = \sigma \omega_i = \frac{1}{2Z_0} \sigma \left| E^i \right|^2. \tag{2.2}$$

If the target scatters the captured power uniformly in all 4π sr of space, the scattering power at the distance R can be expressed as follows:

$$\omega_s = \frac{P}{4\pi R^2} = \frac{\sigma \left| E^i \right|^2}{8\pi Z_0 R^2} \tag{2.3}$$

The scattering power can also be represented as follows:

$$\omega_s = \frac{1}{2Z_0} \left| E^s \right|^2. \tag{2.4}$$

where E^i denotes the electric field intensity of the scattered wave.

We can then solve Eqs. (2.3) and (2.4) for σ and assume that the distance R from the target is large, as follows:

$$\sigma = 4\pi R^2 \frac{\left| \boldsymbol{E}^s \right|^2}{\left| \boldsymbol{E}^i \right|^2}. \tag{2.5}$$

Because the target is illuminated by a plane wave and assumed to be a point scatterer, the distance from the target to the transmitter and receiver should approach infinity. Thus, Eq. (2.5) can be expressed more strictly as follows:

$$\sigma = \lim_{R \to \infty} 4\pi R^2 \frac{\left| \boldsymbol{E}^s \right|^2}{\left| \boldsymbol{E}^i \right|^2} = \lim_{R \to \infty} 4\pi R^2 \frac{\left| \boldsymbol{H}^s \right|^2}{\left| \boldsymbol{H}^i \right|^2} \tag{2.6}$$

Here, Eq. (2.6) is the fundamental theoretical definition of the RCS.

If the vector properties of the EM field are preserved, the vector definition of the RCS can be expressed as follows:

$$\sigma = \frac{1}{\pi}(\lambda \hat{\boldsymbol{e}}_r \cdot \boldsymbol{S})^2 \tag{2.7}$$

where $\hat{\boldsymbol{e}}_r$ denotes the unit vector in the polarization direction of the receiving antenna, and $\boldsymbol{S}$ denotes the vector scattering function:

$$\boldsymbol{S} = k\boldsymbol{R}e^{jkR}\frac{\boldsymbol{E}^s}{|\boldsymbol{E}^i|} \tag{2.8}$$

where $k = 2\pi/\lambda$ denotes the wavenumber in free space, and e^{jkR} denotes the phase shift from the target to the receiving antenna. Consequently, $\boldsymbol{S}$ is a dimensionless vector function.

The parameters R^2 and λ^2 in Eqs. (2.6) and (2.7), respectively, result in the unit for σ being the area. For a 3D geometric structure, the scattered electric field $|\boldsymbol{E}^s|$ (or $\boldsymbol{E}^s$) is inversely proportional to R. Thus, the RCS does not depend on the distance of the target from the transmitter. Although scattering problems of 2D structures are seldom encountered, they are useful in analyzing the corresponding 3D problems.

The scattered field of a 2D structure is inversely proportional to $\sqrt{R}$, so its RCS is defined in a slightly different way. For a 2D structure, the scatterer is an infinitely long cylinder. The symbol σ' can be used instead of σ to represent the RCS. The definitions of the RCS corresponding to Eqs. (2.6) and (2.7) are expressed in Eqs. (2.9) and (2.10), respectively:

$$\sigma' = \lim_{\rho\to\infty} 2\pi\rho\frac{|E^s|^2}{|E^i|^2} = \lim_{\rho\to\infty} 2\pi\rho\frac{|H^s|^2}{|H^i|^2} \tag{2.9}$$

$$\sigma' = \frac{4}{k}|P^2| \tag{2.10}$$

The scattering function P in Eq. (2.10) corresponds to S in Eq. (2.8), where:

$$\mathrm{P} = \sqrt{\frac{\pi k\rho}{2}}\frac{|\boldsymbol{E}^s|}{|\boldsymbol{E}^i|}e^{j\left(k\rho-\frac{\pi}{4}\right)} \tag{2.11}$$

The parameter ρ in Eqs. (2.9)–(2.11) denotes the distance from the cylinder to the receiver, the scattering function being polarization-dependent. There is a simple relationship between the RCS (σ) of a cylinder with a finite length (l) and the RCS (σ') of an infinitely long cylinder, which can be expressed as follows:

$$\sigma = \frac{2l^2}{\lambda}\sigma'(l \gg \lambda) \tag{2.12}$$

where σ denotes a scalar (m^2). It is often expressed in logarithmic form, as follows:

$$\sigma_{\mathrm{dBsm}} = 10\lg\sigma \tag{2.13}$$

The units of σ' are m or dBm.

2.2 Fundamental Theories of Radiation and Receiving of an Antenna

This section introduces several fundamental theories of radiation and receiving of an antenna, which will be used in the derivation of the antenna scattering theory.

2.2.1 *Antenna Radiation Theory*

The EM field radiated from a source described by current densities $(\boldsymbol{J}, \boldsymbol{J}_m)$ in an unbounded homogeneous space characterized by the constitutive parameters (ε, μ) can be calculated using the following equations [2]:

$$E(r) = -j\omega\mu\left(A + \frac{1}{k^2}\nabla\nabla\cdot A\right) - \nabla\times F \tag{2.14}$$

$$H(r) = -j\omega\varepsilon\left(F + \frac{1}{k^2}\nabla\nabla\cdot F\right) - \nabla\times A \tag{2.15}$$

where

$$A(r) = \int_r G\left(\frac{r}{r'}\right)J\left(r'\right)dr' \tag{2.16}$$

$$F(r) = \int_r G\left(\frac{r}{r'}\right)J_m\left(r'\right)dr' \tag{2.17}$$

$$G\left(\frac{r}{r'}\right) = \frac{e^{-jk|r-r'|}}{4\pi|r-r'|} \tag{2.18}$$

$$k = \omega\sqrt{\mu\varepsilon} \tag{2.19}$$

The far field has the following representations:

$$\boldsymbol{E}(\boldsymbol{r}) = \sqrt{Z}\boldsymbol{A}(\boldsymbol{k})\frac{e^{-jkr}}{r} \tag{2.20}$$

$$\boldsymbol{H}(\boldsymbol{r}) = \sqrt{Y}\hat{\boldsymbol{k}} \times \boldsymbol{A}(\boldsymbol{k})\frac{e^{-jkr}}{r} \tag{2.21}$$

where

$$Z = \frac{1}{Y} = \sqrt{\frac{\mu}{\varepsilon}} \tag{2.22}$$

$$\boldsymbol{k} = \hat{\boldsymbol{k}}k = \hat{\boldsymbol{r}}k \tag{2.23}$$

$$A(k) = \frac{k}{4\pi j}\int_{r}\left[\sqrt{Z}\left(\overleftrightarrow{I} - \hat{r}\hat{r}\right)\cdot J\left(r'\right) + \sqrt{Y}J_m\left(r'\right) \times \hat{r}\right]e^{-jk\cdot r'}dr' \tag{2.24}$$

2.2.2 *Antenna Receiving Theory [2]*

1. **Three ideal sources for transmitting antennas**: When the antenna is used for transmitting, a source needs to be connected to the terminal port. Here, we introduce three principle ideal sources as shown in Fig. 2.1.
   (1) Unit amplitude source $\left(a = 1\sqrt{\text{watt}}^{1/2}\right)$: For a unit amplitude source, the internal impedance (Z_s) is equal to the characteristic impedance of the transmission line (Z_c). Thus, when the source is matched to the transmission line, there is no reflection at the source–transmission-line junction. However, the mismatch at the antenna–transmission-line junction may still exist; that is, Z_c may or may not be equal to Z_{in}.
   (2) Unit voltage source ($v = 1$ V): For a unit voltage source, the internal impedance (Z_s) is equal to zero. As a result, there is no internal voltage drop within the source.
   (3) Unit current source ($I = 1$ A): For a unit current source, the internal impedance (Z_s) approaches infinity, so the total source current enters the transmission line.

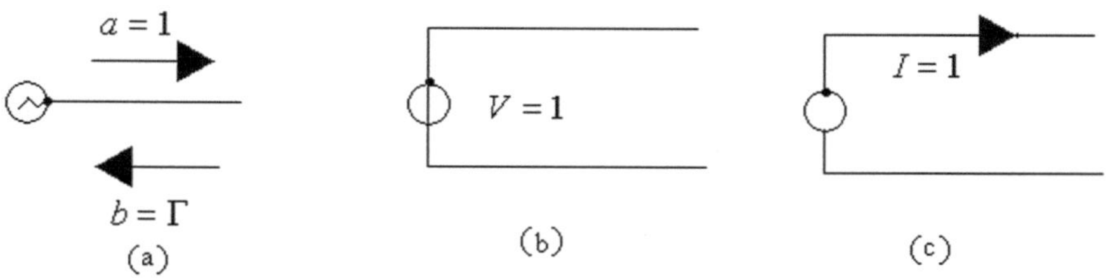

Fig. 2.1 Three ideal sources for transmitting antennas

For transmitting antennas, any of the three ideal sources can be used as the excitation source. The relationships between the parameters (U, I, a, b) and the external radiation quantities (radiated field F and its power P_t) are listed in Table 2.1. Here, F denotes the vector field ($\boldsymbol{E}, \boldsymbol{H}$). The subscripts 1, 2, and 3 are used to identify the unit amplitude, voltage, and current source, respectively. The three fields are related as follows:

$$F_2 = \frac{\sqrt{Y_c}}{1+\Gamma}\frac{V_2}{a_1}F_1 = \frac{1}{2}\sqrt{Y_c}\left(1+\frac{Y_{in}}{Y_c}\right)\frac{V_2}{a_1}F_1 \tag{2.25}$$

$$F_3 = \frac{\sqrt{Z_c}}{1-\Gamma}\frac{I_3}{a_1}F_1 = \frac{1}{2}\sqrt{Z_c}\left(1+\frac{Z_{in}}{Z_c}\right)\frac{I_3}{a_1}F_1 \tag{2.26}$$

where $a = 1\sqrt{\text{watt}}^{1/2}$, $v = 1$ V, and $I = 1$ A.

2. **Three ideal meters for receiving antennas**: When an antenna is used for receiving, the source can be replaced by a receiver or a meter. Here, we introduce three ideal meters as shown in Fig. 2.2.

   (1) Amplitude meter: An amplitude meter can be used to measure the incoming traveling-wave's amplitude (b) at the reference plane $z = 0$. Because its internal impedance (Z_s) is equal to Z_c, the meter can be matched to the

Table 2.1 Relationships for transmitting antennas

Source quantity	Amplitude	Voltage	Current
V (v)	$\sqrt{Z_c}(1+\Gamma)a_1$	$V_2 = 1$	$Z_{\text{in}}I_3$
I (A)	$\sqrt{Y_c}(1-\Gamma)a_1$	$Y_{\text{in}}\ V_2$	$I_3 = 1$
a (W)$^{1/2}$	$a_1 = 1$	$\frac{1}{2}\sqrt{Y_c}(1+\frac{Y_{in}}{Y_c})V_2$	$\frac{1}{2}\sqrt{Z_c}(\frac{Z_{in}}{Z_c}-1)I_3$
b (W)$^{1/2}$	Γa_1	$\frac{1}{2}\sqrt{Y_c}(1-\frac{Y_{in}}{Y_c})V_2$	$\frac{1}{2}\sqrt{Z_c}(\frac{Z_{in}}{Z_c}-1)I_3$
P_t (W)	$(1-\lvert\Gamma\rvert^2)a_1^2$	$V_2^2\text{Re}Y_{in}$	$I_3^2\text{Re}Z_{in}$
Field	F_1	F_2	F_3
Source impedance	Z_c (matched)	0 (short)	∞ (open)

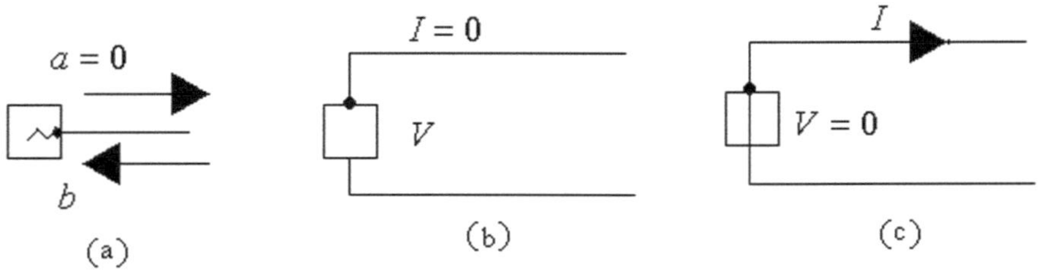

Fig. 2.2 Three ideal meters for receiving antennas

Table 2.2 Relationships for receiving antennas

Meter quantity	Amplitude	Voltmeter	Ammeter
V (v)	$\sqrt{Z_c}b_4$	V_5	0
I (A)	$-\sqrt{Y_c}b_4$	0	I_6
a (W)$^{1/2}$	0	$\frac{1}{2}\sqrt{Y_c}V_5$	$\frac{1}{2}\sqrt{Z_c}I_6$
b (W)$^{1/2}$	b_4	$\frac{1}{2}\sqrt{Y_c}V_5$	$-\frac{1}{2}\sqrt{Z_c}I_6$
Source impedance	Z_c (matched)	∞ (open)	0 (short)

transmission line (Z_{in} may or may not be equal to Z_c). It should be noted that b is a complex number that includes both amplitude and phase information.

(2) Voltmeter: A voltmeter can be used to measure the voltage (U) at the reference plane $z = 0$. Because its internal impedance (Z_s) approaches infinity, U is the open-circuit voltage.

(3) Ammeter: An ammeter can be used to measure the current (I) at the reference plane $z = 0$. Because its internal impedance (Z_s) is zero, I is the short-circuit voltage.

When the antenna is illuminated by a plane wave, we can connect any of the three meters to its transmission line, corresponding to matched-load, open-circuit, and short-circuit situations. The parameters (U, I, a, b) are listed in Table 2.2. The subscripts 4, 5, and 6 are used to identify the use of the amplitude meter, voltmeter, and ammeter, respectively. For the same incident field, the three situations are related as follows:

$$V_5 = \sqrt{Z_c}\frac{2b_4}{1-\Gamma} = b_4\sqrt{Z_c}\left(1 + \frac{Z_{in}}{Z_c}\right) \tag{2.27}$$

$$I_6 = \sqrt{Y_c}\frac{(-2b_4)}{1+\Gamma} = -b_4\sqrt{Y_c}\left(1 + \frac{Y_{in}}{Y_c}\right) \tag{2.28}$$

Commonly, U_5 is known as the open-circuit voltage, and I_6 is the short-circuit current of the receiving antenna.

3. **Reciprocity between antenna transmitting and receiving**: The reciprocity theorem in a circuit is well known; however, its application to antennas is not simple. The transmitting and receiving properties of an antenna can be described by two complex numbers (a, b) and (U, I), respectively. In the free space beyond the antenna, either the radiated field of the transmitting antenna or the incident field on the receiving antenna can be complex. They are vector fields with certain polarizations and spatial variations, which cannot be described by circuit quantities. Thus, the reciprocity of an antenna cannot be stated by the usual exchange of sources and meters.

Here, we simply Introduce the reciprocity relationships for antennas. When an antenna excited by a unit amplitude source with $a_1 = 1\sqrt{\text{watt}}^{1/2}$ is used for transmitting, its radiated field is $F_1 = (\boldsymbol{E}_1, \boldsymbol{H}_1)$. When the same antenna in the receiving situation is connected to an amplitude meter with a matched impedance (so that $a_4 = 0$) and is illuminated by an incident field $F_4 = (\boldsymbol{E}_4, \boldsymbol{H}_4)$, the reciprocity can be expressed as follows:

$$b_4 = \frac{1}{2a_1}\int_s (\vec{E}_4 \times \vec{H}_1 - \vec{E}_1 \times \vec{H}_4) \cdot \hat{n} dS \tag{2.29}$$

where S denotes a closed surface that encloses the antenna but excludes the source of F_4, and $\hat{n}$ denotes the outward unit normal vector of S.

If F_1 and F_4 are known:

$$F_1 = \begin{Bmatrix} \vec{E}_1 \\ \vec{H}_1 \end{Bmatrix} = \begin{Bmatrix} \sqrt{Z}\vec{A}(\vec{k}) \\ \sqrt{Y}\hat{k} \times \vec{A}(\vec{k}) \end{Bmatrix} \frac{e^{-jkr}}{r} \text{ (far-field radiation field)} \tag{2.30}$$

$$F_4 = \begin{Bmatrix} \vec{E}_4 \\ \vec{H}_4 \end{Bmatrix} = C\begin{Bmatrix} \sqrt{Z}\vec{u}^* \\ -\sqrt{Y}\hat{k} \times \vec{u}^* \end{Bmatrix} e^{j\vec{k}\cdot\vec{r}} \text{ (incident plane wave)} \tag{2.31}$$

then reciprocity states:

$$b_4 = -j\lambda C[\vec{A}(\vec{k}) \cdot \vec{u}^*] \tag{2.32}$$

It should be noted that the fields in the states $(\boldsymbol{k}, \boldsymbol{u})$ and $(-\boldsymbol{k}, \boldsymbol{u}^*)$ have the same polarization and different propagation directions.

2.3 Scattering Theory of Antennas

In this section, we introduce the basic theory of antenna scattering, which clearly illustrates the mechanism of scattering. As shown in Fig. 2.3, (a_0, b_0) are the amplitudes of the outgoing and incoming signals on the transmission line at the reference plane (*S1*). The field in the region outside the antenna can be expanded using the vector spherical wave function, as follows:

$$\vec{E} = \sum_i \left(a_i \vec{e}_i^{\,in} + b_i \vec{e}_i^{\,out} \right) \tag{2.33}$$

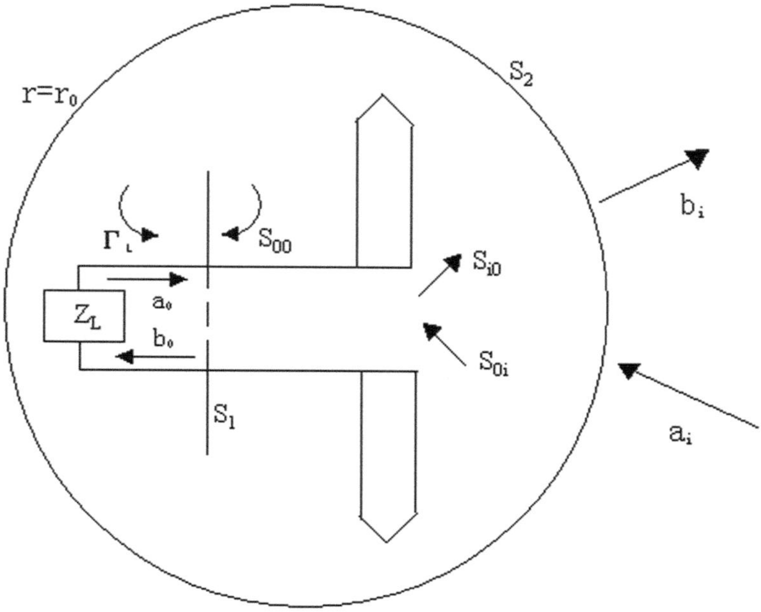

Fig. 2.3 Scattering matrix representation of an antenna

where $(a_i, b_i)(i \neq 0)$ denote the ith complex amplitudes of the incoming and outgoing spherical waves, respectively, $\boldsymbol{e}_i$ denotes the basis function for spherical wave expansion, and the subscripts in and out denotes the incoming and outgoing waves, respectively.

The antenna scattering properties can be described using the scattering matrix S [3], as follows:

$$\begin{bmatrix} b_0 \\ b_1 \\ b_2 \\ \vdots \\ \vdots \end{bmatrix} = \begin{bmatrix} S_{00} & S_{01} & S_{02} & \cdots & \cdots \\ S_{10} & S_{11} & S_{12} & \cdots & \cdots \\ S_{20} & S_{21} & S_{22} & \cdots & \cdots \\ \vdots & \vdots & \vdots & \ddots & \vdots \\ \vdots & \vdots & \vdots & \vdots & \ddots \end{bmatrix} \begin{bmatrix} a_0 \\ a_1 \\ a_2 \\ \vdots \\ \vdots \end{bmatrix} \tag{2.34}$$

For convenience, Eq. (2.34) can be expressed as follows:

$$[b] = [S]\,[a] \tag{2.35}$$

We can obtain b_0 from Eq. (2.34), as follows:

$$b_0 = \sum_{j=0}^{\infty} S_{0j}a_j = S_{00}a_0 + \sum_{j=1}^{\infty} S_{0j}a_j \text{ (reflection/receiving amplitude)} \tag{2.36}$$

where the first and second items denote the reflection amplitude of a_0 owing to the mismatch and the amplitude of the received signals of a_i at the reference plane ($S1$), respectively.

$$b_i = \sum_{j=0}^{\infty} S_{ij}a_j = S_{i0}a_0 + \sum_{j=1}^{\infty} S_{ij}a_j \text{ (radiation/scattering amplitude)} \tag{2.37}$$

where the first and second items represent the amplitude of the field radiated by a_0 and the amplitude of the scattered field of a_i in the region outside the antenna, respectively.

For the pure radiation case, $a_i = 0 (i \neq 0)$, so:

$$b_0 = S_{00}a_0 \text{ (reflection amplitude)} \tag{2.38}$$

$$b_i = S_{i0}a_0 \text{ (radiation amplitude)} \tag{2.39}$$

$$S_{00} = \Gamma_a = \frac{Z_{in} - Z_c}{Z_{in} + Z_c} \text{ (reflection coefficient)} \tag{2.40}$$

For the receiving and scattering case, the outgoing signal (a_0) can be expressed as follows:

$$a_0 = \Gamma_l b_0 \tag{2.41}$$

$$\Gamma_l = \frac{Z_l - Z_c}{Z_l + Z_c} \text{ (reflection coefficient of the receiver load)} \tag{2.42}$$

When the terminal load is matched to the transmission line ($Z_l = Z_c$), $a_0 = 0$. Therefore, the receiving signal (b_0) can be expressed as follows:

$$b_0 \Big|_{a_0=0} = \sum_{j=1}^{\infty} S_{0j}a_j = b_0^m \tag{2.43}$$

We can define b_0^m as the receiving magnitude under matched conditions. The relationship between b_0^m and the receiving signal for the case of an arbitrary terminal load can be described as follows:

$$b_0 = \frac{1}{1 - \Gamma_l S_{00}} \sum_{j=1}^{\infty} S_{0j}a_j = \frac{1}{1 - \Gamma_l S_{00}} b_0^m \tag{2.44}$$

Using Eqs. (2.37) and (2.38), we can obtain the amplitudes of the outgoing spherical waves as follows:

$$b_i = \sum_{j=0}^{\infty} S_{ij}a_j = S_{i0}\Gamma_l b_0 + \sum_{j=1}^{\infty} S_{ij}a_j \tag{2.45}$$

Similarly, we can define b_i^m as the scattering magnitude under matched conditions, expressed as follows:

$$b_i^m = \sum_{j=1}^{\infty} S_{ij}a_j \tag{2.46}$$

Substituting Eqs. (2.40), (2.44) and (2.46) into Eq. (2.45) results in the following:

$$b_i = b_i^m + \frac{\Gamma_l}{1-\Gamma_l\Gamma_a} b_0^m S_{i0} \tag{2.47}$$

It is evident from Eq. (2.33) that the total scattered field is the sum of the coefficients multiplied by the basis function [4]:

$$\vec{E}^s(Z_l) = \vec{E}^s(Z_c) + \frac{\Gamma_l}{1-\Gamma_l\Gamma_a} b_0^m \vec{E}_1^t \tag{2.48}$$

For convenience, Eq. (2.48) can be expressed as follows:

$$\vec{E}^s(Z_l) = \vec{E}^s(Z_c) + \vec{E}^a(Z_l) \tag{2.49}$$

In Eq. (2.49), the first item is known as the structural mode scattering field, which is the field scattered with a matched load. When the transmission line is connected to a matched load, $Z_l = Z_c$, $\Gamma_l = 0$, and all of the signals received by the antenna are absorbed by the load. Hence, the scattered field is the structural mode scattering field. The second item is known as the antenna mode scattering field. $\boldsymbol{E}_1^t$ denotes the radiation field when the antenna is connected to a unit amplitude excitation source. When the antenna is connected to an arbitrary load, the signals reflected by the load are transmitted to the antenna, some of the signals are reradiated to the free space, and some signals are reflected to the load again. This process of multiple reflections continues, and the total signal received by the load can be expressed as follows:

$$b_0^m[1 + \Gamma_l\Gamma_a + (\Gamma_l\Gamma_a)^2 + \cdots] = b_0^m \frac{1}{1-\Gamma_l\Gamma_a} \tag{2.50}$$

After reflection, the received signals become the incoming signals $\left(\frac{\Gamma_l}{1-\Gamma_l\Gamma_a} b_0^m\right)$, which can then be regarded as the excitation of the antenna. Consequently, the radiated field of the antenna is equal to the second item in Eq. (2.48). It is worth noting that the expressions for the antenna scattering field proposed by previous scholars, as well as their tenable conditions, can be derived from Eq. (2.48).

2.4 RCSR Method for Antenna Mode Scattering

Based on the previous theoretical analysis, several antenna RCSR methods can be developed.

1. Reducing the structural mode scattering field ($\boldsymbol{E}^s(Z_c)$): Different antennas are designed for different applications and have various structures. Therefore, different methods are required to reduce the structural mode scattering of different antennas.
2. Reducing the antenna mode scattering field:

$$\vec{E}^a(Z_l) = \frac{\Gamma_l}{1 - \Gamma_l S_{00}} b_0^m \vec{E}_1^t \tag{2.51}$$

Here, the influence of the incident wave polarization on antenna scattering is considered. First, we can define a unitary vector as follows:

$$\vec{U} \cdot \vec{U}^* = 1 \tag{2.52}$$

The signal received from the incident wave at the terminal load can then be expressed as follows:

$$b_0^m = -j\lambda\left(\frac{c}{a_1}\right)[\vec{A}(\vec{k}) \cdot \vec{u}^*] \tag{2.53}$$

The radiation field when an antenna is connected to a unit amplitude excitation can be expressed as follows:

$$\vec{E}_1^t = \sqrt{Z}\vec{A}(\vec{k})\frac{e^{-jkr}}{r} \tag{2.54}$$

where $\boldsymbol{A}(k) = \boldsymbol{u}A(\boldsymbol{k}, \boldsymbol{u}) + \boldsymbol{v}A(\boldsymbol{k}, \boldsymbol{v})$ denotes the amplitude vector. Here, $(\boldsymbol{u}, \boldsymbol{v})$ satisfies the following relationships:

$$\left|\vec{u}\right|^2 = \vec{u} \cdot \vec{u}^* = \left|\vec{v}\right|^2 = 1 \tag{2.55}$$

$$\vec{u}^* \cdot \vec{v} = 0 \tag{2.56}$$

$$\vec{u} \cdot \vec{k} = \vec{v} \cdot \vec{k} = 0 \tag{2.57}$$

So:

$$\vec{H}^t = \sqrt{Y}\hat{r} \times \vec{E}^t = \sqrt{Y}\hat{k} \times \vec{A}(\vec{k})\frac{e^{-jkr}}{r} \tag{2.58}$$

Then, $A(\boldsymbol{k}, \boldsymbol{u})$ and $A(\boldsymbol{k}, \boldsymbol{v})$ can be represented by $\boldsymbol{A}(k)$, as follows:

$$A(\vec{k}, \vec{u}) = \vec{A}(\vec{k}) \cdot \vec{u}^* \tag{2.59}$$

$$A(\vec{k}, \vec{v}) = \vec{A}(\vec{k}) \cdot \vec{v}^* \tag{2.60}$$

Therefore, $\boldsymbol{A}(k)$ can be expressed as:

$$\vec{A}(\vec{k}) = \left|\vec{A}(\vec{k})\right| \vec{U} \tag{2.61}$$

where $\boldsymbol{U} = \frac{1}{|A(k)|}[\boldsymbol{u}A(\boldsymbol{k}, \boldsymbol{u}) + \boldsymbol{v}A(\boldsymbol{k}, \boldsymbol{v})]$.

Thus, the polarization of the scattered field in the direction of $\boldsymbol{k}$ can be represented by the separate vector $\boldsymbol{U}$. The incident wave can be expressed as follows:

$$\vec{E}^i = \sqrt{Z}ce^{j\vec{k}_i \cdot \vec{r}}\vec{u}^* \tag{2.62}$$

Substituting these equations into Eq. (2.53) results in:

$$b_0^m = -j\lambda(\frac{c}{a_1})\left|\vec{A}(\vec{k}_i)\right|(\vec{U} \cdot \vec{u}^*) \tag{2.63}$$

Based on these equations, several methods can be used to reduce the antenna mode scattering.

(1) Keep Z_l as close as possible to 0: When $Z_l = 0$, the antenna mode scattering $\boldsymbol{E}^a$ is 0. Therefore, the total scattering field of the antenna is irrelevant to the antenna mode scattering and is only related to the antenna structure.
(2) Keep b_0^m as close as possible to 0: Eq. (2.63) shows that there are two methods capable of achieving this goal. The first is to make $\boldsymbol{A}(\boldsymbol{k}_i) = 0$. In this case, the antenna radiation pattern has a null in the direction of the incident wave. Therefore, regardless of how strong the excitation is, the field reradiated by the antenna is always 0 in this direction. Accordingly, the antenna RCS can be reduced by adjusting its radiation properties—such as by making the pattern

null in the direction toward the threatened angular domain. Another method is to make the polarization of the incident wave perpendicular to that of the receiving antenna ($(\boldsymbol{U} \cdot \boldsymbol{u}^*) = 0$). In this case, no signals are transmitted to the antenna load.

2.5 Other Expressions for Antenna Scattering

The fundamental formula for antenna scattering is given in Sect. 2.2. All of the formulae for antenna scattering proposed by previous scholars [5–8] as well as their tenable conditions, can be derived from this formula.

When the antenna is terminated with a short circuit, $Z_l = 0$, $\Gamma_l = -1$, (2.48) can be rewritten as follows:

$$\vec{E}^s(0) = \vec{E}^s(Z_c) - \frac{1}{1+\Gamma_a} b_0^m \vec{E}_1^t \qquad \vec{E}^s(Z_c) = \vec{E}^s(0) + \frac{1}{1+\Gamma_a} b_0^m \vec{E}_1^t \tag{2.64}$$

Substituting Eq. (2.64) into Eq. (2.48) results in:

$$\vec{E}^s(Z_l) = \vec{E}^s(0) + \left\{ \frac{\Gamma_l}{1-\Gamma_l\Gamma_a} + \frac{1}{1+\Gamma_a} \right\} b_0^m \vec{E}_1^t \tag{2.65}$$

The relationship between the receiving magnitude under matched conditions $\left(b_0^m\right)$ and the receiving current under short-circuit conditions ($\boldsymbol{I}(0)$) can be expressed as follows:

$$b_0^m = -\frac{1}{2}\sqrt{Z_c} I(0)(1+\Gamma_a) \tag{2.66}$$

The relationship between the fields radiated by an antenna with a unit amplitude excitation $\boldsymbol{E}_1^t$ and unit current excitation $\boldsymbol{E}_3^t$ cam be expressed as follows:

$$\vec{E}_1^t = \frac{1-\Gamma_a}{\sqrt{Z_c}} \vec{E}_3^t \tag{2.67}$$

Substituting Eqs. (2.66) and (2.67) into Eq. (2.65) results in:

$$\vec{E}^s(Z_l) = \vec{E}^s(0) - \frac{1}{2} \frac{(1+\Gamma_l)(1-\Gamma_a)}{(1-\Gamma_l\Gamma_a)} I(0) \vec{E}_3^t \tag{2.68}$$

Substituting $\Gamma_a = \frac{Z_{in}-Z_c}{Z_{in}+Z_c}$ and $\Gamma_l = \frac{Z_l-Z_c}{Z_l+Z_c}$ into Eq. (2.68) results in:

$$\vec{E}^s(Z_l) = \vec{E}^s(0) - \frac{Z_l}{Z_{in}+Z_l} I(0) \vec{E}_3^t \text{ (classical scattering formula 1)} \tag{2.69}$$

This is the classic antenna scattering formula with a short-circuit load proposed by Collin [6]. In Eq. (2.69), we make Z_l equal to Z_c, as follows:

$$\vec{E}^s(Z_{in}) = \vec{E}^s(0) - \frac{1}{2}I(0)\vec{E}_3^t \tag{2.70}$$

This can be rewritten as follows:

$$\vec{E}^s(0) = \vec{E}^s(Z_{in}) + \frac{1}{2}I(0)\vec{E}_3^t \tag{2.71}$$

Substituting Eq. (2.71) into Eq. (2.69) results in:

$$\begin{aligned}\vec{E}^s(Z_l) &= \vec{E}^s(Z_{in}) + \left(\frac{1}{2} - \frac{Z_l}{Z_{in}+Z_l}\right)I(0)\vec{E}_3^t \\ &= E^s(Z_{in}) + \frac{1}{2}I(0)\frac{Z_{in}-Z_l}{Z_{in}+Z_l}\vec{E}_3^t\end{aligned} \tag{2.72}$$

Making $\Gamma = \frac{Z_{in}-Z_l}{Z_{in}+Z_l}$, which denotes the mismatch between the load impedance and the input impedance of the antenna, results in:

$$\vec{E}^s(Z_l) = \vec{E}^s(Z_{in}) + \frac{1}{2}I(0)\Gamma\vec{E}_3^t \tag{2.73}$$

Because the receiving current under short-circuit conditions ($I(0)$) is double the receiving current under matched conditions ($I_m = I(Z_c)$), another classic scattering formula for an antenna with a matched load proposed by Hansen [7] can be derived, as follows:

$$E^s(Z_l) = \vec{E}^s(Z_{in}) + \Gamma I_m\vec{E}_3^t \text{ (classical scattering formula 2)} \tag{2.74}$$

In Eq. (2.69), making $Z_l = Z_{in}^*$ results in:

$$\vec{E}^s(Z_{in}^*) = \vec{E}^s(0) - \frac{Z_{in}^*}{Z_{in}+Z_{in}^*}I(0)\vec{E}_3^t \tag{2.75}$$

Substituting Eq. (2.75) into Eq. (2.69) results in:

$$\vec{E}^s(Z_l) = \vec{E}^s(Z_{in}^*) + \frac{Z_{in}(Z_{in}^*-Z_l)}{(Z_{in}^*+Z_{in})(Z_{in}+Z_l)}I(0)\vec{E}_3^t \tag{2.76}$$

Here, we can define $\tilde{\Gamma}_A = \frac{Z_{in}^*-Z_l}{Z_{in}^*+Z_l}$ and $\tilde{I}_{in} = \frac{Z_{in}}{Z_{in}^*+Z_{in}}I(0)$. Substituting them into Eq. (2.75) results in:

$$\tilde{\Gamma}_A = \frac{Z_{in}^* - Z_l}{Z_{in} + Z_l} \tag{2.77}$$

This is the classic scattering formula for an antenna with a conjugate-matched load proposed by Green [8]. The derivations presented above show that the previous scattering formulae can be derived from our proposed formula, which proves its validity.

2.6 Summary

This chapter presented the fundamental theoretical formula of antenna scattering, from which the previously proposed classic formulae could be derived. This formula clearly illustrated the scattering mechanism of antennas. Additionally, several antenna RCSR methods could be developed based on this formula. Consequently, this research on the fundamental antenna scattering formula offers theoretical innovation and reference value in the field of antenna RCSR.

References

1. Yingzheng Ruan, Radar Cross Section and Stealth Technology, Beijing National Defense industry Publishing House, 1998.
2. Y.T. Lo, S.W. Lee, Antenna Handbook: Theory, Applications, and design, Van Nostrand Reinhold Company, New York, 1988.
3. E.F. Knott, John Shaeffer, Michael Tuley, Radar Cross Section, 2nd edition, New York: Artech House, 1993.
4. Ying Liu, Prediction and Reduction of Antenna Radar Cross Section, Xidian University, Doctoral Dissertation, 2004.
5. Appel-Hansen, J., Accurate Determination of Gain and Radiation Patterns by Radar Cross Section Measurements, IEEE Trans. on AP, Vol. 27, pp. 640–646, September 1979.
6. Collin, R.E., The Receiving Antenna, Antenna Theory, Part1, R.E. Collin and F.J. Zucker(eds.), McGraw-Hill, New York, pp.123–133, 1969.
7. Hansen, R.C., Relationships Between Antennas as Scatters and as Radiators, Pro. IEEE, Vol. 77, pp. 659–662, May 1989.
8. Green, R.B., The General Theory of Antenna Scattering, Report 1223-17, Antenna Laboratory, Ohio State University, November 30, 1963.

Chapter 3
Antenna RCS Analysis

3.1 Antenna RCS Analysis Model

Solving for the antenna RCS using a purely theoretical approach is usually impractical. The structural mode and antenna mode scatterings can sometimes be approximated; however, the phase difference between them can be difficult to simulate. In this section, an analysis model that can be used to solve this problem is proposed. By using the scattering fields of the antenna under open-circuit and short-circuit conditions, the proposed model can obtain the structural mode and antenna mode scatterings, separately. Moreover, the scattering field of an antenna with arbitrary loads can be easily calculated.

A sketch of the analysis model is shown in Fig. 3.1 [1]. First, a reference plane (S_0) is chosen at the transmission line and set to be a short circuit. Thus, $Z_l = 0$, and $\Gamma_l = 1$. Based on the formula (Eq. 2.48) for antenna scattering, we can obtain:

$$\vec{E}^s(0) = \vec{E}^s(Z_c) - \frac{1}{1+\Gamma_a} b_0^m \vec{E}_1^t \tag{3.1}$$

The reference plane can then be shifted by $\lambda/4$ and denoted as S_1. We set S_1 to be a short circuit; thus, S_0 corresponds to an open circuit. Substituting $Z_l = \infty$ and $\Gamma_l = 1$ into Eq. (2.48) results in:

$$\vec{E}^s(\infty) = \vec{E}^s(Z_c) + \frac{1}{1-\Gamma_a} b_0^m \vec{E}_1^t \tag{3.2}$$

From Eqs. (3.1) and (3.2), we can obtain the structural mode scattering as follows:

$$\vec{E}^s(Z_c) = \frac{(1-\Gamma_a)\vec{E}^s(\infty) + (1+\Gamma_a)\vec{E}^s(0)}{2} \tag{3.3}$$

Y. Liu et al., *Antenna Radar Cross Section: Theory and Design*,
https://doi.org/10.1007/978-981-96-1226-0_3

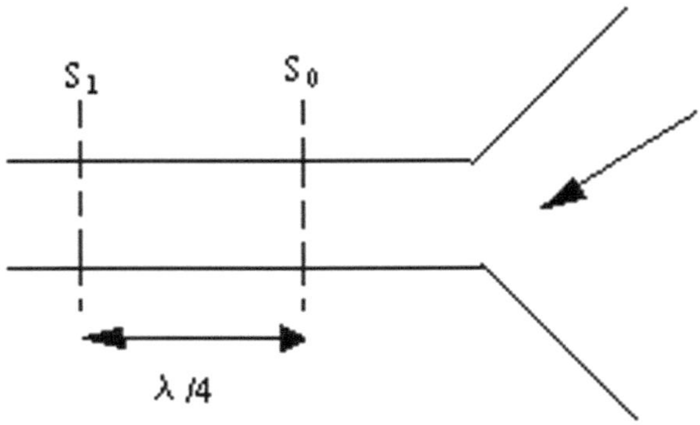

Fig. 3.1 Sketch of the analysis model

By substituting Eq. (3.3) into Eq. (3.1), we can obtain:

$$b_0^m \vec{E}_1^t = \frac{(1-\Gamma_a^2)}{2}[\vec{E}^s(\infty) - \vec{E}^s(0)] \tag{3.4}$$

Then, by substituting Eqs. (3.3) and (3.4) into Eq. (2.48), the scattering field of an antenna with arbitrary loads can be obtained, as follows:

$$\vec{E}^s(Z_l) = \frac{(1-\Gamma_a)\vec{E}^s(\infty) + (1+\Gamma_a)\vec{E}^s(0)}{2} \frac{\Gamma_l}{1-\Gamma_l\Gamma_a} \frac{1-\Gamma_a^2}{2}[\vec{E}^s(\infty) - \vec{E}^s(0)] \tag{3.5}$$

The first component of Eq. (3.5) is the structural mode scattering field, and the second component is the antenna mode scattering field, which is related to both Γ_l and Γ_a. When the transmission line and antenna are matched, Γ_a is small. Thus, the scattering field of an antenna with arbitrary loads can be calculated from the scattering fields under short-circuit and open-circuit conditions.

3.2 Microstrip Antenna RCS Analysis Results

In this section, we investigate the RCS of a microstrip antenna. When investigating the scattering of a microstrip antenna, most previous studies analyzed only the scattering of a patch. However, we believe that the scattering of a patch is not equivalent to the structural mode scattering and that the feeding structure should be considered. When the antenna is connected to a matched load, there is no reradiating field, and the total scattering field is the structural mode scattering. Moreover, when the load is not matched with the antenna, a reflected signal is reradiated by the antenna, resulting in antenna mode scattering. Using the proposed analysis model, we can calculate the structural mode and antenna mode scattering fields separately. Before this analysis, we first briefly review the research status of the microstrip antenna RCS.

3.2.1 Research Status of Microstrip Antenna RCS

Reference [2] analyzed the scattering of a microstrip antenna without a feeding structure and presented the simulated and measured results in the frequency domain. Reference [3] added the feeding structure and presented the calculated results of an antenna with a matched load. A precise feeding model was used in Aberle et al. [4], and the calculated and measured results of a circular microstrip antenna with a matched load and without a load were presented. Scattering results from a finite array of microstrip patches were presented in King and Bow [5]. There were also several other numerical analyses of microstrip antenna scattering, using methods such as the fast multiple method (FMM) [6] and discrete complex image method [7]. The RCSR of microstrip antennas has been studied based on the MoM [8, 9] and the finite element method (FEM) [10]. However, all of these studies analyzed only the scattering of a patch or antenna scattering with a matched load. The phase difference between the structural mode scattering and antenna mode scattering remains unknown. Consequently, in this study we investigate the scattering of the two modes separately using our proposed analysis model.

3.2.2 Microstrip Antenna RCS Analysis Procedure

The analysis model of microstrip antenna scattering is shown in Fig. 3.2. The model was developed using Ansoft HFSS (a 3D full-wave EM field solver for high-frequency and high-speed electronic component designs), which is based on the FEM and can precisely model practical targets. Direct utilization of the software cannot solve the problem of antenna scattering. However, using the analysis mdel proposed in Sect. 3.1, the two components of antenna scattering can be obtained separately.

The coaxial probe can be modeled as a cylinder as it is in practice, a more exact method than the approximate methods that use a linear structure. First, the feed is set to a short circuit and is chosen as reference plane 1. Second, the feeding structure is lengthened by $\lambda/4$. With the scattering fields in these two cases, the structural mode

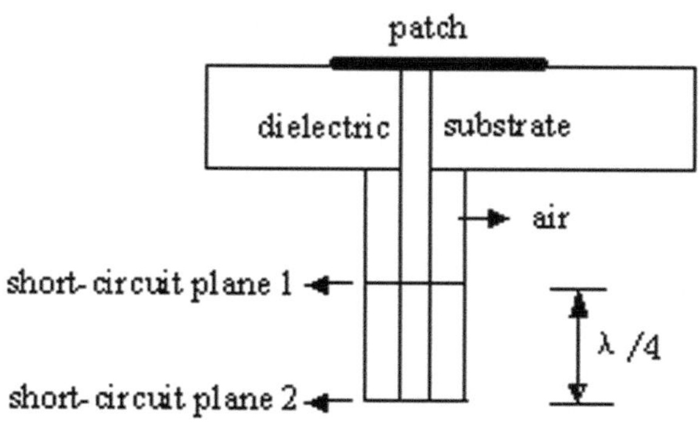

Fig. 3.2 Schematic diagram of a microstrip antenna

and the antenna mode scattering fields of the antenna with a short-circuit load can be obtained. For the case of an arbitrary load, the scattering field can be calculated using Eq. (3.5).

3.2.3 Calculated Results

The sample microstrip antenna used for the calculation is shown in Fig. 3.3, with the antenna parameters being as shown in the figure. The relative dielectric constant is $\varepsilon_r = 2.17$, and the resonant frequency is $f = 3.7$ GHz.

1. **Radiation characteristics**: Fig. 3.4 shows the radiation patterns of the antenna in the planes $\phi = 0°$ and $90°$. In the plane $\phi = 0°$ the ϕ component is the main component and nearly coincides with the total field. By contrast, in the plane $\phi = 90°$, the θ component is the main component.

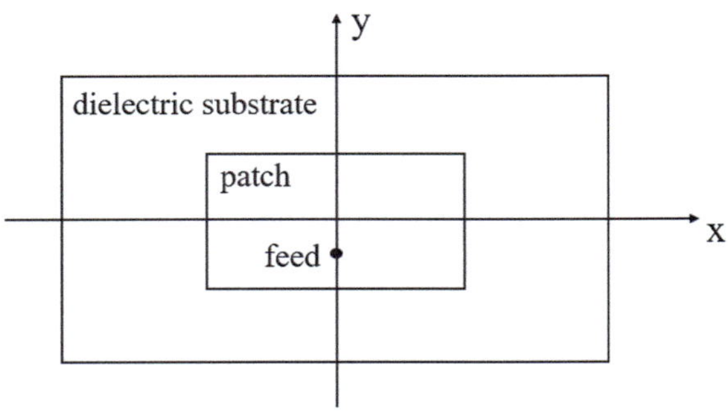

Fig. 3.3 Calculation example of a microstrip antenna

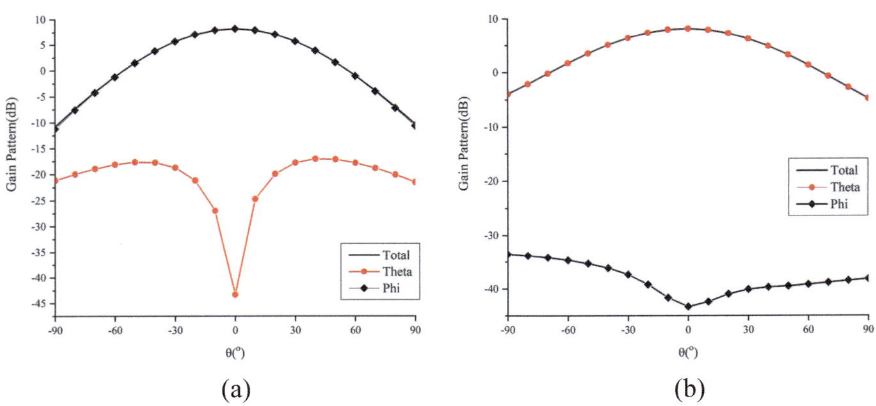

Fig. 3.4 The radiation pattern versus angle on two planes: **a** $\phi = 0°$ and $\phi = 90°$

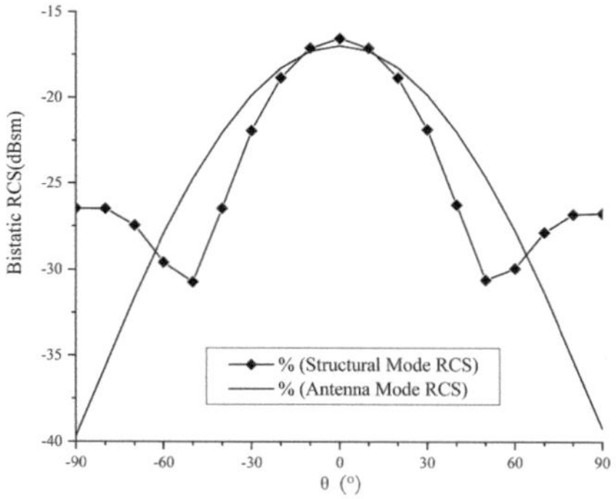

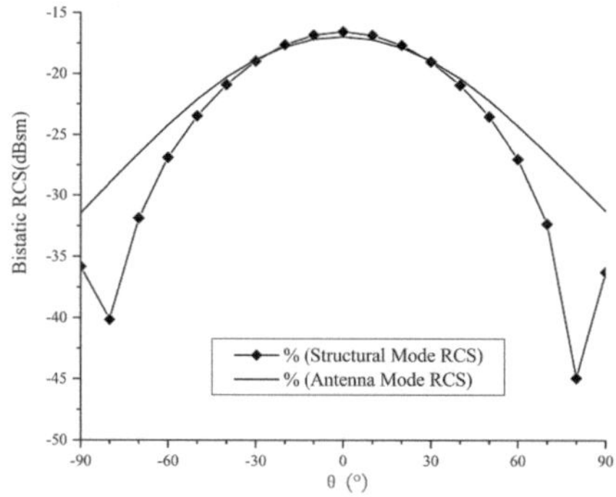

Fig. 3.5 Bistatic RCS under perpendicularly incident waves in two planes: **a** $\phi = 0°$, ϕ-polarized incident wave, **b** $\phi = 90°$, θ-polarized incident wave

2. **Scattering characteristics**: The scattering is a monostatic RCS if the incident direction and scattering direction are at the same point, and it is a bistatic RCS if not. Figure 3.5ba and b show the bistatic RCS when the plane wave is perpendicularly incident on antenna patches with ϕ and θ polarizations, respectively. Two incident waves with different polarizations are used because ϕ and θ polarizations are the primary polarizations of the two planes.

   Figure 3.5b shows that the structural mode RCS is not symmetrical, which is caused by the asymmetric feeding structure. The structural mode RCS given here differs from the RCS of the patch calculated in previous studies. The scattering of the patch ignores the existence of the feeding structure. However, the feeding structure exists in practical antennas and affects the scattering, which is shown through the asymmetry of the structural mode RCS in the plane $\phi = 90°$. Because the antenna is symmetrical in the plane $\phi = 0°$, the structural mode RCS is symmetrical.

   Figure 3.6 shows the monostatic antenna mode RCS and structural mode RCS derived from the analysis model. When the antenna is match-loaded, the incident wave received by the antenna is absorbed completely by the load, and there is no reradiation.

   The RCS in this case is the structural mode RCS. The results show that the structural mode RCS under the matched load and that from the model are nearly coincident. The special point at which the horizontal axis is zero also highlights the results. When the incident wave is perpendicular to the antenna and $\theta = 0°$, $\sigma_{\theta\theta}$ on the $\phi = 0°$ plane should be equal to $\sigma_{\phi\phi}$ on the $\phi = 90°$ plane. The results shown in Fig. 3.6 verify the validity of the computational results. However, at the point at which the horizontal axis is nearly 90°, there is a difference between the two structural modes; this difference is caused by the computational error when the plane wave illuminates the antenna at a grazing angle of incidence.

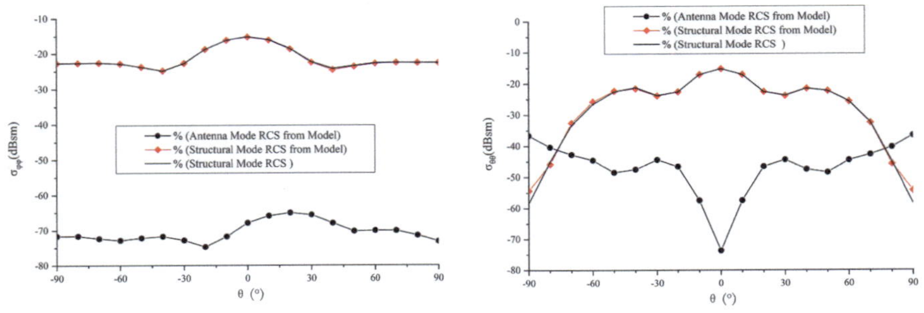

(a) (b)

Fig. 3.6 Monostatic antenna mode RCS and structural mode RCS calculated from the analysis model compared with the structural mode RCS under a matched load simulated using Ansoft HFSS: **a** $\phi = 0°$, θ-polarized incident wave, **b** $\phi = 90°$, ϕ-polarized incident wave

3. **Antenna mode RCS comparison**: We demonstrated the scattering and radiation performance of the antenna above. Next, we investigate the relationship between them. Because the antenna gain is known, the antenna mode RCS can be derived from its radiation characteristics [11]. The comparison between the antenna mode RCS calculated from the antenna gain and the analysis model is as follows. Assume an antenna with an effective area A_e is illuminated by an incident plane wave with a power density S_i. The power received by the antenna can be simply expressed as follows:

$$P_r = S_i A_e. \tag{3.6}$$

When the antenna is terminated with a short circuit, the signals transmitted to the load are completely reflected, the total reradiated power being equal to the received power. The power density of the reradiated energy of an antenna with a gain G at a range R can be expressed as follows:

$$S_r = P_r G / \left(4\pi R^2\right) = S_i A_e G / \left(4\pi R^2\right) \tag{3.7}$$

The definition of the RCS in terms of the power densities can be expressed as follows:

$$\sigma = \lim_{R \to \infty} (4\pi R^2)(S_r / S_i). \tag{3.8}$$

Substituting Eq. (3.7) into Eq. (3.8):

$$\sigma = A_e G \tag{3.9}$$

It is well known that the antenna gain and its effective area are related as follows:

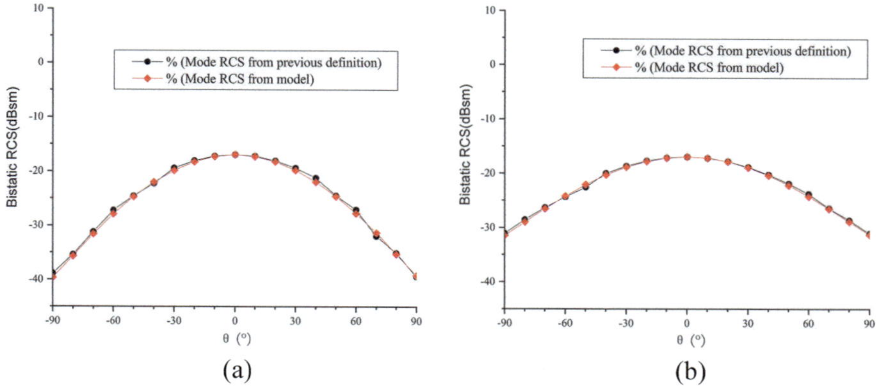

Fig. 3.7 Antenna mode RCS calculated from the model and that calculated from the radiation pattern: **a** $\phi = 0°$, ϕ-polarized incident wave, **b** $\phi = 90°$, θ-polarized incident wave

$$G = 4\pi A_e/\lambda^2 \tag{3.10}$$

Thus, the antenna mode RCS calculated from the antenna gain can be expressed as follows:

$$\sigma = G^2\lambda^2/(4\pi) \tag{3.11}$$

After this process, the antenna mode RCS, which is consistent with the previous definition, can be obtained and compared with the results of the respective component calculated from the analysis model, as shown in Fig. 3.7. The results show that there are minor differences between them. However, the model can realize the structural mode RCS and antenna mode RCS separately; thus, the scattering field with an arbitrary load can be easily obtained.

3.3 Horn Antenna RCS Analysis Results

Many kinds of antennas can be analyzed using the antenna RCS analysis model. Here, we analyze the scattering of a pyramidal horn as an example. Pyramidal horns are commonly used as the feed for reflector antennas. Analyzing them can assist in the analysis of the scattering performance of an aperture array. When the incident wave is paraxial, there are two main structural mode scattering components [12, 13]. One component comes from the diffraction along the aperture edge of the horn, the magnitude of the field reflected from the edge parallel to the polarization direction of the incident wave being approximately L^2/π, where L denotes the length of the edge. The other component is caused by the higher-order modes of the horn. Because these cannot be transmitted to the antenna feed, the higher-order modes radiate at the aperture of the horn. For the case in which the incident wave is misaligned from the axis of the pyramidal horn, the waves reflected by the outer surface become the main contributor to the total RCS. The analysis of a pyramidal horn antenna is given below.

Figure 3.8 shows the structure of a pyramidal horn, whose radiation and scattering performance will be calculated. In the previous section, the reference plane was chosen to be on the feeding coaxial when calculating the RCS of the microstrip antenna. Here, we set the port of the waveguide to be the reference plane. Moreover, the waveguide's dominant mode occurs at 10 GHz.

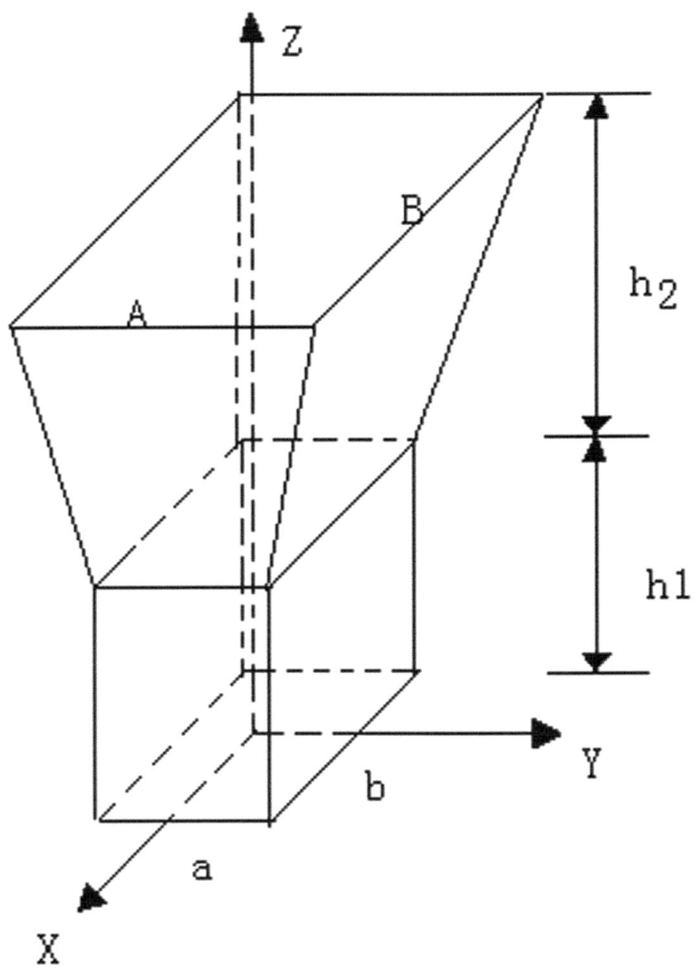

Fig. 3.8 Structure of the pyramidal horn

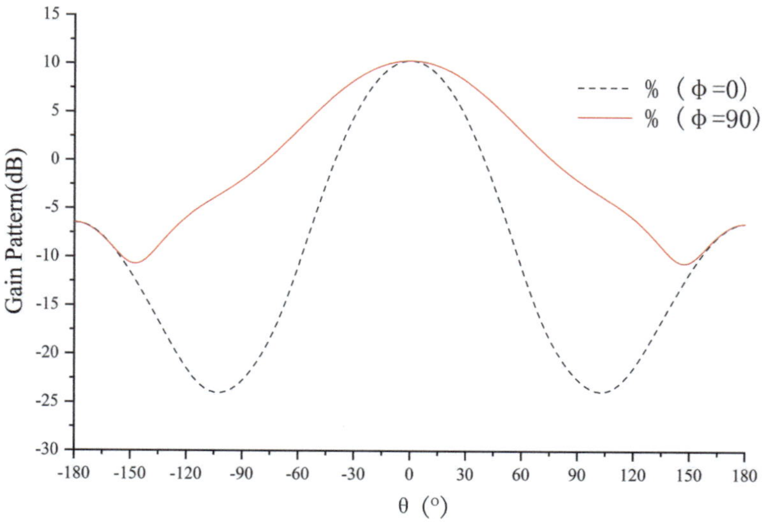

Fig. 3.9 Calculated radiation pattern of the horn antenna

1. **Radiation pattern**: The copolarization of the horn antenna is ϕ polarization on the $\phi = 0°$ plane and θ polarization at $\phi = 90°$. The radiation pattern of the horn antenna is shown in Fig. 3.9, and the peak gain G = 10.37 dB.
2. **Scattering characteristics**: In the previous section, the analysis method and calculated results of the antenna mode RCS and structural mode RCS were discussed in detail when analyzing a microstrip antenna. Here, we show the calculated RCS at $\phi = 0°$ plane as a practical example of the proposed RCS analysis model applied to a horn antenna. Figure 3.10 shows a comparison of the monostatic RCS of the horn antenna illuminated by ϕ-polarized incident waves. The S_{11} of the antenna was considered in the calculation. The results show that the structural mode RCS calculated using the proposed analysis model is consistent with the RCS when the antenna is terminated with a matched load.

   The main lobe of the scattered pattern is caused by the aperture of the horn antenna. When the incident wave is misaligned from the axis of the horn antenna, the structural mode RCS has an upward trend. This occurs because the outer surface of the horn antenna contributes considerably to the total RCS. An understanding of the scattering mechanism of the horn antenna can help to reduce its RCS. For example, to reduce the RCS contribution of the outer surface, a microwave absorber can be used to coat the outer surface. Additionally, the RCS analysis model can be applied to other antennas with irregular shapes.
3. **Rapid calculation method of antenna mode RCS based on reciprocal theory**: When the RCS analysis model introduced in Sect. 3.1 is used, two reference planes should be selected and set to short and open circuits, respectively, so that the antenna mode RCS and structural mode RCS can be separated. However, this procedure is inconvenient when dealing with complex antennas or array

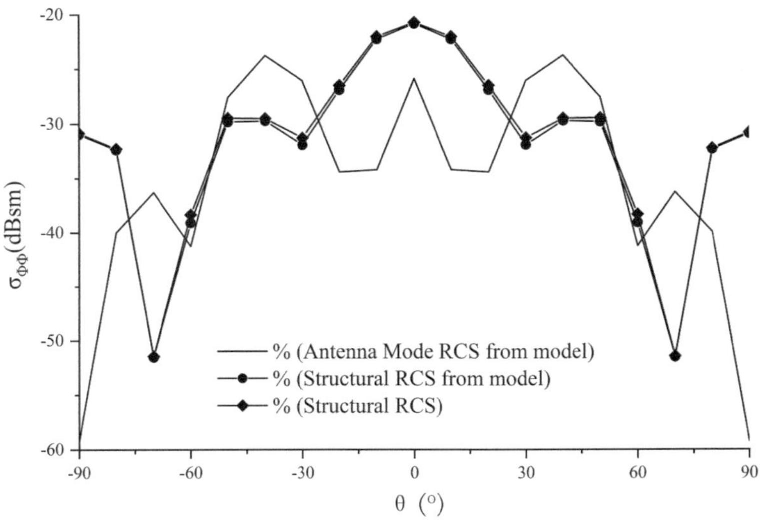

Fig. 3.10 Monostatic RCS of the horn antenna

antennas. Consequently, in this section, we focus on another calculation method for the antenna mode RCS based on the reciprocal theory. By using the radiation characteristics and reflection coefficient measured at the receiver load, this method can calculate the antenna mode RCS and be used to analyze the antenna scattering fields.

When an antenna excited by a unit amplitude source with $a_1 = 1\sqrt{watt}^{1/2}$ is used for transmitting, its radiated field can be expressed as follows [14]:

$$F_1 = \begin{Bmatrix} \vec{E}_1 \\ \vec{H}_1 \end{Bmatrix} = \begin{Bmatrix} \sqrt{Z_0}\vec{A}(\vec{k}) \\ \sqrt{Y_0}\hat{k} \times \vec{A}(\vec{k}) \end{Bmatrix} \frac{e^{-jkr}}{r} = \begin{Bmatrix} \sqrt{Z_0}\left|\vec{A}(\vec{k})\right| \vec{u} \\ \sqrt{Y_0}\left|\vec{A}(\vec{k})\right| \hat{k} \times \vec{u} \end{Bmatrix} \frac{e^{-jkr}}{r} \tag{3.12}$$

where

$$\vec{A}(\vec{k}) = \frac{k}{4\pi j} \int_v \left(\sqrt{Z_0}(\overline{\overline{I}} - \hat{r}\hat{r})\, \vec{J}(\vec{r}') + \sqrt{Y_0}(\vec{J}_m(\vec{r}') \times \hat{r}) \right) e^{jkr} dv' \tag{3.13}$$

(Z_0 *denotes the wave impedance in free space*)

Assume that the antenna is illuminated by an incident field, then:

$$\vec{F}_4 = \begin{Bmatrix} \vec{E}_4 \\ \vec{H}_4 \end{Bmatrix} = C \begin{Bmatrix} \sqrt{Z_0}\vec{u}' \\ -\sqrt{Y_0}\hat{k} \times \vec{u}' \end{Bmatrix} e^{j\vec{k}\cdot\vec{r}} \tag{3.14}$$

(It is worth noting that F_4 denotes not only the incident wave but also the scattering field after the incident wave illuminates the antenna.)

As shown in Fig. 3.11, S' is composed of S_a and the virtual boundary that is close to the antenna conductor wall, and S is a sphere surrounding the antenna. The area surrounded by S' and S is passive. According to the reciprocity theorem:

$$\oint_{s} \left(\vec{E}_4 \times \vec{H}_1 - \vec{E}_1 \times \vec{H}_4 \right).(\hat{n})ds = \oint_{s'} \left(\vec{E}_4 \times \vec{H}_1 - \vec{E}_1 \times \vec{H}_4 \right) \cdot (\hat{n})ds. \tag{3.15}$$

On the conductor surface of the antenna, $\hat{n} \times \vec{E}_1 = 0, \hat{n} \times \vec{E}_4 = 0$. Thus, we can obtain the following:

$$\oint_{s'} \left(\vec{E}_4 \times \vec{H}_1 - \vec{E}_1 \times \vec{H}_4 \right) \cdot (\hat{n})ds = \oint_{s_a} \left(\vec{E}_4 \times \vec{H}_1 + \vec{H}_4 \times \vec{E}_1 \right) \cdot (\hat{n})ds \tag{3.16}$$

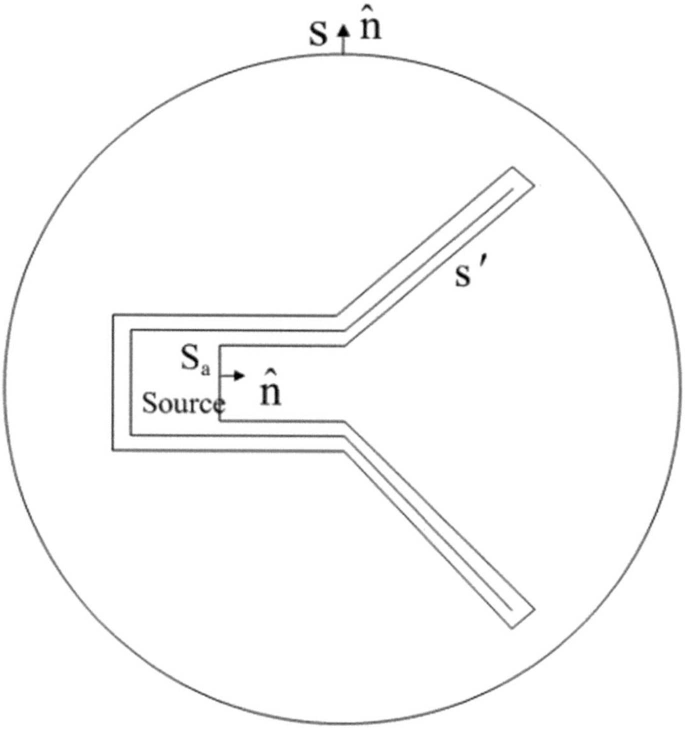

Fig. 3.11 Schematic diagram of an antenna

Because only the normalized residence mode can propagate on S_a, we record the normalized electric field and magnetic field of the principal mode as $\vec{e}$, $\vec{h}$; then:

$$\begin{cases} \underset{1}{\vec{E}} = \vec{e}(a_1 + b_1) \\ \underset{1}{\vec{H}} = \vec{h}(a_1 - b_1) \end{cases}, \begin{matrix} \underset{4}{\vec{E}} = \vec{e}(a_4 + b_4) \\ \underset{4}{\vec{H}} = \vec{h}(a_4 - b_4) \end{matrix} \begin{cases} \underset{1}{\vec{E}} = \vec{e}(a_1 + b_1) \\ \underset{1}{\vec{H}} = \vec{h}(a_1 - b_1) \end{cases}, \begin{matrix} \underset{4}{\vec{E}} = \vec{e}(a_4 + b_4) \\ \underset{4}{\vec{H}} = \vec{h}(a_4 - b_4) \end{matrix}. \tag{3.17}$$

where a_1, b_1 denote the incident and reflected amplitudes on S_a, respectively, when the antenna is radiating, and a_4, b_4 denote the incident and reflected amplitudes on S_a, respectively, when the antenna is receiving. If the antenna is well matched, $a_4 = 0$. Therefore:

$$\vec{E}_4 \times \vec{H}_1 = \vec{e} \times \vec{h}(a_1 b_4 - b_1 b_4), \vec{H}_4 \times \vec{E}_1 = -\vec{e} \times \vec{h}(-a_1 b_4 - b_1 b_4) \tag{3.18}$$

By substituting this formula into $\oint_{S_a} \left(\vec{e} \times \vec{h}\right) \cdot (\hat{n}) ds = 1$, we obtain:

$$\oint_{(s_a)} \left(\underset{4}{\vec{E}} \times \underset{1}{\vec{H}} + \underset{4}{\vec{H}} \times \underset{1}{\vec{E}}\right) \cdot (\hat{n}) ds = 2 a_1 b_4 \tag{3.19}$$

Then, substituting Eq. (3.19) into Eq. (3.15) gives $\oint_s \left(\underset{4}{\vec{E}} \times \underset{1}{\vec{H}} - \underset{1}{\vec{E}} \times \underset{4}{\vec{H}}\right) \cdot (\hat{n}) ds = 2 a_1 b_4$, such that:

$$\text{i.e, } b_4 = \frac{1}{2a_1} \oint_s \left(\underset{4}{\vec{E}} \times \underset{1}{\vec{H}} - \underset{1}{\vec{E}} \times \underset{4}{\vec{H}}\right) \cdot (\hat{n}) ds \tag{3.20}$$

Combining Eqs. (3.12) and (3.13) and performing the proper integral transformation:

$$b_4 = \frac{-c}{2a_1} \frac{4\pi j}{k} \left[\vec{A}\left(\vec{k}\right) \cdot \vec{u'}\right] = -j\lambda \frac{c}{a_1} \left[\vec{A}\left(\vec{k}\right) \cdot \vec{u'}\right] = -j\lambda \frac{c}{a_1} \left[\left|\vec{A}\left(\vec{k}\right)\right| \vec{u} \cdot \vec{u'}\right] \tag{3.21}$$

It should be noted that b_4 denotes the matched receiving amplitude, which corresponds to b_0^m in Eq. (2.30).

When the antenna is excited by a unit amplitude ($a_1 = 1$), the total radiation field of the antenna at a radius r in the $\hat{k}$ direction can be calculated as follows:

$$\vec{E_0^{1t}}\left(\hat{k}, r\right) = E_0\left(\hat{k}, r\right) \vec{u}\left(\hat{k}, r\right) \tag{3.22}$$

Thus,

$$\left|\vec{A}\left(\vec{k}\right)\right| = E_0\left(\frac{e^{-jkr}}{r}\right)^{-1} / \sqrt{Z_0} \tag{3.23}$$

We can obtain $c = 1/\sqrt{Z_0}$ from Eq. (3.14). For a $\vec{u}'$-polarized incident wave propagating in the $-\hat{k}$ direction with a unit amplitude, we can substitute these formulae into Eq. (3.21), as follows:

$$b_o^m = -j\lambda \frac{E_0\left(\vec{k}, r\right)}{Z_0}\left(\frac{e^{-jkr}}{r}\right)^{-1}\left(\vec{u} \,.\, \vec{E}'\right) \tag{3.24}$$

(The incident waves $\left(\hat{k}, \vec{u}\right)$ and $\left(-\hat{k}, \vec{u}^*\right)$ denote plane waves with the same polarization and opposite propagation directions in free space. When the incident wave is $\left(-\hat{k}, \vec{u}^*\right)$ then $\vec{u} \cdot \vec{u}' = \vec{u} \,.\, \vec{u}^* = 1$).

If the receiver loads are mismatched, we can obtain the antenna mode scattering fields by substituting Eq. (3.23) into the scattering expression of the antenna mode terms, as follows:

$$\vec{E}_s\left(\hat{k}^s, r\right) = \left[-j\lambda \frac{\Gamma_l}{1 - \Gamma_l\Gamma_a} \frac{E_0(\vec{k}, r)}{Z_0}\left(\frac{e^{-jkr}}{r}\right)^{-1}\left(\vec{u} \cdot \vec{u}'\right)\right]\vec{E}_0^{1t}\left(\hat{k}^s, r\right) \tag{3.25}$$

Equation (3.24) shows how to calculate the antenna mode scattering under plane wave incidence using the reciprocity theorem on the premise that the radiation field is known.

We can verify the separation method by using the microstrip antenna shown in Fig. 3.12 as an example [15]. The antenna operates at 3 GHz with linear polarization. The permittivity of the substrate is 2.64, and the thickness of the substrate is 2 mm. The widths along the x-axis and y-axis are 38.4 and 28.66 mm, respectively. The antenna is fed by a coaxial line, and the feeding point is on the y-axis 5.98 mm from the coordinate origin. The impedance of the coaxial line is 50 Ω. The radii of the inner conductor and outer conductor are 0.5 and 1.51 mm, respectively. We can assume that the middle medium is air and use the cross section of the coaxial that is 10 mm from the antenna ground as the reference plane S_a.

The bistatic RCS of the antenna mode calculated using the reciprocity theorem under the conditions of a plane wave incident at $\theta = 0°$, $\phi = 0°$ with $\hat{\phi}$ polarization and a unit amplitude is shown in Fig. 3.13a. Here, $\Gamma_l = -1$ and $\Gamma_a = 0$. The bistatic RCS of the antenna mode in the $\phi = 0°$ plane separated by the method discussed in Sect. 3.2is also shown in Fig. 3.13. The resultant curves are in good agreement. The bistatic RCS of the antenna mode under a plane wave incident at $\theta = 45°$, $\phi = 45°$ with $\hat{\phi}$ polarization and a unit amplitude is shown in Fig. 3.13b. Here, $\Gamma_l = 0.5$

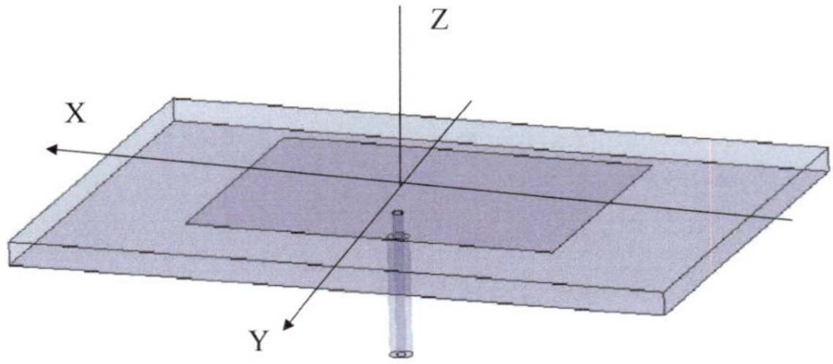

Fig. 3.12 HFSS model of the microstrip antenna

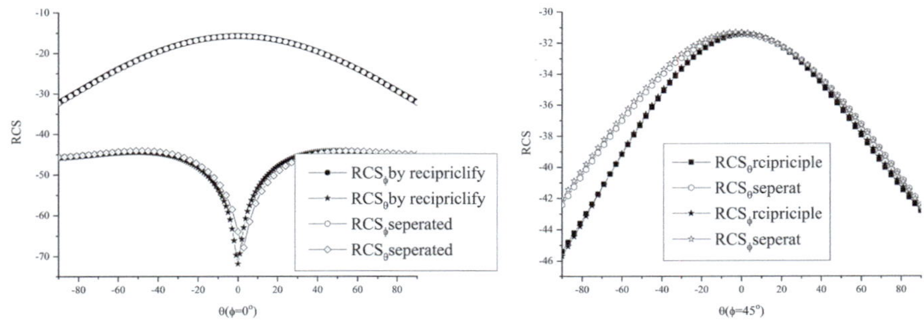

(a) (b)

Fig. 3.13 Scattering patterns of antenna mode scattering

and $\Gamma_a = 0$. These results also match those separated by the method discussed in Sect. 3.2 in the $\phi = 45°$ plane. These results demonstrate the feasibility of the reciprocity theorem in RCS prediction of the antenna mode.

3.4 Summary

Based on the fundamental theoretical formula of antenna scattering, we proposed an analysis model to calculate the antenna mode RCS and structural mode RCS separately, solving the problem of obtaining the phase difference between the two components. Based on the results calculated using the analysis model, the structural mode RCS and antenna mode RCS with arbitrary loads could be conveniently derived. The results calculated using the proposed model and the results presented in previous studies were compared, and their agreement verified the validity and feasibility of the proposed model. The RCSs of a microstrip antenna and a pyramidal horn antenna were calculated using this analysis model. Finally, a rapid calculation method of the antenna mode RCS and a practical example were presented.

References

1. Ying Liu, De-Min Fu, Shu-Xi Gong, A Novel Model for Analyzing the RCS of Microstrip Antenna, Journal of Electromagnetic Waves and Application, Vol. 17, No. 9, 1301–1310, 2003.
2. D.M. Pozar, Radiation and Scattering from a Microstrip Patch on a Uniaxial Substrate, IEEE Trans. on AP, Vol. 35, No. 6, pp. 613–621, 1987.
3. Edward H. Newman, David Forrai, Scattering from a Microstrip Patch, IEEE Trans. on AP, Vol. 35, No. 3, March 1987.
4. James T. Aberle, David M. Poazr, Craig R. Birtcher, Evaluation of Input Impedance and Radar Cross Section of Probe-Fed Microstrip Patch Elements Using an Accurate Feed Model, IEEE Trans. on AP, Vol. 39, No. 12, pp. 1691–1696, December 1991.
5. Adrian S. King, Wallace J. Bow, Scattering from a Finite Array of Microstrip Patches, IEEE Trans. on AP, Vol. 40, No. 7, pp. 770–774, July 1992.
6. Vikram Jandhyala, Eric Michielssen, Balasubramaniam Shanker, Weng C. Chew, A Fast Algorithm for the Analysis of Radiation and Scattering from Microstrip Arrays on Finite Substrates, Microwave and Optical Technology Letters, Vol. 23, No. 5, pp. 306–310, December 5 1999.
7. Feng Ling, Jian-Ming Jin, Scattering and Radiation Analysis of Microstrip Antennas Using Discrete Complex Image Method and Reciprocity Theorem, Microwave and Optical Technology Letters, Vol. 16, No. 4, pp. 212–216, November 1997.
8. R.V. McGahan, B. Rama Rao, Electromagnetic Backscatter from Microstrip arrays: Theory and Measurement, 1985 Phased Arrays Symp. Proc., RADC-TR-85-171.
9. David M. Pozar, RCS Reduction for a Microstrip Antenna Using a Normally Biased Ferrite Substrate, IEEE Microwave and Guided Wave Letters, Vol. 2, No. 5, pp. 196–198, 1992.
10. Jin-Ming Jin, John L. Volakis, A Hybrid Finite Element Method for Scattering and Radiation by Microstrip Patch Antennas and Arrays Residing in a Cavity, IEEE Trans. on AP, Vol. 39, No. 11, pp. 1598–1604, November 1991.
11. Appel-Hansen, J., Accurate Determination of Gain and Radiation Patterns by Radar Cross Section Measurements, IEEE Trans. on AP, Vol. 27, pp. 640–646, September 1979.
12. Jefferson F. Lindsey, Radar Cross-Section Effects Relating to a Horn Antenna, IEEE Trans. on AP, Vol.37, No.2, pp.257-260, February 1989.
13. Li-Chung T. Chang, Walter D. Burnside, An Ultrawide-Bandwidth Tapered Resistive TEM Horn Antenna, IEEE Trans. on AP, Vol. 48, No. 12, pp. 1848–1857, December 2000.
14. Y.T. Lo, S.W. Lee, Antenna Handbook: Theory, Applications, and design, Van Nostrand Reinhold Company, New York, 1988.
15. Pengfei Zhang, Prediction and Control of Radar Cross Section in Stealth Technology, Xidian University, Doctoral Dissertation, 2008.

Chapter 4
Array Antenna Scattering Theory

4.1 Array Antenna Analysis Model

The radiation and scattering of an array antenna are the result of the contributions of all of its elements. Two factors increase the complexity of array antenna analysis—that is, the complex bidirectional reflection of discontinuous points on the antenna system caused by the feeding network and the mutual coupling between the array elements. The array antenna can be studied using a network analysis model, which is presented below.

As shown in Fig. 4.1, the array antenna system can be assumed to comprise a transmitter/receiver, feeding network, and antenna array. The single mode transmission area between the transmitter/receiver and the feeding network can be taken to be the reference plane 0 (the characteristic impedance of the transmission line is $\boldsymbol{Z}_0$). The port on the reference plane can be viewed as port 0. Moreover, the port reference plane of the element is the cross section taken from the single mode transmission area. This area connects the feeding network and each element of the antenna array, and can be viewed as reference plane 1, as shown in Fig. 4.1. The corresponding ports of the r array elements on the reference plane are denoted as ports $1 \sim r$.

The feeding network is an $r + 1$-port network. The complex amplitude of the incident mode from port 0 to the feeding network is $\boldsymbol{a}_0$, and the exiting mode's amplitude is $\boldsymbol{b}_0$. The other r ports are related to the antenna elements. The complex amplitude of the incident mode whose direction is from the ith antenna element's port among the r ports to the feeding network is $a_{i'}$, and the exit mode's complex amplitude is $\boldsymbol{b}_i'$. Consequently, the S parameter equation of the feeding network can be expressed as follows:

Y. Liu et al., *Antenna Radar Cross Section: Theory and Design*,
https://doi.org/10.1007/978-981-96-1226-0_4

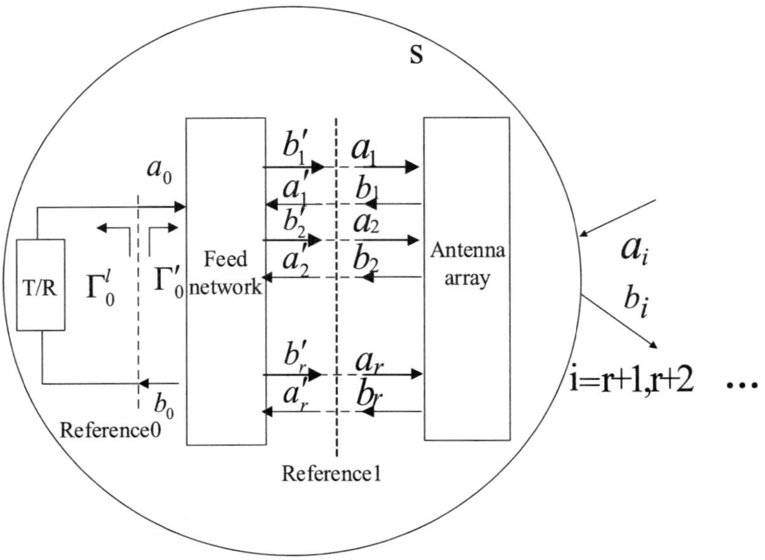

Fig. 4.1 Array antenna system

$$\begin{bmatrix} b_0 \\ \hline b_1' \\ b_2' \\ \vdots \\ b_r' \end{bmatrix} = \left[\begin{array}{c|cccc} S_{00}' & S_{01}' & S_{02}' & \cdots & S_{0r}' \\ \hline S_{10}' & S_{11}' & S_{12}' & \cdots & S_{2r}' \\ S_{20}' & S_{21}' & S_{22}' & \cdots & S_{3r}' \\ \vdots & \vdots & \vdots & \ddots & \vdots \\ S_{r0}' & S_{r2}' & S_{r3}' & \cdots & S_{rr}' \end{array}\right] \begin{bmatrix} a_0 \\ \hline a_1' \\ a_2' \\ \vdots \\ a_r' \end{bmatrix} \tag{4.1}$$

Let

$$[b']_r = [b_1' \; b_2' \; .... \; b_r']^T; \; [a']_r = [a_1' \; a_2' \; .... \; a_r']^T$$

$$[S']_{0r} = [S_{01}' \; S_{02}' \; .... \; S_{0r}']; \; [S']_{r0} = [S_{10}' \; S_{20}' \; .... \; S_{r0}']^T$$

$$[S']_{rr} = \begin{bmatrix} S_{11}' \; S_{12}' \; ... \; S_{1r}' \\ S_{21}' \; S_{22}' \; ... \; S_{2r}' \\ ... \; ... \; ... \; ... \\ S_{r1}' \; S_{r2}' \; ... \; S_{rr}' \end{bmatrix} \tag{4.2}$$

Thus, Eq. (4.1) can be expressed as follows:

$$\begin{bmatrix} b_0 \\ [b']_r \end{bmatrix} = \begin{bmatrix} S_{00}' & [s']_{0r} \\ [s']_{r0} & [s']_{rr} \end{bmatrix} \begin{bmatrix} a_0 \\ [a']_r \end{bmatrix} \tag{4.3}$$

The antenna array can also be analyzed using spherical wave function expansion. Consider a spherical surface surrounding an antenna array in the radiation space of the antenna. The radiation and total scattering field of the array antenna in the outer space of S can be expanded using the spherical wave function, as follows:

$$\vec{E} = \sum_{i=r+1}^{\infty} \left(a_i \vec{e}_i^{\,in} + b_i \vec{e}_i^{\,out} \right) \tag{4.4}$$

where (a_i, b_i) $(i = r + 1, r + 2, \ldots)$ denotes the complex amplitudes of the ith incident spherical wave e_i^{in} and the exiting spherical wave e_i^{out} in the spherical wave function expansion. The characteristic of the antenna array can be completely expressed by the scattering matrix, as follows:

$$\begin{bmatrix} b_1 \\ b_2 \\ \vdots \\ b_r \\ \hline b_{r+1} \\ \vdots \end{bmatrix} = \left[\begin{array}{ccccc|cc} s_{11} & s_{12} & \cdots & s_{1,r} & s_{1,r+1} & \cdots \\ s_{21} & s_{22} & \cdots & s_{2,r} & s_{2,r+1} & \cdots \\ \vdots & \vdots & \ddots & \vdots & \vdots & \cdots \\ s_{r,1} & s_{r,2} & \cdots & s_{r,r} & s_{r,r+1} & \cdots \\ \hline s_{r+1,1} & s_{r+1,2} & \cdots & s_{r+1,r} & s_{r+1,r+1} & \cdots \\ \vdots & \vdots & \vdots & \vdots & \vdots & \ddots \end{array} \right] \begin{bmatrix} a_1 \\ a_2 \\ \vdots \\ a_r \\ \hline a_{r+1} \\ \vdots \end{bmatrix} \tag{4.5}$$

We can use the block matrix to express $\begin{bmatrix} [b]_r \\ [b]_s \end{bmatrix} = \begin{bmatrix} [s]_{rr} & [s]_{rs} \\ [s]_{sr} & [s]_{ss} \end{bmatrix} \begin{bmatrix} [a]_r \\ [a]_s \end{bmatrix}$, as follows:

$$[b]_r = [s]_{rr}[a]_r + [s]_{rs}[a]_s \text{ reflection/acceptance amplitude vector} \tag{4.6}$$

$$[b]_s = [s]_{sr}[a]_r + [s]_{ss}[a]_s \text{ radiation/scattering amplitude vector} \tag{4.7}$$

The feeding network is connected to r ports of the array antenna; then

$$\left[b'\right]_r = [a]_r, \left[a'\right]_r = [b]_r \tag{4.8}$$

4.2 Analysis of Mutual Coupling of the Array Antenna Elements

Internal circuit coupling is generated because the array antenna's elements are connected by the feeding system. Additionally, external field coupling occurs because of the transmitting and receiving between the elements. $\left[S'\right]_{rr}$ is analyzed in Eq. (4.2); when $i \neq j$, $S'_{j,i}$ characterizes the internal circuit coupling generated by the feeding system connecting the ports for element i and element j. When $i = j$, $S'_{i,i} = \Gamma'_i$,

$\Gamma_i' = \frac{Z_i'-Z_c}{Z_i'+Z_c}$ expresses the reflection coefficient of port i, while the other ports of the feeding network are matched. $[S]_{rr}$ is analyzed in Eq. (4.6); when $i \neq j$, $S_{j,i}$ expresses the matched receiving amplitude of element j when element i is excited by a unit amplitude, and the other elements are matched. This parameter reflects the intensity of the external field coupling between the ports of elements i and j. When $i = j$, $S_{i,i} = \Gamma_i^a$, $\Gamma_i^a = \frac{Z_i^a-Z_c}{Z_i^a+Z_c}$ expresses the antenna element reflection coefficient of port i, and all of the other ports are matched except for port i in the antenna array ports $1 \sim r$.

Not all field couplings among the elements pass through the ports. The scattering field of element j is excited by the radiating field of element i. A portion of the scattering field is an expression of element j's radiation field because of the reflected signal caused by its mismatched receiver load. The scattering generated by element j is called antenna mode scattering. The field coupling passes through the ports. The other portion of the scattering field is independent of the matching state of the receiver load of element j, and is introduced when element j is connected to the matched receiver load. The scattering caused by element j is called structural mode scattering. The field coupling does not pass through the feeding ports of the elements. Both scattering effects should be considered in the array antenna analysis.

4.3 Array Antenna Scattering Analysis

4.3.1 Scattering Theory Analysis of an Array Antenna

To examine the problem of receiving and scattering, consider the receiver resistance Z_l. $\Gamma_0^l = \frac{Z_l-Z_0}{Z_l+Z_0}$ denotes the reflection coefficient of the receiver load for port 0. Substituting $a_0 = \Gamma_0^l b_0$ into Eq. (4.3)

$$[b']_r = [s]_{rr}^l [a']_r, \; [s]_{rr}^l = [S']_{rr} + \frac{\Gamma_0^l}{1 - S'_{00}\Gamma_0^l}[S']_{r0}[S']_{0r} \tag{4.9}$$

Combining this with Eq. (4.8) results in:

$$[a]_r = [s]_{rr}^l [b]_r \tag{4.10}$$

Substituting Eq. (4.10) into Eqs. (4.6) and (4.7) results in:

$$[b]_r = \left\{[I]_{rr} - [s]_{rr}[s]_{rr}^l\right\}^{-1}[s]_{rs}[a]_s \tag{4.11}$$

$$[b]_s = [s]_{sr}[s]_{rr}^l [b]_r + [s]_{ss}[a]_s \tag{4.12}$$

If all ports 1 ~ r of the array antenna are matched, $[a]r = [0]$, and Eq. (4.6) can be written as follows:

$$[b]_r|_{\Gamma_i=0,(i=1,2\ldots r)} = [s]_{rs}[a]_s = [b]_r^m \tag{4.13}$$

Equation (4.13) is called the matched receiver amplitude vector of the array. When all of the ports are matched, b_k^m indicates the receiving amplitude of port k when the antenna is excited by an incident wave. If the antenna is illuminated by a plane wave whose incident direction is $\vec{k}_i$ and whose polarization is $\vec{u}^*$, the antenna reciprocity can be expressed as follows:

$$b_k^m = -j\lambda C[\vec{A}_k(\vec{k}_i)\cdot\vec{u}^*] \tag{4.14}$$

where $A_k(k_i)$ denotes the magnitude of the vector corresponding to the radiation fields E_k^{1t} of the antenna element.

When $[a]_r=[0]$, Eq. (4.7) can be expressed as follows:

$$[b]_s|_{\Gamma_i=0,(i=1,2\ldots r)} = [s]_{ss}[a]_s = [b]_s^m \tag{4.15}$$

which is called the matched scattering amplitude vector.

Substituting Eqs. (4.11), (4.13) and (4.15) into Eq. (4.12) results in:

$$[b]_s = [b]_s^m + [s]_{sr}[s]_{rr}^l\left\{[I]_{rr} - [s]_{rr}[s]_{rr}^l\right\}^{-1}[b]_r^m \tag{4.16}$$

Let $[C] = \left\{[I]_{\mathrm{rr}} - [S]_{\mathrm{rr}}[S]_{\mathrm{rr}}^{\mathrm{l}}\right\}^{-1}$, and the scattering amplitude vector Eq. (4.16) can be expressed as follows:

$$b_i = b_i^m + \sum_{k=1}^{r} s_{ik}\sum_{p=1}^{r} s_{kp}^l\sum_{q=1}^{r} C_{pq}b_q^m,\ (i = r+1, r+2\ldots) \tag{4.17}$$

where S_{ik} denotes the element of $[S]_{\mathrm{sr}}$, and $S_{\mathrm{kp}}^{\mathrm{l}}$ denotes the element of $[S]_{\mathrm{rr}}^{\mathrm{l}}$. The ith peripheral spherical wave function $e_{\mathrm{i}}^{\mathrm{out}}$ can be multiplied by the two sides of this equation, and the subscripts i ranging from $r+1$ to ∞ can be summed as follows:

$$\sum_{i=r+1}^{\infty} b_i\vec{e}_i^{\,out} = \sum_{i=r+1}^{\infty} b_i^m\vec{e}_i^{\,out} + \sum_{k=1}^{r}\sum_{p=1}^{r}\sum_{q=1}^{r}\left(\sum_{i=r+1}^{\infty}\vec{e}_i^{\,out}s_{ik}\right)s_{kp}^l C_{pq}b_q^m \tag{4.18}$$

We can examine Eq. (4.18) as follows:

(1) Let $E^{\mathrm{s}}=\sum_{\mathrm{i=r+1}}^{\infty}\mathrm{b_i e_i^{out}}$, indicating that the spatial scattering field consists of outer scattering wave superposition.

(2) Let $E_0^{\mathrm{s}}=\sum_{\mathrm{i=r+1}}^{\infty}\mathrm{b_i e_i^{out}}$, indicating that all of the ports of the antenna from 1 to r are matched. Thus, $[\mathrm{a}] = [0]$, and the outer scattered wave is excited by the

matched scattering amplitude vector and is composed of the scattering fields in space—that known as the structural mode scattering of the array antenna.

(3) Let

$$\vec{E}_a^s = \sum_{k=1}^{r}\sum_{p=1}^{r}\sum_{q=1}^{r}\left(\sum_{i=r+1}^{\infty} \vec{e}_i^{\,out} s_{ik}\right) s_{kp}^l C_{pq} b_q^m \tag{4.19}$$

and

$$\sum_{i=r+1}^{\infty} \vec{e}_i^{\,out} s_{ik} = \vec{E}_k^{1t},\ \vec{E}_k^{1t} = \vec{E}_k^{0t} \exp(j\vec{k}\cdot\vec{\rho}_k)$$

Moreover, let

$$V_k = \sum_{p=1}^{r}\sum_{q=1}^{r} s_{kp}^l C_{pq} b_q^m$$

Equation (4.19) can then be expressed as follows:

$$\vec{E}_a^s = \sum_{k=1}^{r} V_k \exp(j\vec{k}\cdot\vec{\rho}_k)\vec{E}_k^{0t} \tag{4.20}$$

This equation can be considered to be the antenna mode scattering of the array antenna. The model maintains the form of the superposition theory. If the difference between the units is ignored, then $E_{\mathrm{k}}^{0\mathrm{t}} \approx E_0^{0\mathrm{t}}$ with order $S^{\mathrm{a}} = \sum_{k=1}^{r} V_k \exp(jk \cdot \rho_k)$. Equation (4.18) can be transformed into the equation $E_{\mathrm{a}}^{\mathrm{s}} = E_0^{0\mathrm{t}} \times S^{\mathrm{a}}$, which is the expression of the product theorem mode of the array antenna scattering fields. In summary, the total spatial scattering fields can be represented as follows:

$$\vec{E}^s = \vec{E}_0^s + \vec{E}_a^s \tag{4.21}$$

where the spatial scattering fields are composed of structural mode scattering and antenna mode scattering.

4.3.2 Array Antenna Scattering Theory Analysis

To explore the main features of antenna mode scattering, several approximations can be presented, as follows: (1) ignore the mutual coupling between the antenna elements; and (2) ignore the difference between the antenna elements. Then, $E_{\mathrm{k}}^{0t} \approx E_0^{0t}$, $\Gamma_{\mathrm{k}}^{\mathrm{a}} = \Gamma^{\mathrm{a}}$, $\Gamma_{\mathrm{k}}^{\mathrm{l}} = \Gamma^{\mathrm{l}}$, $[S]_{\mathrm{rr}} = \Gamma^{\mathrm{a}}[I]$, $[S]_{\mathrm{rr}}^{\mathrm{l}} = \Gamma^{\mathrm{l}}[I]$, and the antenna mode scattering fields can be expressed as follows:

$$\vec{E}_a^s = \vec{E}_0^{0t} \frac{\Gamma^l}{1 - \Gamma^l \Gamma^a} \sum_{k=1}^{r} b_k^m \exp(j\vec{k} \cdot \vec{\rho}_k) \tag{4.22}$$

Considering the phase difference of the array elements excited by the incident plane wave, (which is oblique illumination), we can assume that the reference is element 0, which is at the center of the array. Then, $b_k^m \approx b_0^m \exp\left[-jk^i \cdot \rho_k\right]$, which can be substituted into Eq. 4.22, as follows:

$$\vec{E}_a^s = \vec{E}_0^0 \frac{\Gamma^l}{1 - \Gamma^l \Gamma^a} b_0^m \sum_{j=1}^{r} \exp[j(\vec{k} - \vec{k}^i) \cdot \vec{\rho}_j] \tag{4.23}$$

The maximum superposition direction of Eq. (4.23) is located in the specular reflection direction in which the array plane is the reflector, and the incoming wave direction is the incident direction. Because this scattering is related to the array factor, a strong scattering peak forms in this direction.

Structural mode scattering of an array antenna contains two distinct components. One component is the scattering field with superposition properties. Most of the elements in a large array have the same scattering characteristics under the illumination of plane waves—that is, regardless of the characteristic changes of the elements on the edge, the scattering fields of all the cells can be considered to have the same form, and they are affected by the phase difference of the incoming wave excitation and the spatial difference of the scattering field. This superposition forms a peak value in the specular reflection direction. The other component is nonsuperimposed scattering, which consists of the edge diffractions of the array structure and other scatterings, such as tip diffractions. The analysis of the characteristics of the structural mode scattering field of the array antenna provide theoretical guidance for the RCS control of the array.

4.4 Active Phased Array Scattering Analysis

4.4.1 Introduction to Phased Array Radar

Phased array radar [1] was first designed in the 1960s. Many technologies—such as very large-scale integrated circuits, solid-state power devices, digital phase shifters, and embedded computers—are becoming increasingly mature. Nonetheless, with the continuous development of digital beam forming, adaptive technology, and ultralow sidelobe technology, phased array radar has been the primary choice for modern advanced radar systems because of its high performance. Current research on phased array radar has focused on the following features:

(1) Miniaturization of the T/R components of the active phased array radar (APAR);
(2) Low sidelobe and ultralow sidelobe phased array design;

(3) Multi-beamforming technology;
(4) Wideband phased array radar technology; and
(5) Conformal antenna array technology.

Phased array radar systems are complicated. Moreover, their radiation and scattering characteristics are affected by the feeding system. Before presenting the research on the EM scattering of a phased array, the composition of a phased array radar system is introduced. As shown in Fig. 4.2, a phased array radar system comprises an antenna array, T/R module, multichannel receiving system, central processing system and feed control system. The following sections introduce the antenna array and the T/R module.

1. **Antenna array**: The antenna's radiating element can be a microstrip printed dipole, umbrella dipole, coupling slotted dipole, waveguide slot, or waveguide horn. The arrangement of the array can be rectangular, although triangular grid arrays are most commonly used.
2. **T/R module**: This component is one of the main differences between an APAR and a common phased array radar. The T/R module is usually an integrated part of the smaller unit circuit and comprises a phase shifter, circulator, low-noise amplifier (LNA), high-power amplifier (HPA), and circuit protection device. The T/R module is one of the core components of an APAR. Figure 4.3 shows the T/R module schematics of the transceiver circuit division and transceiver circuit sharing.

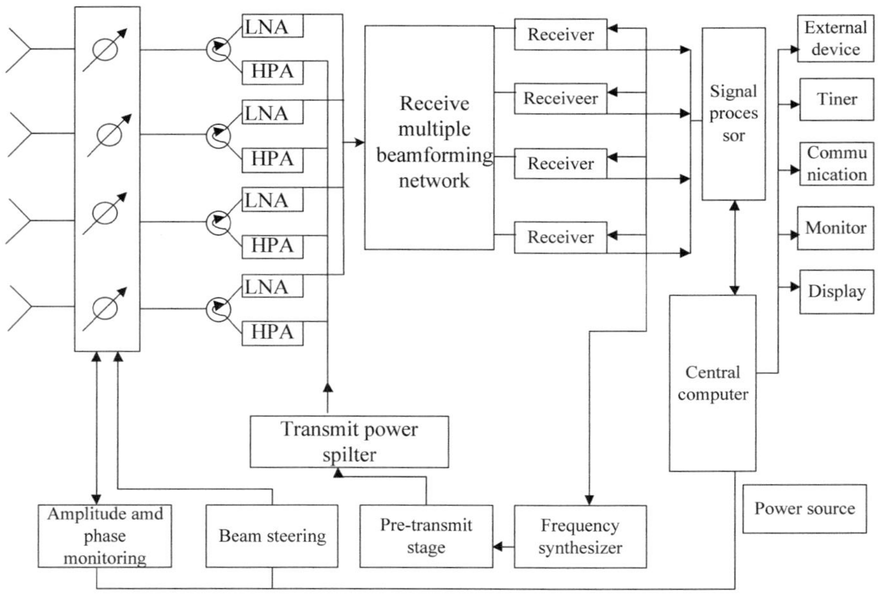

Fig. 4.2 Schematic of a phased array radar system

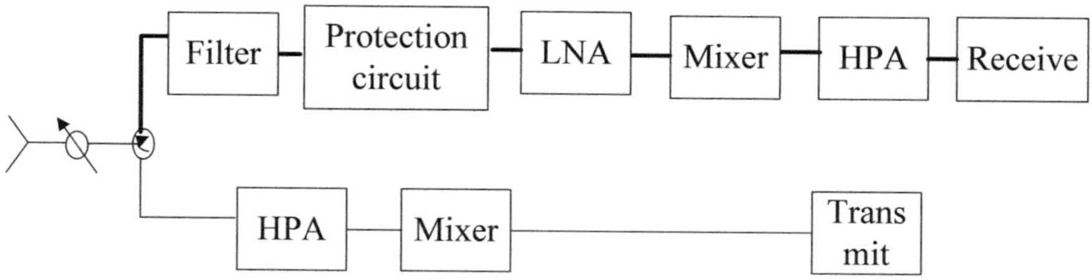

(a) T/R module schematic of the transceiver circuit split

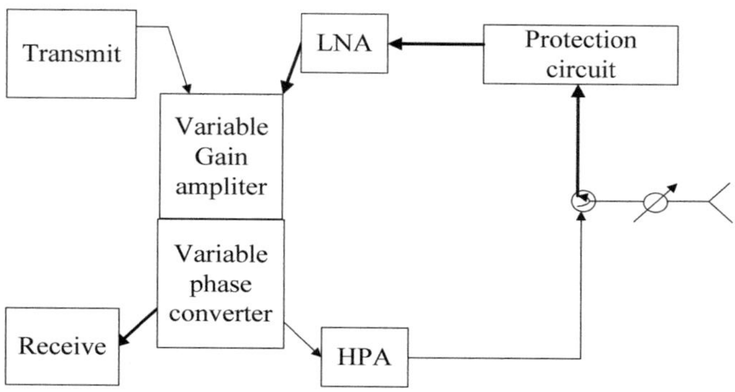

(b) T/R module schematic of a common transceiver circuit

Fig. 4.3 T/R module schematics of a transceiver circuit

4.4.2 Feeding System Analysis of an Active Phased Array

The feeding system structure of an active phased array is complicated, and its parameters change constantly with the phase state of the beam steering direction. Consequently, the scattering analysis of a phased array system is based on an analysis of the feeding system. As shown in Fig. 4.4, the antenna element module of an active phased array comprises a radiating antenna element, a phase shifter, the signal processing circuit of the T/R module, and a feed and receiving system connected to the T/R module.

The signal-port active module comprises a signal processing circuit, the feed and the circulator connecting them in the T/R module on the left-hand side of the T/R reference plane. This module comprises an LNA and HPA. It can also be analyzed using circuit simulations and graph theory. For the scattering problem, the parameter corresponding to the module can be considered—that is, the reflection coefficient Γ_m of the port that is connected only to the phase shifter. The T/R component is shown in Fig. 4.3b. When the loop comprises the transceiver circuit, the LNA, the public module and the HPA are under special conditions, and $|\Gamma_m| > 1$.

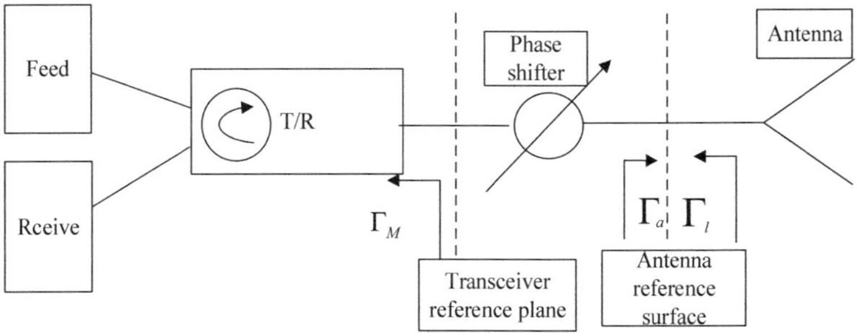

Fig. 4.4 Phased array antenna elements

The phase shifter can be analyzed using the S-parameters. Here, we assume that the S-parameters of the phase shifter are $s = \begin{bmatrix} s_{11}^p & s_{12}^p \\ s_{21}^p & s_{22}^p \end{bmatrix}$ (the S-parameters are normalized by the characteristic impedance Z to the reference plane), such that:

$$S_{12}^p = \left|T_p'\right| e^{j\psi_k'} \; S_{21}^p = \left|T_p\right| e^{j\psi_k} \tag{4.24}$$

where $|T_P|$ denotes the loss, ψ_K denotes the phase shift of element k, and S_{11}^P and S_{22}^P denote the reflection parameters of the two ports of the phase shifter. The reflection coefficient of the reference antenna plane can be expressed as follows:

$$\Gamma_l = S_{11}^p + \frac{S_{12}^p S_{21}^p \Gamma_M}{1 - S_{22}^p \Gamma_M} \tag{4.25}$$

4.4.3 Scattering Analysis of an Active Phased Array Antenna

To analyze the main characteristics of the scattering field and the scattering of the array affected by the feeding system of an active phased array, the following assumptions can be made:

(1) The differences between the elements are ignored:

$$\vec{E}_k^{0t} = \vec{E}_0^{0t}, \Gamma_k^a = \Gamma^a$$

(2) The mutual coupling between the elements is ignored:

$$[s]_{rr} = \Gamma^a[\mathrm{I}], [s]_{rr}^l = \Gamma^l[\mathrm{I}]$$

(3) Under oblique incidence of the plane wave and considering the phase difference of the array elements excited by the incident wave, we can assume $b_k^m \approx b_0^m \exp\left[-jk^i \cdot \rho_k\right]$, where b_0^m denotes the matched receiving amplitude.
(4) The two ports of the phase shifter corresponding to the receiving and transmitting reflection coefficients can be assumed to be matched well. S_{11}^P and S_{22}^P are small, so higher-order $S_{22}^P\Gamma_m$ can be ignored.
(5) $S_{21}^P = \left|T_P\right|e^{j\psi_k}$; if the working conditions of the phase shifter are linear, the following equation can be obtained:

$$S_{12}^P = \left|T_p'\right|e^{j\psi_k'} = \left|T_p'\right|e^{jt\psi_k} \tag{4.26}$$

When the two S-parameters are reciprocal, t = 1. The parameter ψ_k, which is included in the exponential factors of S_{21}^P and S_{12}^P, can be controlled by the working conditions of the phased array, the other parameters being constant. Thus, the reflection coefficient expression (Eq. (4.25)) can be simplified as follows:

$$\Gamma_k^l = S_{11}^P + S_{12}^P S_{21}^P \Gamma_M = \Gamma_{const}^l + \Gamma_{MP}^l e^{j(\psi_k+\psi_k')} = \Gamma_{const}^l + \Gamma_{MP}^l e^{j(1+t)\psi_k}$$

where

$$\begin{cases} \Gamma_{const}^l = S_{11}^P \\ \Gamma_{MP}^l = \left|T_p\right|\left|T_p'\right|\Gamma_M \end{cases} \tag{4.27}$$

Based on these approximations, the following equation can be obtained from the scattering field expression of the antenna array mode:

$$\begin{aligned} \vec{E}_a^s &= \vec{E}_0^{0t} b_0^m \Gamma_{const}^l \sum_{k=1}^{r} \exp[j(\vec{k} - \vec{k}^i) \cdot \vec{\rho}_k] \\ &+ \vec{E}_0^{0t} b_0^m \sum_{k=1}^{r} \Gamma_{MP}^l e^{j(1+t)\psi_k} \exp[j(\vec{k} - \vec{k}^i) \cdot \vec{\rho}_k] \end{aligned} \tag{4.28}$$

Next, let:

$$\vec{E}_a^{rs} = \vec{E}_0^{0t} b_0^m \Gamma_{const}^l \sum_{k=1}^{r} \exp[j(\vec{k} - \vec{k}^i) \cdot \vec{\rho}_k] \tag{4.29}$$

where E_a^{rs} denotes the reflection scattering field of the antenna mode.

The scattering state is not affected by the element phase shift of the active phased array. The maximum value of the scattering field is in the mirror reflection direction, where the array plane is the reflection plane and the incoming wave direction is the incident direction. Because this scattering field corresponds to the array factor, there will be a strong scattering peak in the reflected direction.

Let:

$$\vec{E}_{\mathrm{a}}^{\mathrm{ps}} = E_0^0 b_0^m \sum_{k=1}^{r} \Gamma_{MP}^l e^{j(1+\mathrm{t})\psi_k} \exp[j(\vec{k} - \vec{k}^{\,i}) \cdot \vec{\rho}_k]$$

$$= \vec{E}_0^0 b_0^m \Gamma_{MP}^l \sum_{k=1}^{r} \exp\{j[(\vec{k} - \vec{k}^{\,i}) \cdot \vec{\rho}_k + ((1+\mathrm{t})\psi_k)\} \tag{4.30}$$

where E_a^{ps} denotes the phased array scattering field of the antenna mode. It is affected by ψ_k, and the scattering field changes with the phase shift of each element of the antenna system. If the element spacing is equal and the feeding increases uniformly, the stacked maximum scattering directivity occurs in the k direction, which can be confirmed by $\left(k - k^i\right) \cdot \rho_k + ((1+t)\psi_k) = 0$. For a one-dimensional array consisting of antenna elements along the x-axis, $\rho_k = k\Delta x\hat{x}$. If the feeding phase of the element increases uniformly, $\psi_k = k\Delta\psi$, which can be substituted into $\left(k - k^i\right) \cdot \rho_k + ((1+t)\psi_k) = 0$, the result of which is as follows:

$$k_x = \frac{-(1+\mathrm{t})\Delta\psi}{\Delta x} + k_x^i \tag{4.31}$$

The phased array scattering field (Eq. (4.30)) of the antenna mode contains the array factor. After superpositioning, there will be a strong scattering peak. It should be noted that in the above expression, if $\Delta\psi = 0$, the phased array scattering field of the antenna mode that the array adopts is not zero. However, the maximum directivity of the scattering field coincides with the reflection scattering field of the antenna mode. Consequently, the phased array scattering field of the antenna mode can be considered to be the reflection scattering field of the antenna mode—that is, the antenna mode scattering field of the common antenna array.

4.4.4 Validation of Structural Mode and Antenna Mode Scattering Analysis of an Active Phased Array

As shown in Fig. 4.5, the results of the analysis of scattering theory of a phased array can be verified using a one-dimensional array antenna that comprises nine linearly polarized microstrip patch antenna elements, where the spacing between two elements is half the wavelength. The feeding phase can be controlled by a delay line. The length of the coaxial cable for each element increases in sequence to obtain an increase in the sequence of the feeding phase. The phase increment ($\Delta\psi$) can be controlled by the coaxial cable increment (ΔL). When the increment is zero, the antenna array can be fed uniformly. Moreover, the feeding port of each element added to the delay line is the transmit-receive reference plane. The copolarized radiating

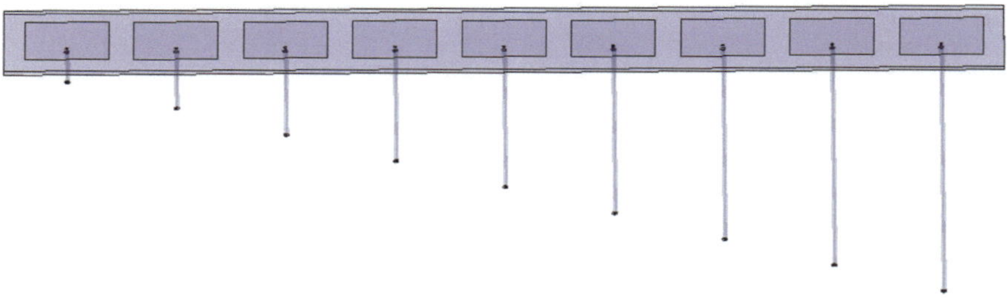

Fig. 4.5 Geometry of the one-dimensional antenna array

field (E_ϕ) can be studied on the plane comprising the array orientation x and the normal direction of the element z.

When $\Gamma_i^l = 0$, the total scattering field is the antenna structural mode scattering field. Ordering $\Delta L = 0$, which is the reference plane (as shown in Fig. 4.5, the feeding port cross-sections of all of the elements with $\Delta L = 0$), with the matching load obtains the structural mode scattering field of the array. The antenna mode scattering field can be obtained by subtracting the structural mode scattering field from the total scattering field of the array.

Figure 4.6 shows the antenna mode scattering field illuminated normally by a plane wave with $\Delta L = 0$, $\Delta\psi = 0$. The antenna mode reflection scattering field can be separated under the conditions that all of the ports are shorted (the antenna mode phased scattering field being merged into the antenna mode reflection scattering field).

Figure 4.7 shows the antenna radiation pattern with uniform excitation. The results show that the antenna mode scattering field is consistent with the radiation pattern, verifying the scattering analysis and separation method to be correct.

To observe changes in the peak value of the array antenna mode reflection field with the incident angle with $\Delta L = 0$, $\Delta\psi = 0$ (the antenna mode phased scattering field being merged into the antenna mode reflection scattering field), an excited array

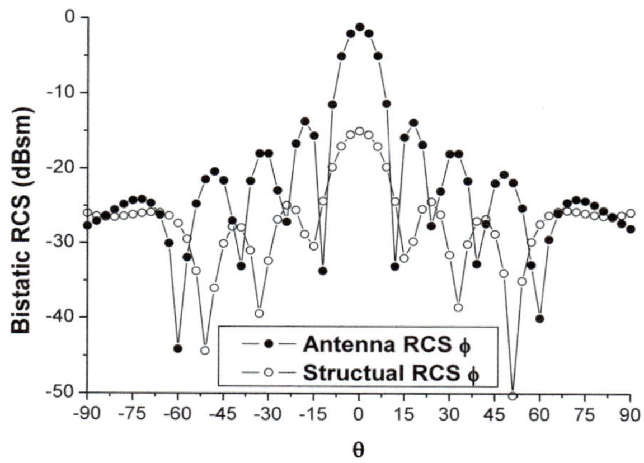

Fig. 4.6 The structure mode scattering and antenna mode scattering (0° plane illumination) of the array antenna

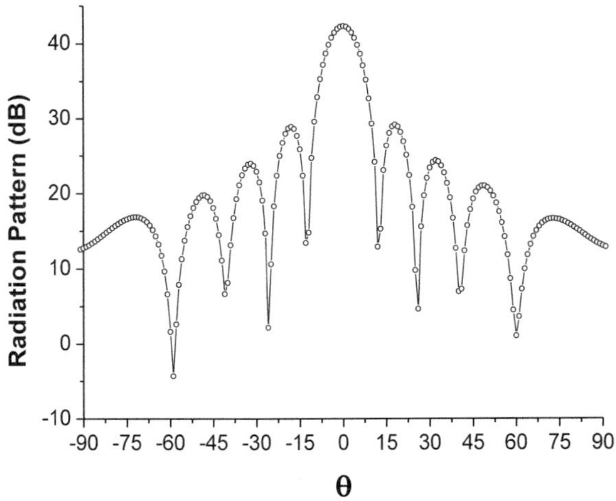

Fig. 4.7 Antenna array radiation pattern

with 45° oblique incidence is adopted. Figure 4.8 shows the separated antenna mode reflection scattering field. The desired peak value of the antenna mode reflection scattering field appears at $-45°$.

To observe the antenna mode phased scattering field of the array, $\Delta L = 0.15\lambda$, $\psi_k = k \times 54°$. The delay line and the reference plane have the same characteristic impedance, $S_{11}^p = 0$ and $\Gamma_{const}^l = S_{11}^p = 0$ can be substituted into Eq. (4.29), the antenna mode reflection scattering field being zero. Here, the separated antenna mode scattering field is the antenna mode phased scattering field. The plane wave that is perpendicular to the array orientation illuminates the antenna array, and $k^i = 0$. It can be assumed that $t = 1$. By substituting $\Delta x = 0.5\lambda$, $\Delta\psi = 54°$ into Eq. (4.31), the antenna mode phased scattering field corresponding to k can be obtained. The maximum radiating direction of the phased scattering field is at $\theta = -36.87°$.

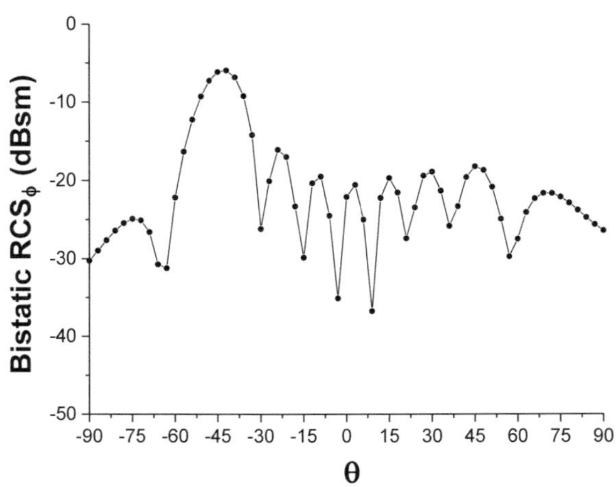

Fig. 4.8 Antenna mode reflection scattering field

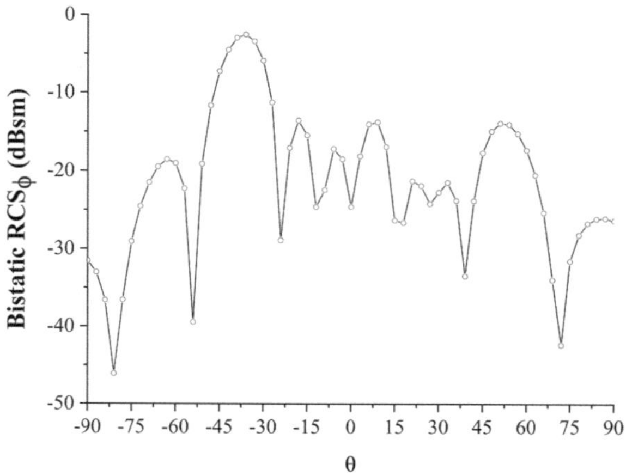

Fig. 4.9 Antenna mode phased scattering field

Figure 4.9 shows the separated antenna mode scattering field. The peak antenna mode scattering value at $\theta = -36.87^\circ$ confirms the validity of the theoretical analysis and the existence of the antenna mode phased scattering field.

4.5 RCS Control of a Shipborne Active Phased Array Antenna

The scattering field of a phased array antenna can be divided into structural mode scattering and antenna mode scattering. Based on the theoretical analysis of antenna scattering, structural mode scattering can be divided into two scattering fields which have superposition and nonsuperposition properties. However, the antenna mode scattering field can be divided into the antenna mode reflection scattering field and the antenna mode phased scattering field. Based on the four components of the scattering field present in studying antenna array RCS control (as presented above) and analyzing the corresponding ship-based antenna RCS control strategy, the strong scattering threat of the antenna can be eliminated during the design stage.

4.5.1 Structural Mode RCS Control of an Active Phased Array Antenna

The scattering mechanism of the antenna's structural mode is similar to the scattering mechanism of a common scatterer. However, the structural mode RCS control of a ship-based array antenna must consider the array scattering characteristics and interactions between the antenna and the surrounding environment.

1. **Superposition scattering field of the structural mode**: As discussed in Sect. 4.4, the scattering field with superposition properties in the structural mode scattering of an array antenna will be a superposed peak in the mirror reflection direction of the antenna array. An effective method to control the threat associated with the scattering peak is an oblique installation, as shown in Fig. 4.10. Because the main beam of a phased array antenna is controlled freely by the feeding phase, the antenna front's normal direction does not point in the detection direction as it does for a common array antenna. For a surface ship, if the installation direction makes the front tilt upwards, as shown in Fig. 4.10, the superimposed items of the array antenna's structural mode that are excited by a horizontal incoming wave also tilt upwards to avoid a detection threat from the mirror reflection peak value aimed at the detection radar. However, for the detection beam of the enemy, the antenna can be adjusted to the detection direction by controlling the feeding phase, although this will sacrifice the antenna's effective aperture area, gain, and feeding phase complexity.

The bow direction region is where the detection threat will appear. After the installation of a phased array antenna, the incident angle of the detection incoming wave (marked as $\hat{k}_i$) will be fixed for the incident angle of the phased array front. It should be noted that the scattering is in the direction $\hat{k} = -\hat{k}_i$. In addition to the oblique installation, the superposition of the other scattering fields should be reduced as much as possible to reduce echoes. This is discussed further by means of an example in later sections.

Superposition scattering of the array antenna's structural mode superposes a strong scattering wave in the mirror reflection direction. In addition to the direct scattering,

Fig. 4.10 Oblique installation of a ship-based phased array antenna

corner reflectors may be formed by the array front and other perpendicular conductor surfaces. On the ship, the front of the phased array antenna and other objects—such as the deck and building structures perpendicular to the deck—can form corner reflectors. To avoid strong scattering effects, the antenna front and the sea surface should be prevented from forming corner reflectors. For example, the array antenna can be obliquely installed on existing stealth ships, preventing the formation of antenna and sea-surface induced corner reflectors. Additionally, the antenna array and other large planar surfaces on the ship should be prevented from forming corner reflection structures.

Element structural scattering is the primary reason for the superposition scattering within array structure scattering. Consequently, reducing the element RCS is a key component of array antenna RCS control. With the development of stealth antenna technology [2–14], antenna miniaturization has become the main focus of antenna RCSR. Using a microstrip patch antenna as an example, based on the assumptions that a zero dielectric resonant antenna adopts a left-hand circuit and a fractal antenna can satisfy the antenna radiating performance, the surface reflection and scattering of an antenna element can be reduced considerably by reducing its physical size. Additionally, because of variations of the antenna structure, the total array antenna RCS can sometimes be reduced by using various antenna coupling effects.

Antenna miniaturization can reduce the stacked antenna element structure scattering and provide a larger arrangement space (and additional choices of the array form) for antenna array design. Through a reasonable arrangement space—noting that an asymmetrical array changes the array resonant characteristics—the use of fractal multiresonant characteristics will become an important antenna stealth technology research field.

2. **Nonsuperposed structural mode scattering of an array antenna**: Nonsuperposed structural mode scattering of an array antenna is caused primarily by edge diffractions, spire diffractions, and the corresponding multiple scattering. Consequently, the array design should employ jagged edges and absorbing coating material to reduce edge and spire diffractions.

4.5.2 Antenna Mode RCS Control of an Active Phased Array Antenna

The antenna mode scattering control technology can be divided into in-band and out-of-band control technology. The former focuses on controlling the scattering peak corresponding to the antenna radiation characteristics. Out-of-band control technology focuses on the resonant point scattering peak (including element-level and array-level resonance), which is composed of a periodic element arrangement.

Theoretical analysis of antenna mode scattering shows that antenna mode out-of-band control of an active phased array antenna can be considered from two perspectives—that is, reflection scattering control or phased scattering control of the antenna mode.

The reflection scattering field of the antenna mode is similar to the stacked scattering field of the structural mode in the antenna array. The array front is the reflecting plane, and its peak value occurs in the mirror reflection direction of the incident direction (which is the incoming wave direction). To achieve stealth, the oblique installation shown in Fig. 4.10 can be adopted to cause the reflection scattering field of the antenna mode to change the incoming wave direction. It should be noted that the side lobe of the array is the main contributor of antenna mode scattering in an antenna array that is obliquely placed. Consequently, for both radar detection performance and stealth, the side lobe of the antenna array should be suppressed.

For the phased scattering field of the antenna mode, the value of the main scattering peak changes with the state of the feeding phase, which can be difficult to control. However, if the pencil beam of the phased array antenna is relatively narrow and the array has a rapid scanning speed, the direction of the beam's peak also changes continuously. Thus, the enemy radar cannot receive a continuous echo signal, and the probability that the antenna will be detected diminishes. For phased scattering field RCS control of the antenna mode, if an oblique installation is adopted, the objective of the reflection scattering field of the antenna mode is to reduce the side lobe of the array antenna.

The out-of-band resonant peak value control of an antenna can be studied based on the element and the array. For out-of-band RCS peak value control (illustrated using the example of a microstrip patch antenna), the basic theory—including the photonic bandgap (PBG) structure, loading, slotting, improving the antenna structure, and aperture coupling—can be used to reduce the high-order resonant coupling, absorb the high-order mode energy, and destroy the high-order resonant requirements to control the higher-order resonant radiation of the antenna. Based on the array factor in the scattering field expression when the frequency increases several times:

$$f_s = pf_0 \tag{4.32}$$

This equation means that the scattering wave frequency fs is p times the antenna's array operating frequency f_0. The element number can be denoted as n, and the array factor of the antenna scattering field can be expressed as follows:

$$\begin{aligned} S_a &= \sum_{n=1}^{N} V_k \exp(j\vec{k}_s\vec{\rho}_n) \approx b_0^m \sum_{n=1}^{N} \exp\left[j\left(\vec{k}_s - \vec{k}_s^i\right)\vec{\rho}_n\right] \\ &= b_0^m \sum_{n=1}^{N} \exp[jp\left(\vec{k} - \vec{k}^i\right)\vec{\rho}_n] \end{aligned} \tag{4.33}$$

Assuming a one-dimensional array along the x-axis $\rho_n = n \times 0.5\lambda\hat{x}$, then:

$$S_a = b_0^m \sum_{n=1}^{N} \exp[jpnk\frac{\lambda}{2}(\hat{k} - \hat{k}^i)\hat{x}]$$

$$
\begin{aligned}
&= b_0^m \sum_{n=1}^{N} \exp[jpn\frac{2\pi}{\lambda}\frac{\lambda}{2}(\hat{k} - \hat{k}^i)\hat{x}] \\
&= b_0^m \sum_{n=1}^{N} \exp\{jn\pi\left[p\left(k_x - k_x^i\right)\right]\}
\end{aligned} \tag{4.34}
$$

Because $0 \le \left|k_x - k_x^i\right| \le 1$, for the operating frequency point, $p = 1$, the peak value only appears in the direction $k_x - k_x^i = 0$. However, for $\mathrm{p} > 1$, owing to the periodicity of the exponential function, the maximum value appears at a series of k_x when $p\left(k_x - k_x^i\right) = 2$. Moreover, when the frequency shifts, a mismatch between the antenna and receiver loads can occur, which leads to a greater reflection coefficient. Additionally, the array antenna also resonates easily at the frequency $f_x = p \times f$ to radiate energy. Consequently, the element radiating field, the receiver load reflection, and the array factor make the scattering field generate several peak values. From another perspective, the illuminating wave shifts to the dominant frequency. The electrical size between the elements changes, which causes the virtual space grating lobe to convert into a real space grating lobe to generate two or more radiating lobes whose gains are nearly equal. These lobes can produce a series of strong scattering lobes. If the side lobe of the antenna array is weakly controlled, the density of the side lobes is high, and the grating lobes are large, the echo wave's peak value will be difficult to avoid in the incoming wave direction. Even if an oblique installation is adopted, it can be difficult to suppress backscattering.

To control the out-of-band RCS of a phased array antenna, the frequency selection surface can be used to avoid the threat of a peak scattering value from the incident wave, after which out-of-band RCS control of the antenna element can then be used to eliminate the resonant hidden danger.

Based on this analysis, if an oblique installation is chosen (the most common installation method for invisible phased array antennas), a key requirement is side lobe control of the array antenna. To ensure the desired antenna radiation characteristics, the research focus aims to control the side lobe and suppress the grating lobe for invisible phased array antennas.

4.5.3 *Density Tapered Sparse Array*

Several methods can be used to control the side lobe and suppress the grating lobe for an array antenna [15, 16], such as using a nonuniform array, feeding amplitude, phase weighting, and feeding phase compensation methods. Many analyses and numerical optimizations of such methods have been performed; the analytical methods have included the Fourier transform method [17], Semyon Kuno J's synthetic method [18], the Woodward synthetic method [19], the Chebyshev synthetic method [20], and the Taylor synthetic method [21, 22]; the numerical optimization methods have included the genetic algorithm and particle swarm optimization.

For the Chebyshev and Taylor synthetic methods, the zero of the radiation pattern can be controlled by controlling the level of the side lobe. The Chebyshev synthetic method uses a Chebyshev polynomial as the synthetic goal of the array radiation pattern. The corresponding feed weighting amplitude can be obtained using the narrowest lobe generated by the designated side lobe level. One-dimensional Chebyshev synthetic parameters include the number of elements (N) and the voltage ratio (r) between the main lobe and the side lobe. The feeding amplitude can be obtained using the formula in [16]. However, the Chebyshev synthetic method forces all of the side lobe levels to remain consistent. With an increase in array size, more side lobes consume more energy, which leads to a decrease in array efficiency. Moreover, the main lobe is limited.

The Taylor synthetic method adopts the null of the Chebyshev radiation pattern polynomial in the near region. In the far region, the polynomial null of a uniform feeding array can be adopted to control the side lobe and the array efficiency. In this method, the expansion coefficient in the far region can be used to ensure that the boundary nulls between the far region and the near region are coincident. The Taylor synthetic method can eliminate the number of elements (N) and the voltage ratio (r) between the main lobe and the side lobe. It can also add the number of boundary nulls (n). The feeding amplitude of the Taylor synthetic method can be obtained using the formula in [22]. The feed weighting is tapered from the center to the edge. In this section, the side lobe is controlled using the Taylor synthetic method.

For the scattering of the array, the feed source cannot be weighted as the side lobe control because the energy comes from a uniform plane wave (the far region of the enemy radar) for the scattering of the phased array antenna. All of the element scatterings are excited equally. Consequently, the scattering problem of an array antenna cannot use the feeding amplitude weight to reduce the side lobe.

Based on the idea of weight, the radiation and scattering side lobe of a phased array antenna can also be controlled using a tapered density. Skolink et al. [21] proposed a statistically sparse technology. The weighted tapered feeding amplitude of a nonsparse common array is proportional to the element density of a large array. Based on the probability corresponding to the weighted amplitude, some elements can be eliminated, and the remainder can be fed uniformly. The feeding of the array remains weighted based on the statistical significance. Consequently, good control of the side lobe can be obtained using the Chebyshev and Taylor weighted feed and an array obtained using this method is called a density tapered sparse array.

To reduce the complexity of the feeding system, a density tapered sparse array was proposed. Here, the remainder of the elements use a uniform feed, which can considerably reduce the complexity of the feeding system without weighted amplitudes. The scattering of each remaining element illuminated by a plane wave is excited equally. Because of the rarefaction of arrays, the superposed scattering of the structural mode of an array element and the reflection scattering excitement of the antenna mode are weighted equally. Stronger scattering is generated in the mirror reflection direction, with low side lobes being generated in the other directions. For an oblique installation, the echo scattering of the incoming wave direction can be reduced considerably. Owing to the sparse array, the sharper phased scanning beam

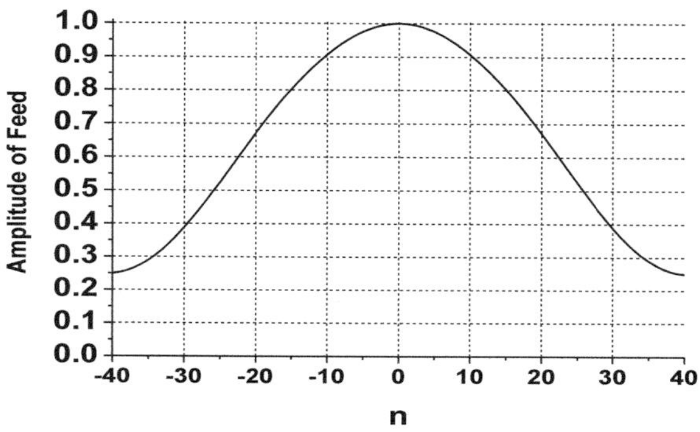

Fig. 4.11 Taylor synthetic tapered feeding amplitudes

of the antenna mode not only performs space scanning but can also prevent the radar from obtaining a continuous signal. Moreover, the detection probability can also be reduced. The design of a density tapered sparse antenna array is examined in the following section. Here, the circular array comprises a triangular grid array on the *xoy* plane. There are 81 elements along the diameter of the array and the design process is described below.

Taylor synthetic feeding amplitude weighted calculation: the goal of the normalized side lobe level is −30 dB. The Taylor feeding weighted coefficient of the 81 one-dimensional elements (marked as −40 ~ 40) can be calculated, the feeding amplitude distribution of which is shown in Fig. 4.11.

Verifying the decrease of the side lobe of the Taylor feed: as shown in Fig. 4.12, the uniform triangular grid array contains 5,815 antenna elements on the *xoy* plane (the element field can be assumed to be the antenna radiation field, as shown in Fig. 3.12). The far-field pattern of the array can be obtained from the Taylor amplitude weighted feed, from which the radiation pattern of the array with an equal amplitude can be calculated. After the Taylor weighted feed, the main lobe level is 92.36068 dB. Compared with the main lobe of the array, which has an equal amplitude of 98.26258 dB, the level decreases by 5.902 dB. All of the side lobes in the radiation pattern decrease substantially; the maximum normalized side lobe decreases from −24.32 to −38.804 dB, which satisfies the design objective. The side lobe level is lower than −30 dB because of the change in the element radiation pattern and the weights in both the *x* and *y* directions.

Array sparsity: the sparse array shown in Fig. 4.13 can be designed from the circular array shown in Fig. 4.12, the circular array being dispersed with the probabilities corresponding to the Taylor feeding tapered amplitudes shown in Fig. 4.11 along the radius direction of the array. The element density along the radius in Fig. 4.13 is consistent with the curve in Fig. 4.11. The number of sparse elements is 2,941. The radiation pattern can be obtained using the array element fed with the equal amplitude with sparsity (Fig. 4.14).

Array optimization: the size of the array is limited owing to the sparsity process based on the random sparsity of the probability distribution. The randomness of the radiation pattern is greater after the sparsity process. We can use random ergodicity

Fig. 4.12 The uniform array

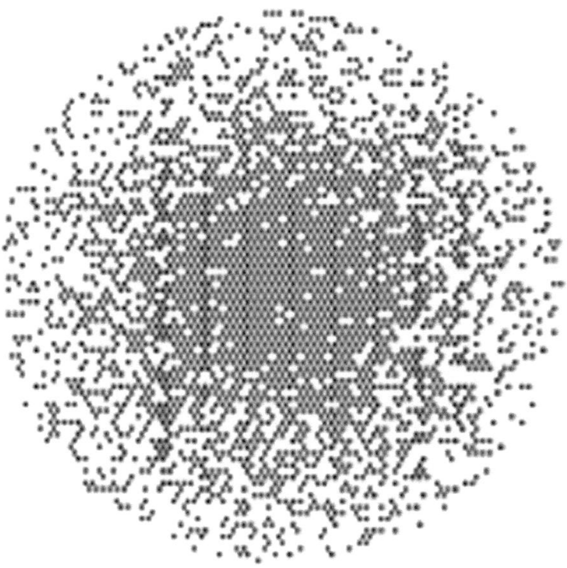

Fig. 4.13 The sparse array

to optimize the radiation pattern. Figure 4.15 shows the curve of the normalized side lobe level in the 1,000-step random ergodic process. After optimization, the relative average value of the side lobe level is −31.5 dB, which is 6 dB lower than before the optimization.

Result verification: the calculated result of the optimum sparse array after the optimization is shown in Fig. 4.16. The main lobe level is 92.3875 dB, and the

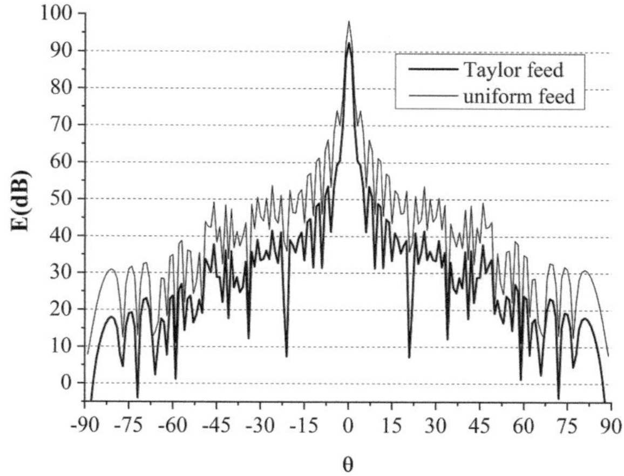

Fig. 4.14 Comparison between the Taylor feed and the uniform feed for the circular array

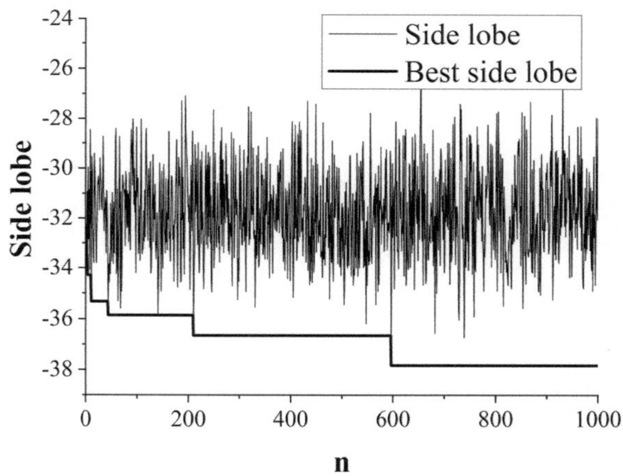

Fig. 4.15 Curve of the side lobe level in the optimization process

maximum normalized side lobe level is −37.82983 dB. The characteristics of the sparse array are similar to those of the nonsparse array with the Taylor feeding weights. The results therefore achieve the design objective.

For stealth, because of the decrease in the side lobe level, the threat from reflection scattering of the antenna mode of the array antenna diminishes. As the ratio between the main lobe and the side lobe increases, the probability that the phased scattering of the antenna mode will be detected can be reduced using the high-quality pencil beam. Additionally, the detection accuracy of the radar can be improved using the high-quality pencil beam. The interception probability also decreases. From an enemy detection and stealth perspective, this method is advisable. The design and manufacturing costs of the sparse array are less than those of the nonsparse array, and the complexity and cost of the feeding system can also be reduced by means of the uniform feed.

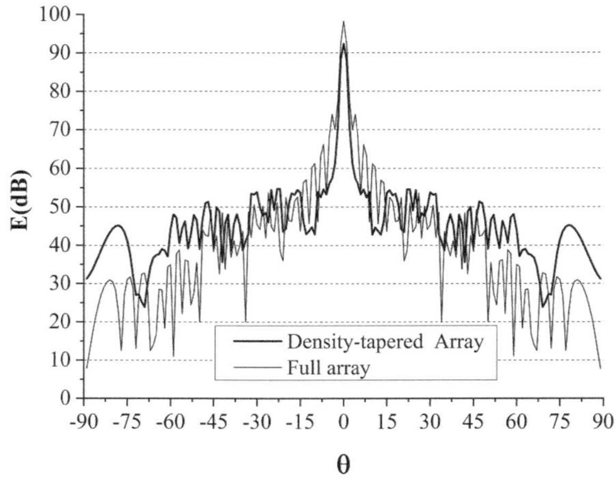

Fig. 4.16 Radiation patterns of the sparse array and the nonsparse array

The antenna array can be assumed to rotate 5° along the x-axis in the xoy plane, which is an oblique placement. The array elements that have a spatial phase difference are excited by the incident wave. The excitement of an incident wave along the z-axis is equivalent to the excitement of an incident wave at an angle of 5° with the z-axis before rotation. Because of the computer computational limitations, the scattering field of the oblique array with the plane wave was not calculated directly. To observe the peak value of the scattering, an array whose excitement is equal to the spatial phase difference above was adopted (the spacing between the antenna being 0.5 λ, and $\Delta\psi_x = k \times \sin(5°) \times \lambda/2$).

Figure 4.17 shows the radiation pattern on the xoz plane. The obvious peak value is in the mirror direction of −5°. Similar to the discussion in Sect. 4.5.2, the superposition is in the direction $\hat{k} = -\hat{k}_i$, and the corresponding radiation pattern is also optimized, the difference from the main lobe's peak value being −62 dB. Consequently, the results suggest that the superposed scattering of the antenna's structural mode and the reflection scattering of the antenna mode can be suppressed well in this direction. The figure also shows the radiation pattern under the conditions that the array is fed with different phases and equal amplitudes. The minimum value of the adjacent side lobe in the incoming direction of −5° is higher than the side lobe level of −62 dB by approximately 20 dB. The results also show that random discrete values can be used to optimize the echo in the fixed direction. The calculations verify that density tapered sparse technology can control the RCS peak of an antenna array with an oblique placement.

Reference [23] showed that the RCSR method could destroy the radiation characteristics of the antenna, regardless of the type of antenna. The low RCS corresponds to a low antenna gain. Moreover, the radiation and scattering characteristics of the antenna must both be considered and controlled quantitatively in the antenna design. Using the weighted feed, the characteristics of a low-RCS antenna can be used properly. Additionally, low-RCS (low-gain) antennas that adopt the weighted tapered idea at various levels can be adopted successively one-by-one. With decreases in the

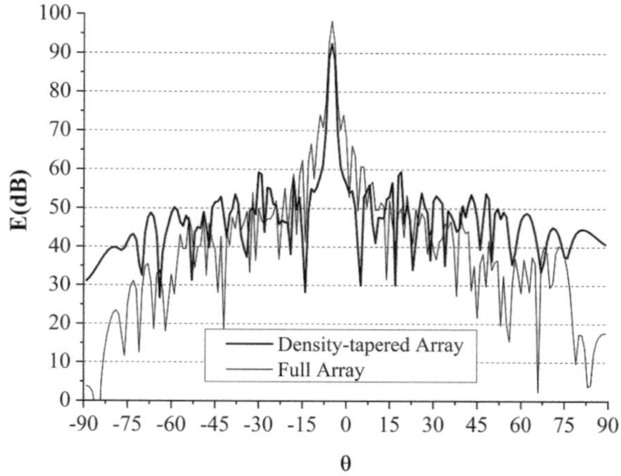

Fig. 4.17 Radiation patterns of the sparse array with oblique placement and the nonsparse array

gain of the edge elements, the tapered radiation field can realize low side lobes of the radiation pattern. Additionally, a low side lobe of the superposed scattering field can be obtained. Consequently, the stealth threat from the reflection scattering of the antenna mode and the scattering side lobe level of the antenna's structural mode can be reduced.

4.6 Summary

In this chapter, the mutual coupling of array elements was analyzed using the S-parameters of an array antenna. Spherical wave function expansion and a generalized scattering matrix were used in the radiation and scattering analysis of the array antenna, and the basic theoretical formula of array antenna radiation and scattering could be derived. Moreover, the scattering field characteristics and the direction of the peak value of the array antenna were analyzed using the approximate calculations of the basic theoretical formula of array antenna radiation and scattering. Based on a brief introduction to active phased array antennas, the element feeding method in the analysis of active phased array antenna scattering was presented. Additionally, based on this analysis, the scattering characteristics of an active phased array antenna were analyzed. Finally, the theoretical analysis results were verified using an example. The scattering RCS control methods for shipborne phased array antennas were discussed from a scattering characteristics of active phased array antennas perspective. The radiation and scattering characteristics of an array antenna with an oblique installation were analyzed, and the requirements of the antenna side lobe level for the RCS control were discussed. Taylor synthetic and density tapered sparse technologies were used to design and optimize an array antenna with a low side lobe level.

References

1. Guangyi Zhang, Phased Array Radar Technology, 2nd edition, Beijing National Defense industry Publishing House, 2006.
2. Yingzheng Ruan, et al. Radar Cross Section and Stealth Technology, Beijing National Defense industry Publishing House, 1998.
3. D.M. Pozar. Radiation and Scattering from a Microstrip Patch on a Uniaxial Substrate, IEEE Trans. on AP, Vol. 35, No. 6, pp. 613–621, 1987.
4. J.L. Volakis, et al., Radar Cross Section Analysis and Control of Microstrip Patch Antennas, 1992 PIEEE AP-S International Symposium, pp. 2225–2228, 1992.
5. J.L. Volakis, et al., Broadband RCS Reduction of Rectangular Patch by Using Distributed Loading, Electronic Letters, Vol. 28, No. 25, pp. 2322–2325, 1992.
6. Yingzheng Ruan, Lin Feng, et al., Research on Antenna RCS Reduction Technology, Collection of Achievements in the Electronic Science Research Institute of the Ministry of Mechanical and Electronic Industry, pp. 16–66, 1990.
7. Yingzheng Rua, Scattering Mechanism of Antenna and Reduction of Radar Cross Section, Journal of Astronautics, Vol.11, No.4, pp. 94–100， 1990.
8. Ying Liu, Shuxi Gon, Demin Fu, Application of Fractal Antenna in Radar Cross Section Reduction, Journal of Microwaves, Vol. 19, No. 2, pp. 28–31, 2003.
9. J.T. Aberle, et al., Scattering and Radiation Properties of Varactor-Tuned Microstrip Antennas, 1992 PIEEE AP-S International Symposium, pp. 2229–2232, 1992.
10. Ying Liu, Shuxi Gong, Demin Fu, Analysis and Optimization Design of Radar Cross Section Reduction Using Resistance Bars, Chinese Journal of Radio Science, Vol. 18, No. 3, pp. 301–304, 2003.
11. Ying Liu, Shuxi Gong, Demin Fu, Research Progress of Fractal Antennas, Chinese Journal of Radio Science, Vol. 17, No. 1, pp. 54–58, 2002.
12. Ying Liu, Shuxi Gong, Demin Fu, Fractal and Array Antennas, Chinese Journal of Radio Science, Vol. 16, No. 9, pp. 147–149, 2001.
13. D.R. Jackson, The RCS of a Rectangular Microstrip Patch in a Substrate-Superstrate Geometry, IEEE Trans. on AP, Vol. 38, No. 1, pp. 2–8, 1990.
14. H.Y. Yang, et al., Mulifunctional Antennas with Low RCS, 1992 PIEEE AP-S International Symposium, pp. 2240–2243, 1992.
15. Jun Lu, Jinsong Gao, Lianchun Sun, Frequency Selective Surface and Its Application in Stealth Technology, Ome Information, No. 9, pp. 1–4, 2003.
16. T.K. Wu. Frequency Selective Surface and Grid Array. John Wiley and Sons, Inc., New York, 1995.
17. Ben A. Munk. Frequency Selective Surfaces, Theory and Design. First edition New York: John Wiley and Sons, Inc., 2000.
18. Xiangming Xu, Yong Yue, Yingzheng Ruan, 雷达舱和天线罩的电磁散射, Chinese Journal of Radio Science, Vol. 15, No. 2, pp. 239–243, 2000.
19. Skolnik M. J W Sherman, F C Org, Jr. Statistically Designed Destiny-Tapered Arrays. IEEE Trans on Antenna and Propagation, 1964, 12(4):408–471.
20. Robert J. Mailloux, Phased Array Antenna Handbook, Systems Engineering and Electronics, 2007.
21. Galindo-Israel, V. Shung-Wu Lee Mittra, R. Synthesis of a laterally displaced cluster feed for a reflector antenna with application to multiple beams and contoured patterns, IEEE Transactions on Antennas and Propagation, Vol. 26, No. 2, pp. 220–228, 1978.
22. Thomas, D. T. Multiple beam synthesis of low sidelobe patterns in lens fed arrays, IEEE Transactions on Antennas and Propagation, No. 26, pp. 883–886, 1978.
23. Kahn, W. Kurss, H. The uniqueness of the lossless feed network for a multibeam array, IEEE Transactions on Antennas and Propagation, Vol. 10, No. 1, pp. 100–101, 1962.

Chapter 5
Numerical Calculation of a Microstrip Antenna and Array RCS

5.1 Method of Moments Analysis of Microstrip Antenna RCS

5.1.1 Introduction

Lo proposed the cavity model to analyze microstrip antennas, the theory of which assumes a thin microstrip antenna [1, 2]. The space between the microstrip patch and the ground sheet can be considered to be a resonance cavity surrounded by magnetic and electric walls. The radiation field can be determined from the equivalent magnetic current around the cavity. The model provides an analytical solution for microstrip antennas with simple regular shapes and can be used as a guide for actual projects. Moreover, owing to its simplicity and limited number of calculations, rapid results can be achieved. However, because of the simplified analysis process, larger errors in the results may be produced when solving complex problems.

Recently, several researchers have studied the full-wave model for accurate analyses of complicated microstrip antennas. The so-called full-wave model considers the effects of dielectric loss, conductor loss, space wave radiation, surface waves, and external coupling. Moreover, the method of moments (MoM) has become the most commonly used method for analyzing microstrip antennas owing to its unique advantages. The MoM for analyzing microstrip antennas can be divided into two categories. The first is the moment method based on the free-space Green's function, which does not require Green's function and offers a simple solution. However, it requires more unknowns and greater computational overhead because the media substrate, ground plane, and the metal patch must be divided together. Another common method used is the MoM based on the layered-media Green's function. This method only requires the antenna structure to be divided, and it has the advantages of fewer unknowns and faster calculation.

Y. Liu et al., *Antenna Radar Cross Section: Theory and Design*,
https://doi.org/10.1007/978-981-96-1226-0_5

The MoM based on the layered-media Green's function can itself be divided into the spectral domain MoM and the spatial domain MoM. The spectral domain MoM requires that the basis functions have the form of analytic Fourier transforms. Moreover, it needs to solve the time-consuming dual infinite integral. Although [3] proposed progressive wave extraction technology to speed up the calculation of the impedance matrix element, the calculation time is still lengthy. The basis function can be arbitrarily selected when using the spatial domain MoM, which is a more versatile method than the spectral domain MoM. However, the spatial domain Green's function is actually a Sommerfeld integral with strong oscillation and slow convergence, whose solution can be time-consuming. Consequently, the efficiency of the spatial domain Green's function is dependent on the speed of its solution. Based on the spatial domain Green's function, a fast and precise solution for the electrical performance of microstrip antennas can be obtained using the spatial domain moment method.

In this section, Michalski's C-type Green's function is combined with the mixed-bit integral equation. Additionally, the spatial domain MoM based on the Rao–Wilton–Glisson (RWG) basis function is used to solve the electrical performance of microstrip antennas. Because of the dielectric loss, surface waves and other effects have already been considered in the Green's function and a microstrip structure with an arbitrary shape can be analyzed using the RWG basis function; consequently this method can be used to accurately analyze the performance of microstrip antennas. The spatial domain MoM based on the mixed-bit integral equation is then examined, and the effectiveness of the algorithm is verified by means of an example that solves the scattering performance of microstrip antennas. A solution for microstrip antennas fed with microstrip lines is examined in the third part of this section, and the analysis method for microstrip antennas fed with coaxial lines is studied in the fourth sub-section.

5.1.2 Spatial Domain MoM Based on the Mixed-Bit Integral Equation

When we use the MoM to analyze the electrical characteristics of microstrip structures, the first step is to combine Michalski's C-type Green's function with a mixed-bit integral equation to represent the electric field created by the current source in the microstrip structure, in which the spatial domain Green's function is solved using the discrete complex image method. The next step is to use the spatial domain MoM to solve the mixed-bit integral equation, the RWG basis function to expand the current distribution, and the Gagarin method to match the obtained matrix equation. The final step is to solve the matrix equation to obtain the current distribution before determining the other electrical characteristics of the antennas.

1. **Mixed-bit integral equation**: The electric field created from a current source placed in multi-layered media in Fig. 5.1 can be expressed as follows:

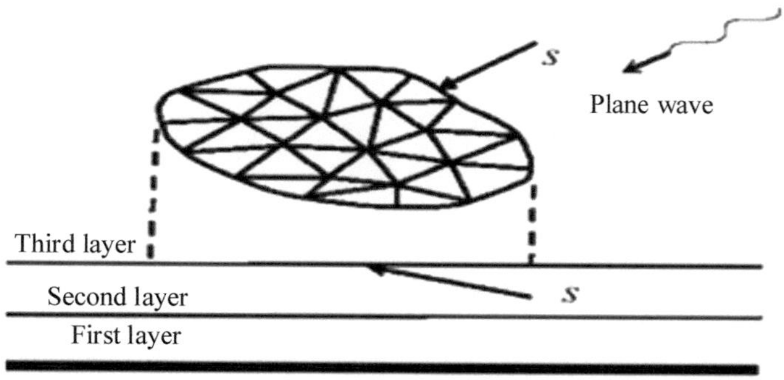

Fig. 5.1 Schematic of the microstrip patch structure

$$E = -jw\mu_0\langle G^A; J\rangle + \frac{1}{jw\varepsilon_0}\nabla\left\langle G_\varphi; \nabla' \cdot J\right\rangle \tag{5.1}$$

The mixed-bit integral equation that the microstrip antenna satisfies can be obtained based on the boundary condition of the tangential current field being zero on the metal patch's edge, as follows:

$$-jw\mu_0 \int_S G_{xx}^A(r, r')J_s(\rho')ds' + \frac{1}{jw\varepsilon_0}\nabla \int_S G_\varphi(r, r')\nabla_{s'} \cdot J_s(\rho')ds' = -E_s^i(r) \tag{5.2}$$

where $E_s^i(r)$ denotes the electric field [3] of the original position excited by a plane wave after removing the patch structure (Fig. 5.1), (θ^i, φ^i) denotes the incident direction of the plane wave, E_θ^{inc} and E_φ^{inc} denote the incident intensity, and G_{xx}^A and G_φ denote the spatial domain Green's function.

2. RWG basis function: The selection of the basis function is a vital part of the MoM. Here, we adopt the triangular element basis function (known as the RWG basis function) presented by Rao [4]. Based on this method, the current distribution of patch antennas with arbitrary shapes can be simulated. The first step is to divide the patch into triangles, after which a basis function can be defined by a pair of triangles that share one edge. As shown in Fig. 5.2, the RWG basis function related to the nth public edge can be defined as follows:

$$f_n(r) = \begin{cases} \frac{l_n}{2A_n^+}\rho_n^+ , & r \in T_n^+ \\ \frac{l_n}{2A_n^-}\rho_n^- , & r \in T_n^- \\ 0 , & \text{other} \end{cases} \tag{5.3}$$

The divergence of the RWG basis function can be expressed as follows:

Fig. 5.2 Schematic of the triangular element basis function

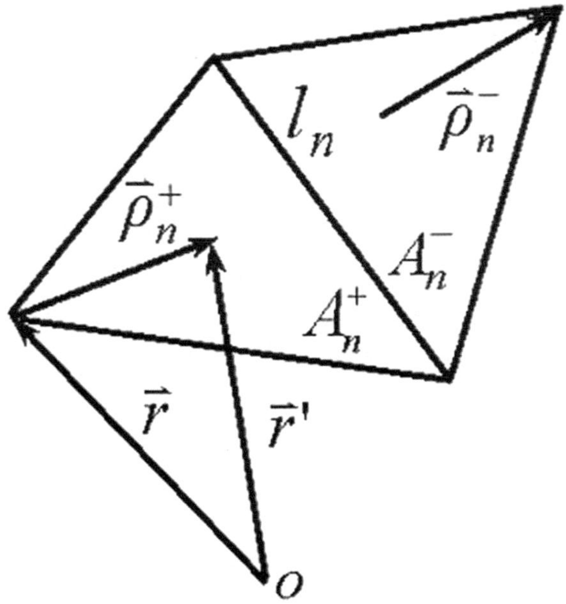

$$\nabla \cdot f_n(r) = \begin{cases} \frac{l_n}{A_n^+}, & r \in T_n^+ \\ -\frac{l_n}{A_n^-}, & r \in T_n^- \\ 0, & \text{other} \end{cases} \quad (5.4)$$

where $A_n^{\pm}$ denotes the area of triangle $T_n^{\pm}$, and l_n denotes the length of the public edge. In the calculation, the number of basis functions is the number of public edges. After solving the expansion coefficient of each basis function, the current inside any triangle can be represented by the linear superposition of the current corresponding to the public side of this triangle.

3. MoM based on the mixed binary integral equation: Based on the RWG triangular element basis function presented above, the current distribution on the patch can be expressed as follows:

$$J_s(\rho') = \sum_{n=1}^{N} I_n f_n(r') \quad (5.5)$$

By substituting Eq. (5.5) into Eq. (5.2) and exchanging the sequence of integral sums, the following can be obtained:

$$- jw\mu_0 \sum_{n=1}^{N} I_n \int_{Sn} G_{xx}^A(r, r') f_n(r') ds'$$

$$+\frac{1}{jw\varepsilon_0}\sum_{n=1}^{N} I_n \nabla \int_{S_n} G_\varphi(r, r')\nabla\prime \cdot f_n(r')ds' = -E_s^i(r) \tag{5.6}$$

By using the Galerkin method to test this formula, the impedance matrix equation can be obtained, as follows:

$$[Z][I] = [V] \tag{5.7}$$

where the expression of the impendence element Z_{mm} is,

$$Z_{mm} = l_m\left[jw\mu_0\left(A^{+}_{mm}\cdot\frac{\rho_m^{c^+}}{2} + A^{-}_{mm}\cdot\frac{\rho_m^{c^-}}{2}\right) - \frac{1}{jw\varepsilon_0}(\varphi^{-}_{mm} - \varphi^{+}_{mm})\right] \tag{5.8}$$

where

$$A^{+}_{mm} = \int_{T_n^{\pm}} G^A_{xx}(r_m^{c\pm}, r')f_n(r')ds',\ \varphi^{\pm}_{mm} = \int_{T_n^{\pm}} G_\varphi(r_m^{c\pm}, r')\nabla_{s\prime}\cdot f_n(r')ds' \tag{5.9}$$

where $r_m^{c\pm}$ denotes the position vector of the barycenter of the triangle $T^{\pm}_{mm}$ related to the mth common side. After solving the spatial domain Green's function (G^A_{xx} and G_φ) using the discrete complex mirror method, the impedance matrix element can be easily obtained. Additionally, the treatment of the singularity encountered in solving the impedance element is the same as that in [5, 6].

When calculating the impendence element, if we directly compute the interaction of a pair of common sides, the interaction between two triangles would be repeatedly computed nine times. Consequently, in performing the calculation, the interaction between each pair of triangles is computed first, after which the interaction is then put into the corresponding impedance elements. In the calculation of the interaction between the surface elements, the integral field is transformed from the triangular area to the rectangular area using coordinate transformation. The computational efficiency can then be improved using the Gaussian boson integral [7]. Moreover, the method of matrix interpolation [8–10] can be used in the algorithm when calculating the electrical characteristics of the antenna across the entire band. This method needs only to calculate the impedance matrixes of a few sampling points. The computational speed can be further improved by using the method of interpolation to calculate the impedance matrixes for the other frequencies.

4. **Far-field calculation**: After obtaining the current distribution by solving the matrix (Eq. (5.7)), the scattered field can be obtained using the reciprocity theorem, as follows:

$$\iint E_{sc}\cdot J_t ds = \iint_s E_t \cdot J_{pat} ds \tag{5.10}$$

where E_{sc} denotes the unknown scattered field, J_t denotes the testing current source introduced to solve the scattered field, E_t denotes the electric field in the patch created by the testing current source, and J_{pat} denotes the electric field created in the patch by an incident plane wave. If we assume that the current source introduced is,

$$J_t = \hat{v}\left(\frac{-j\omega\mu_0}{4\pi r}e^{-jkr}\right)^{-1} E_0\delta(r - r') \tag{5.11}$$

where E_{sc} in the patch created by the current source is equivalent to the field created by the plane wave with an intensity of E_0 and a polarization direction of $\hat{v}$. By substituting Eq. (5.11) into Eq. (5.10), the following equation can be obtained,

$$E_{sc} = \hat{v}\left(\frac{-j\omega\mu_0}{4\pi r}e^{-jkr}\right)\iint_s E_t \cdot J_{pat} ds \tag{5.12}$$

If the direction of the incident plane wave is $\hat{u}$ and the direction of the test current source is $\hat{v}$, E_{vu}^{sc} can be obtained. The RCS can be expressed as follows:

$$\sigma_{vu} = \lim_{r\to\infty} 4\pi r^2 \frac{\left|E_{vu}^{sc}\right|^2}{\left|E_u^i\right|^2} \tag{5.13}$$

If $\hat{u} = \hat{0}$, $\hat{v} = \hat{0}$, $\sigma_{\theta\theta}$ can be obtained. If $\hat{u} = \hat{\theta}$, and $\hat{v} = \hat{\varphi}$, $\sigma_{\varphi\theta}$ can be solved.

5. Numerical results: The first example is presented to calculate the scattering characteristics of a rectangular patch supported by a single-layer grounded dielectric substrate. The related parameters were as follows: the length and width of the patch were both 2 cm, the relative dielectric permittivity of the substrate was $\varepsilon_r = 2.33$, and the relative magnetic permeability was $\mu_r = 1.0$. The substrate thickness was $d = 0.07874$ cm, the incident angles of the plane wave were $\theta_i = 60°$ and $\varphi_i = 60°$, and the incident intensity was 1 V/m. The monostatic scattering characteristics as a function of the frequency of the rectangular patch placed on a single-layer dielectric substrate are shown in Fig. 5.3, with good agreement being evident between the calculation results and those reported in [3].

The second example is provided to calculate the scattering characteristics of a circular patch placed on a single-layer grounded dielectric substrate. The related parameters are as follows: the radius of the patch is 7.1 mm, the relative dielectric permittivity is $\varepsilon_r = 2.2$, the relative magnetic permeability is $\mu_r = 1.0$, the thickness of the medium is $d = 0.0787$ mm, the incident angles of the plane wave are $\theta_i = 60°$ and the incident intensity is 1 V/m. The curve of the monostatic RCS versus the frequency of the circular patch placed on the single-layer medium is shown in Fig. 5.4, with good agreement being achieved between the calculation results and those reported in [11].

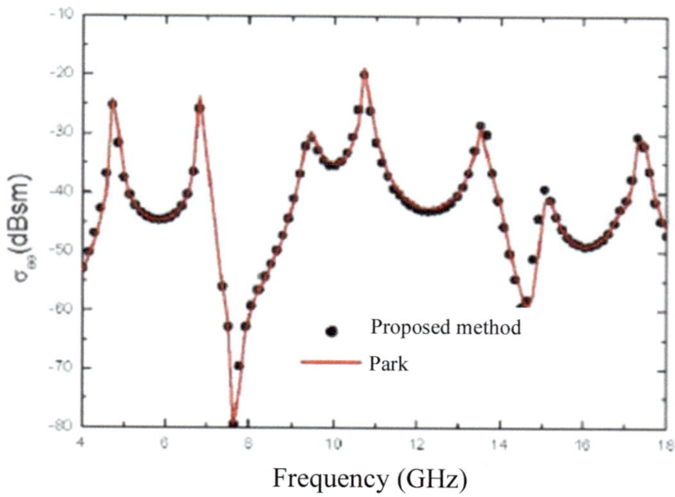

Fig. 5.3 Scattering characteristics of a rectangular patch placed on a single-layer substrate

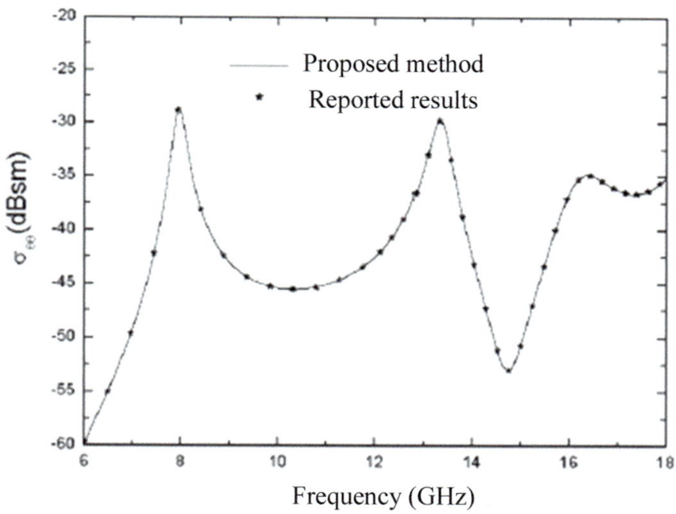

Fig. 5.4 Scattering characteristics of a circular patch placed on a single-layer substrate

The third example is provided to calculate the scattering characteristics of a rectangular patch placed on a dual-layer grounded dielectric substrate. The related parameters are as follows: the relative dielectric permittivity, relative magnetic permeability of the first layer, and the thickness of the medium are $\varepsilon_r = 2.1$, $\mu_r = 1.0$, and $d = 0.07874$ mm, respectively; the relative dielectric permittivity, relative magnetic permeability of the second layer, and the thickness of the medium are $\varepsilon_{r2} = 12.5$, $\mu_r = 1.0$, and $d = 0.03$ mm, respectively, the third layer being the airspace. The side

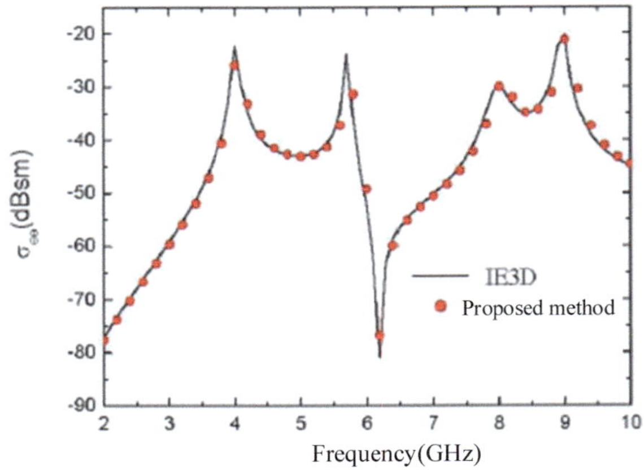

Fig. 5.5 Scattering characteristics of a rectangular patch placed on a dual-layer substrate

length of the rectangular patch is 20 mm, the incident angles of the plane wave are $\theta_i = 60°$ and $\varphi_i = 45°$, and the incident intensity is 1 V/m. The curve of the monostatic RCS versus frequency of the rectangular patch placed on the dual-layer medium is shown in Fig. 5.5, with good agreement being achieved between the calculation results and those calculated using the commercially-available IE3D software.

Figures 5.3, 5.4 and 5.5 show that the results of the calculation method are in good agreement with those from previous studies and commercial software. Moreover, the step approximation error from using the roof basis function in [3] is avoided by using triangles to divide the patch. Additionally, this method can be used to analyze the scattering performance for patches of arbitrary shape. The spatial domain MoM is used in this book. The required spatial domain Green's function can be obtained through simple calculations when computing each matrix element, after which we can obtain the impedance matrix elements. However, the calculation time can be lengthy when using the spectral domain MoM because each matrix element requires time-consuming dual infinite integrals. Nonetheless, using a computer with a 2.4 G Pentium IV processor, the calculation time for the scattering of each frequency point was 26 s when analyzing the scattering characteristics of the rectangular patch (Fig. 5.3) using the proposed method. Moreover, the time was 70 s using the method presented in [3]—that is, the results show the proposed calculation method to be faster than that of [3]. Additionally, the method proposed in [3] could only calculate the scattering characteristics of a single-layer medium. The spatial domain MoM used in this book is more general and can calculate the electrical characteristics of patch antennas with a multi-layer medium.

5.1.3 Calculation of Microstrip Antennas Fed with a Microstrip Line

Microstrip antennas fed with a microstrip line are widely used in engineering, making it important to analyze the performance of such antennas. Using the MoM based on the mixed-bit integral equation studied in the previous section, this section examines the solution for a microstrip antenna fed with microstrip lines or near coupling.

1. Settings of the excitation source: Using a simple excitation source model that is consistent with the physical meaning can simplify the process of analyzing a microstrip antenna using the MoM and further improve the calculation accuracy. The delta-slot voltage model is usually used for the microstrip line feeding structure. As shown in Fig. 5.6, when using a microstrip line feeding structure, the core of the coaxial connector is connected to the microstrip lines, and the skin of the coaxial connector is connected to the ground. In areas far from the microstrip discontinuities and the coaxial connector, the microstrip line can only transmit a quasi-transverse EM (TEM) wave, so the transmission line model can be used to analyze this feeding structure.

The equivalent model of an antenna fed with a microstrip line is shown in Fig. 5.7. The transmission connector can be regarded as an impedance terminal, the transmission line being excited by a total voltage source. In Fig. 5.7, Z_A denotes the input impedance of the antenna, Z_{in} denotes the impedance obtained from the reference sheet, β denotes the transmission constant of the microstrip transmission lines, and Z_c denotes the characteristic impedance of the microstrip lines.

In analysis using the MoM, the voltage matrix can be expressed as follows [13]

$$V_m] = [\delta_{im}] \tag{5.14}$$

in which

Fig. 5.6 Schematic of a microstrip feeding line

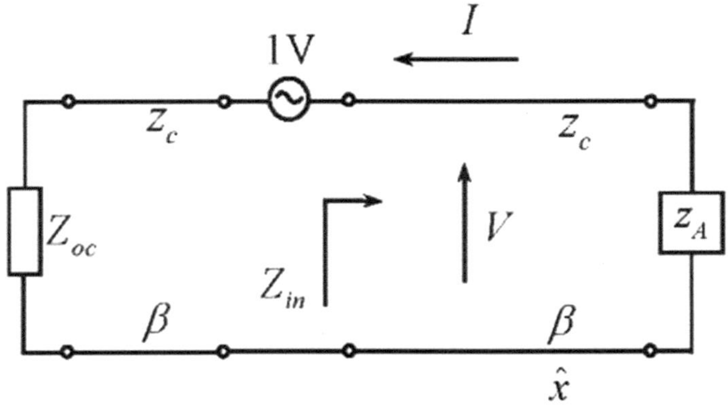

Fig. 5.7 Equivalent circuit diagram of the microstrip feed structure

$$\delta_{im} = \begin{cases} 1, & \text{Location of voltage source } (x_i = x_m) \\ 0, & other(x_i \neq x_m) \end{cases}$$

2. Extraction of the *S* parameter: In the MoM, the impedance matrix can be obtained using Eq. (5.8), the voltage matrix can be obtained using Eq. (5.14), and the *S* parameters can be obtained from the current distribution by solving the matrix equation. This process is called de-embedding. The matching load method [14, 15] and open circuit method [16] can be used to calculate the *S* parameters, and the latter is used in this book. In general, *N* linear independent excitations are required in the extraction of the *S* parameter of an *N*-port microstrip circuit using the open circuit method. A schematic of the multiport microstrip structure is shown in Fig. 5.8.

Assuming that the *i*th port is excited, if the voltage source excitation is added a quarter of a wavelength away from the reference of the *i*th port, the current distribution along the microstrip line of each port can be expressed as follows:

Fig. 5.8 Schematic of the multi-port microstrip structure

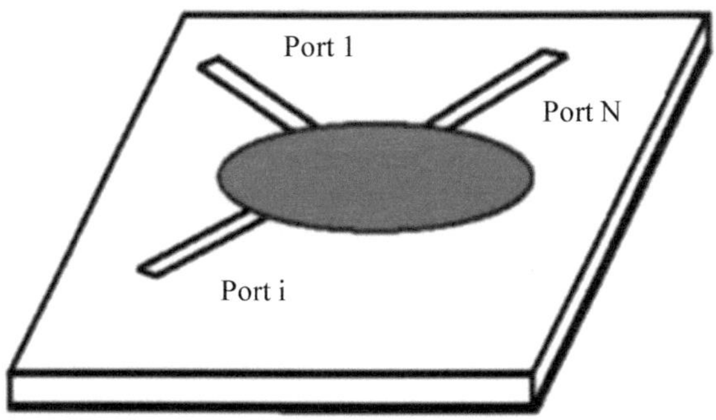

$$I_i = \frac{1}{\sqrt{Z_{oi}}}\left(A_i e^{-j\beta_i x} - B_i e^{j\beta_i x}\right) \tag{5.15}$$

where Z_{oi} denotes the characteristic impedance of the ith port, β_i denotes the transmission constant of the ith port transmission line [17], and A_i and B_i denote the incident wave coefficient and reflected wave coefficient, respectively.

This analysis assumes that the direction of the ith port is in the x-direction. Additionally, we select two consecutive sampling points away from discontinuity and the voltage source of the ith port. The currents at these two points can be expressed as follows [18]:

$$I_i^m = \frac{1}{\sqrt{Z_{oi}}}\left(A_i e^{-j\beta_i x_i^m} - B_i e^{-j\beta_i x_i^m}\right) \tag{5.16}$$

$$I_i^{m+l} = \frac{1}{\sqrt{Z_{oi}}}\left(A_i e^{-j\beta_i x_i^{m+l}} - B_i e^{-j\beta_i x_i^{m+l}}\right) \tag{5.17}$$

The incident wave coefficient and reflected wave coefficient can be obtained by solving these two equations. To obtain a more accurate value, more sampling points and the least squares method can be used to obtain the coefficients. After the analysis of the coefficients of each port, we can use the following equation to obtain the S parameter of the ith port:

$$B_i = \sum_{j=1}^{N} s_{ij} A_i;\ i = 1, N \tag{5.18}$$

By repeating this process for the other $N - 1$ ports, the S parameter of all N ports of the microstrip structure can be obtained. The symmetry and reciprocity theorem are usually used in actual applications to reduce the number of S parameters required for the analysis. For a microstrip antenna fed with a microstrip line, the input impedance of the antenna can be obtained after determining S_{11}.

3. Numerical results: First, the radiation characteristics of a microstrip patch antenna array including four cells fed with a microstrip line were analyzed, the structure of the antenna array being as shown in Fig. 5.9. The dielectric thickness was 1.5748 mm, relative permittivity was 2.1, and operating frequency was 9.42 GHz. Figure 5.10 shows the E-plane radiation pattern calculated using this method, which is consistent with the calculation results of [19]. Deviations occur at 20° and −80°, caused primarily by faulty antenna manufacturing and testing errors.

For the microstrip antenna array, analysis using the MoM requires a greater storage volume and computational overhead. In the calculation of this example, the microstrip antenna was divided into 1,252 triangles with 1,730 basis functions. The storage required was 45.668 Mb, and the computational overhead was large. The adaptive integration method (AIM), which is a fast algorithm, can be applied to this example to reduce the storage capacity and computational resources required.

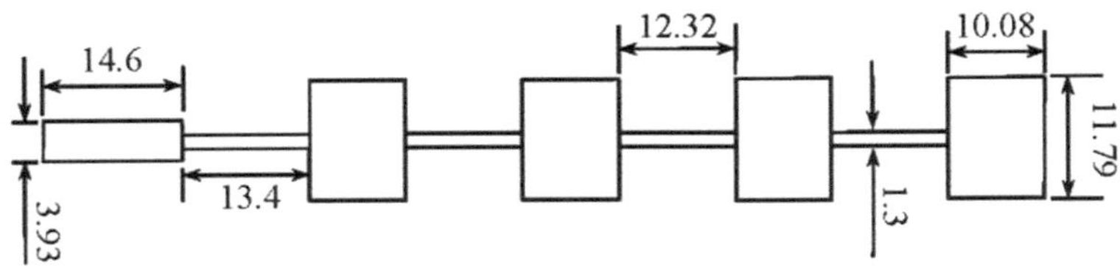

Fig. 5.9 Structural diagram of the microstrip antenna array fed with a microstrip line

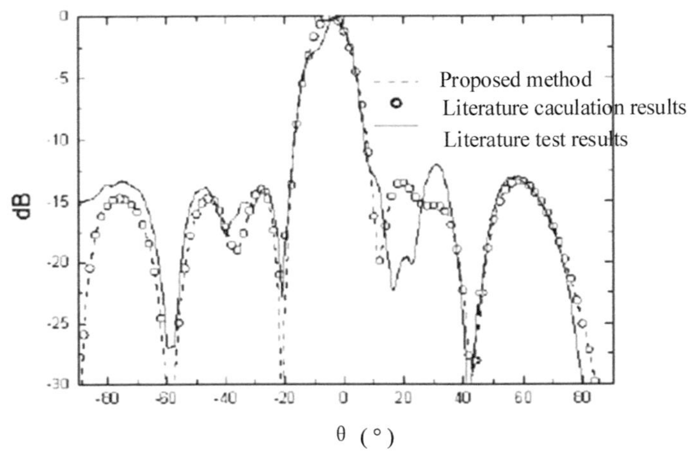

Fig. 5.10 E-plane radiation patterns of the microstrip antenna array fed with a microstrip line

The second example is presented to solve the input impedance of a nearby coupled circular microstrip antenna. The structure of the antenna is shown in Fig. 5.11. The radius of the circular patch is 17.5 mm, the width of the microstrip line is 4.372 mm, and its length is 140 mm. The thickness of the two substrates is 1.59 mm, and the two relative dielectric constants are 2.62. The reference plane is located on the microstrip line 79 mm from the center of the patch. The input impedance on the reference plane was calculated, the calculation frequencies being 2.8–3.15 GHz, and the frequency step being 50 MHz.

Figure 5.12 shows the calculated results, which agree well with the experimental results of [12].

5.1.4 Calculation of a Microstrip Antenna Fed with a Coaxial Line

Microstrip antennas fed with coaxial lines are widely used in real-world applications. Consequently, the MoM for microstrip antennas fed with coaxial lines is examined

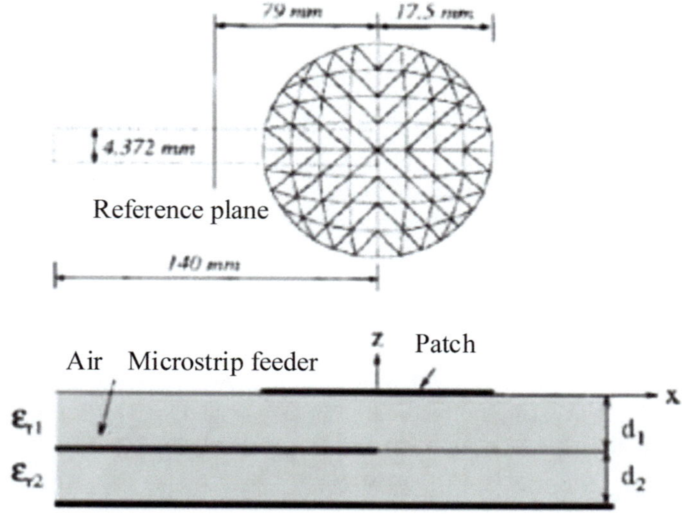

Fig. 5.11 Structure of the nearly coupled microstrip antenna

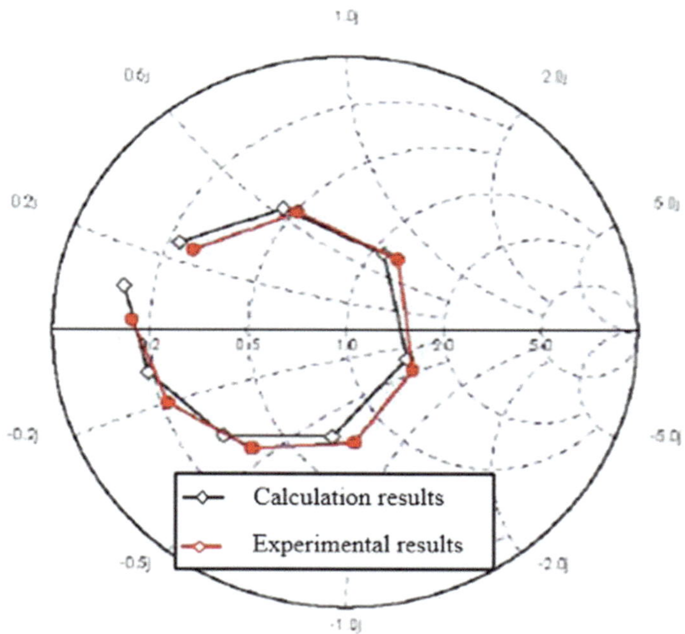

Fig. 5.12 Input impendence of the nearly coupled microstrip antenna

and analyzed in this section. A simplified additional mode is introduced in the analysis to simulate the current distribution at the junction between the probe and the patch. Combined with the discrete complex image method of the spatial domain Green's function, a rapid analysis of microstrip antennas fed with coaxial lines can be achieved.

1. Calculation of the input impedance: Many methods can be used to analyze the input impedance of a microstrip patch antenna, such as the transmission line model, cavity model, and MoM. The transmission line and cavity models neglect the influence of surface waves, and the cavity model assumes that the probe current is uniformly distributed. Consequently, these models can only be used to analyze the input impedance of thin dielectrics. The MoM selects the proper basis and testing functions to solve the electric field integral equation. Moreover, the changing current in the feeding probe is considered. This method can also solve the input impedance of microstrip antennas of any thickness.

The input impedance of a microstrip antenna of arbitrary shape fed with a coaxial line has been computed in [20]. This method uses the numerical integral method to solve the spatial domain Green's function. Because the spatial domain Green's function contains complex Sommerfeld integration, it can be time-consuming to directly adopt the numerical integral method. In [21], the discrete complex image method (DCIM) and roof basis function are used to solve the electrical characteristics of a rectangular microstrip antenna fed with a probe. This method is not universal, so it cannot be used to analyze a microstrip antenna of arbitrary shape, and the excitation source model used by this method is too simple to be used to analyze a microstrip antenna with a thick medium. The spatial domain MoM based on the mixed-bit integral equation is used here to solve the input impedance of a microstrip antenna fed with a probe. In the patch-probe connection, the simplified additional model basis function can be used to simulate the current distribution, ensuring calculation accuracy and convenient computation. The MoM can be used to analyze the input impedance of a microstrip patch antenna fed with a coaxial line, as shown in Fig. 5.13.

The following equations can be obtained based on the boundary condition that the metal surface's tangential electric field is zero:

$$
\begin{aligned}
& j\omega\mu_0 \int_s G_{xx}^{11}(\vec{r},\vec{r}\prime)\vec{J}_s(\vec{r}\prime)ds' - \frac{1}{j\omega\varepsilon_0}\nabla_s \int_s G_{\varphi}^{11}(\vec{r},\vec{r}\prime)\nabla_{s'}\cdot\vec{J}_s(\vec{r}\prime)ds' \\
& + j\omega\mu_0 \int_0^h [\hat{x}G_{xz}^{12}(\vec{r},\vec{r}') + \hat{y}G_{yz}^{12}(\vec{r},\vec{r}')]I(z')dz' \\
& - \frac{1}{j\omega\varepsilon_0}\nabla_s \int_0^h G_{\varphi}^{12}(\vec{r},\vec{r}\prime)\frac{dI(z')}{dz'}dz' = \vec{E}_s^i(\vec{r})\vec{r} \in S
\end{aligned}
\tag{5.19}
$$

Fig. 5.13 Schematic of a microstrip patch antenna fed with a coaxial line

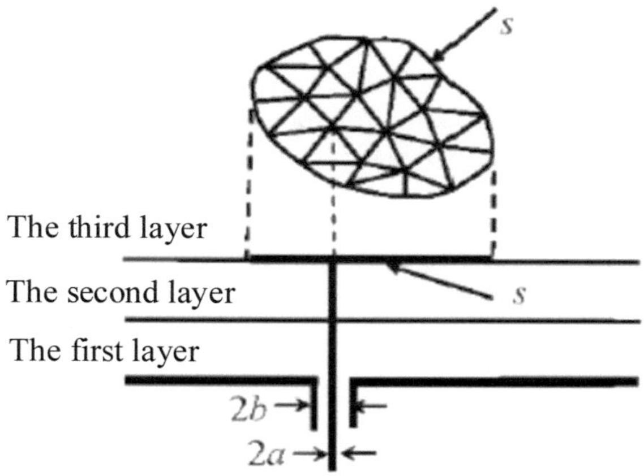

$$
\begin{aligned}
& j\omega\mu_0 \int_s [\hat{x}G_{zx}^{21}(z,\vec{r}') + \hat{y}G_{zy}^{21}(z,\vec{r}')]\vec{J}_s(\vec{r}')ds' \\
& -\frac{1}{j\omega\varepsilon_0}\frac{d}{dz}\int_s G_\varphi^{21}(z,\vec{r}\prime)\nabla_{s\prime}\cdot\vec{J}_s(\vec{r}')ds' \\
& +j\omega\mu_0\int_0^h G_{zz}^{22}(z,z')I(z')dz' \\
& -\frac{1}{j\omega\varepsilon_0}\frac{d}{dz}\int_0^h G_\varphi^{22}(z,z')\frac{dI(z')}{dz'}dz' = \vec{E}_z^i(z) 0 < \mathrm{z} < \mathrm{h}
\end{aligned}
\tag{5.20}
$$

where $G^{mm}(r, r')$ denotes the spatial domain Green's function for the field in the mth layer and the source in the nth layer.

We can assume that there is a TEM mode field in the aperture of the coaxial line and that the voltage between the core and skin of the coaxial line is U_l. The equivalent magnetic current ring can be used to express the coaxial line aperture in the calculation and analysis. Additionally, the electric fields $E_S^i(r)$ and $E_z^i(z)$ excited by the magnetic ring on the conductor surface can be expressed as follows:

$$
\begin{aligned}
E_{x,y}^i(\vec{r}) &= 2\pi K\frac{d}{dx,y}S_0\{[J_0(k_\rho a) - J_0(k_\rho b)]U_v^e/k_\rho^2\}, \\
K &= \frac{U_L}{\ln(b/a)} \\
E_z^i(\vec{r}) &= \frac{2\pi K}{j\omega\varepsilon_0\varepsilon_r}S_0\{[J_0(k_\rho a) - J_0(k_\rho b)]I_v^e\},
\end{aligned}
\tag{5.21}
$$

$$K = \frac{U_L}{\ln(b/a)} \tag{5.22}$$

where S_0 denotes the zero-order Sommerfeld integral, and U_v^e and I_v^e denote the voltage and current in the equivalent transmission line, respectively.

In the calculation using the spatial domain MoM, the current distribution on the patch can be expanded using the RWG vector basis function (f_n^p), and the current on the probe can be expanded using trigonometric functions. As shown in Fig. 5.14, the probe can be divided into N_w segments. The triangle function at two adjacent segments can be defined as follows:

$$\vec{f}_n^{w}(z) = \begin{cases} \frac{z-z_{n-1}}{\Delta n_-}\hat{z},\ z_{n-1} \le z \le z_n \\ \frac{z_{n+1}-z}{\Delta n_+}\hat{z},\ z_n \le z \le z_{n+1} \end{cases} \tag{5.23}$$

At the junction between the probe and the patch, the current requires special handling as it changes rapidly. An additional mode basis function can be introduced to accurately simulate the current distribution at the junction. Wu proposed that the additional mode basis function at the junction between the line and surface [22] could be used to analyze the input impedance of microstrip antennas fed with a probe. However, owing to the singularity of the basis function, this method is complicated. This book adopts the simplified additional model basis function without a singularity proposed in [11] to simulate the current distribution at the junction between the line

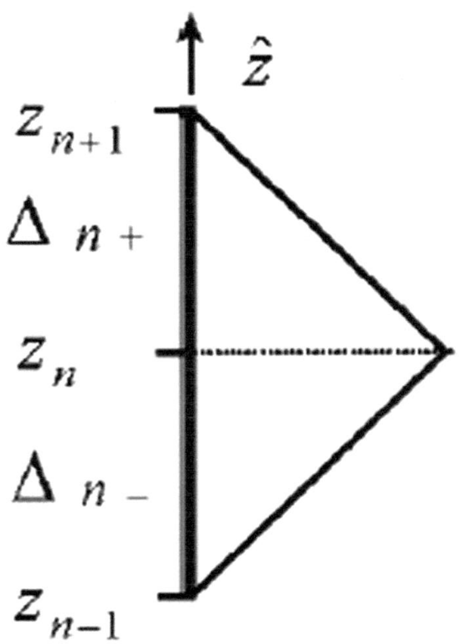

Fig. 5.14 Schematic of the triangle function

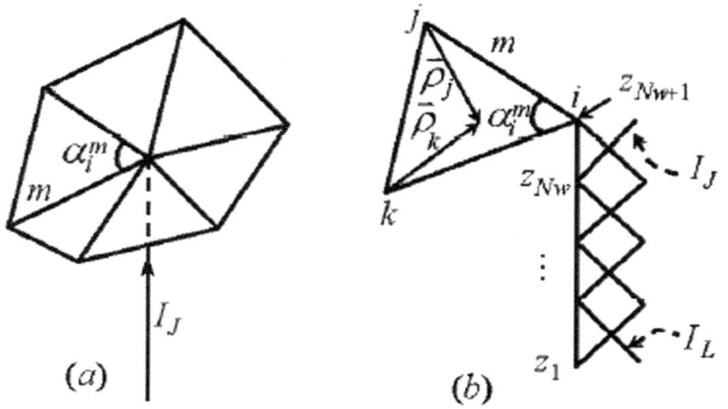

Fig. 5.15 Schematic of the junction between a line and a surface

and surface, which increases the calculation accuracy and reduces the calculation complexity.

As shown in Fig. 5.15a, the number of triangles connected to the probe in the patch is N_t, the N_t triangle being called the node triangle.

The current in the mth node triangle excited by the additional mode basis function can be expressed as follows:

$$\vec{J}_s = I_J \vec{\Pi}_m = -I_J \frac{\alpha_i^m}{2\alpha} \left(\vec{\Lambda}_j^m + \vec{\Lambda}_k^m \right) \tag{5.24}$$

where α denotes the sum of the vertex angles of all N_t node triangles at the nodes. In general, $\alpha = 2\pi$, i, j, and k denote the vertex angles of the mth triangle, and Λ_j^m and Λ_k^m denotes vector functions related to the mth node triangle, which can be defined as follows:

$$\vec{\Lambda}_j^m = \frac{\vec{\rho}_j}{2A}; \ \vec{\Lambda}_k^m = \frac{\vec{\rho}_k}{2A} \tag{5.25}$$

where A denotes the area of mth mode triangle.

The current at the other node triangle created by the additional mode basis function can be obtained similarly. As shown in Fig. 5.15b, the additional mode function can only be applied to the N_{wth} line segment that is connected to node i. The current created by the additional node in this segment can be expressed as follows:

$$\vec{J}_w = \hat{z} I_J f_{Nw}^w(z) = \hat{z} I_J \frac{z - z_{Nw}}{\Delta_{Nw+}} \tag{5.26}$$

where $f_{Nw}^w(z)$ denotes the half-triangle basis function of the N_w^{th} line segment connected to the patch (as shown in Fig. 5.15b). The expansion coefficient of this half-triangle basis function is I_j. The current distribution in the microstrip and the probe can be expanded using the flat triangle RWG vector basis function, the piecewise triangle basis function Eq. (5.25), and the basis function at the junction of the line and the surface Eqs. (5.24) and (5.26).

$$\vec{J}\left(\vec{r}\right) = \sum_{n=1}^{N_P} I_n^P \vec{f}_n^P\left(\vec{r}\right) + I_J \sum_{m=1}^{N_T} \vec{\prod_m}\left(\vec{r}\right), \tag{5.27}$$

$\vec{r}$ is on the conductor surface

$$\vec{I}\left(\vec{r}\right) = \sum_{n=1}^{N_W} I_n^W \vec{f}_n^w\left(\vec{r}\right) + I_J \vec{f}_{Nw}^w\left(\vec{r}\right), \ \vec{r} \text{ is on the probe} \tag{5.28}$$

where N_p and N_w denote the numbers of RWG basis functions and segment triangles, respectively, and I_n^P, I_n^w, and I_J denote the current expansion coefficients on the surface, antenna, and the junction between the line and surface, respectively. By substituting Eqs. (5.27) and (5.28) into Eq. (5.20), using the Galerkin method, and solving the inner product of the two sides of Eqs. (5.19) and (5.20), the impedance matrix equation can be obtained, as follows:

$$\begin{bmatrix} Z_{PP} & Z_{Pw} & Z_{PJ} \\ Z_{wp} & Z_{ww} & Z_{wJ} \\ Z_{JP} & Z_{Jw} & Z_{JJ} \end{bmatrix} \begin{bmatrix} I_P \\ I_w \\ I_J \end{bmatrix} = \begin{bmatrix} V_P \\ V_w \\ V_J \end{bmatrix} \tag{5.29}$$

After solving the matrix equation, the current expansion coefficient can be obtained. By determining the current expansion coefficient I_L of the half-triangle basis function that is defined in the first segment of the probe, the input impedance of the microstrip antenna can be calculated, as follows:

$$Z_{in} = \frac{V_L}{I_L} \tag{5.30}$$

2. Calculation of the scattering properties: Under an incident plane wave, the scattering field of the microstrip antennas **can be** divided into two parts—**that is**, the **EM** field radiated by the current excited by the incident wave on the antenna surface and the **EM** field radiated into space by the excitation of the antenna due to the unmatched load connected to the feeding port. As shown in Fig. 5.16, a plane wave irradiates a microstrip antenna fed with a coaxial line with a load $\boldsymbol{Z_l}$. In the analysis, the intensity of the incident wave is given, and the coaxial feed port is connected to the load $\boldsymbol{Z_l}$ without excitation. It is evident that there are several similarities between using the MoM analysis to analyze the scattering characteristics of the microstrip antenna and using it to analyze the input impedance of the microstrip antenna.

The following two equations can be obtained based on the boundary condition that the tangential electric field on the metal surface is zero:

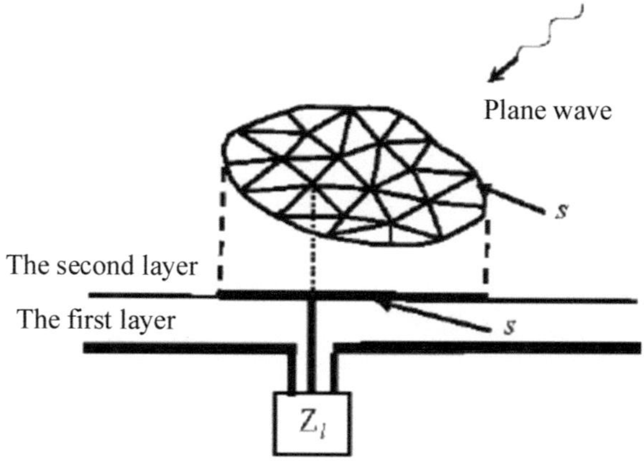

Fig. 5.16 Schematic of a microstrip antenna fed with a coaxial line

$$
\begin{aligned}
& j\omega\mu_0 \int_S G_{xx}^{11}\left(\vec{r}, \vec{r}\,'\right) \vec{J}_S\left(\vec{r}\,'\right) ds' - \frac{1}{j\omega\varepsilon_0} \nabla_s \int_S G_\phi^{11}\left(\vec{r}, \vec{r}\,'\right) \nabla'_s \cdot \vec{J}_S\left(\vec{r}\,'\right) ds' \\
& + j\omega\mu_0 \int_0^h \left[\hat{x} G_{xz}^{12}\left(\vec{r}, \vec{r}\,'\right) + \hat{y} G_{yz}^{12}\left(\vec{r}, \vec{r}\,'\right)\right] I(z') dz' - \frac{1}{j\omega\varepsilon_0} \nabla_s \int_0^h G_\phi^{12}\left(\vec{r}, \vec{r}\,'\right) \frac{dI(z')}{dz'} dz' \\
& - \hat{x} E_x^{mag}\left(\vec{r}, \vec{r}\,'\right) - \hat{y} E_y^{mag}\left(\vec{r}, \vec{r}\,'\right) = \vec{E}_s^{\,i}\left(\vec{r}\right) \quad \vec{r} \in S
\end{aligned}
\tag{5.31}
$$

$$
\begin{aligned}
& j\omega\mu_0 \int_S \left[G_{zx}^{21}\left(z, \vec{r}\,'\right) + \hat{y} G_{zy}^{21}\left(z, \vec{r}\,'\right)\right] \vec{J}_S\left(\vec{r}\,'\right) ds' - \frac{1}{j\omega\varepsilon_0} \frac{d}{dz} \int_S G_\phi^{21}\left(z, \vec{r}\,'\right) \nabla'_s \cdot \vec{J}_S\left(\vec{r}\,'\right) ds' \\
& + j\omega\mu_0 \int_0^h G_{zz}^{22}(z, z') I(z') dz' - \frac{1}{j\omega\varepsilon_0} \frac{d}{dz} \int_0^h G_\phi^{12}(z, z') \frac{dI(z')}{dz'} dz' \\
& - \hat{x} E_x^{mag}\left(\vec{r}, \vec{r}\,'\right) - \hat{y} E_y^{mag}\left(\vec{r}, \vec{r}\,'\right) = \vec{E}_z^{\,i}(z) 0 < z < h
\end{aligned}
\tag{5.32}
$$

where r denotes the field position vector, r' denotes the source position vector, $J_s(r')$ denotes the current distribution on the patch, $I(z')$ denotes the current distribution on the probe, and E_x^{mag}, E_y^{mag}, and E_z^{mag} denote the electric field components generated by the equivalent magnetic current loop at the coaxial aperture.

After removing the patch and the probe (leaving only the dielectric and the ground plane), the total field generated by the incident plane wave at the original position of the patch is $E_s^i(r)$, and the total field generated at the original position of the probe is $E_s^i(z)$.

Next, the coaxial line's aperture can be expressed using an equivalent magnetic current ring. Assuming that there is a TEM mode field in the coaxial line's aperture and that the voltage between the core and skin of the coaxial line is U_L, the electric field excited by the equivalent magnetic ring in Eqs. (5.31) and (5.32) can be expressed as follows:

$$\begin{aligned} E_t^{mag}(\vec{r}) &= 2\pi K \frac{d}{dt} S_0\{[J_0(k_\rho a) - J_0(k_\rho b)] U_v^e / k_\rho^2\}, \\ K &= \frac{U_L}{\ln(b/a)}, t = x, y \end{aligned} \tag{5.33}$$

$$\begin{aligned} E_z^{mag}(\vec{r}) &= \frac{2\pi K}{j\omega\varepsilon_0\varepsilon_r} S_0\{[J_0(k_\rho a) - J_0(k_\rho b)] I_v^e\}, \\ K &= \frac{U_L}{\ln(b/a)} \end{aligned} \tag{5.34}$$

where a denotes the internal diameter of the coaxial line, b denotes the outside diameter of the coaxial line, $U_L = Z_L I_L$, Z_L denotes the load impedance, and I_L denotes the current on the load. Additionally, the current together with the currents in the patch and probe (Eqs. (5.31) and (5.32)) can be solved using the MoM.

We can use the spatial domain MoM to select the basis function that is the same as that used in the analysis of the input impedance of microstrip antennas. Additionally, the Galerkin method can be used to solve the inner product of the two sides of the mixed-bit integral Eqs. (5.31) and (5.32) and then obtain the impedance matrix equation. The scattering field of the microstrip antenna can be solved after obtaining the current distribution by calculating the impedance matrix equation.

3. Numerical results: The following example is presented to solve the input impedance and scattering characteristics of microstrip antennas fed with a coaxial line using the MoM. First, we calculated the input impedance of a rectangular microstrip patch antenna fed with a coaxial line. As shown in Fig. 5.17, the structural parameters of the rectangular microstrip patch antennas were as follows: length $L = 20.1$ mm and width $W = 20.1$ mm. We assumed that the center of the rectangular patch was the coordinate origin, and the feeding positions were $x = 8.75$ mm and $y = 0.05$ mm. The radius of the coaxial line core was $\boldsymbol{r}_0 = 0.635$ mm. The relative dielectric constant was $\boldsymbol{\varepsilon}_r = 2.55$, the dielectric thickness was $d = 1.59$ mm, the working frequency was 4.15–4.65 GHz, and the frequency step was 2.5 MHz.

The rectangle was divided into 98 triangles in the calculation. The calculation results are shown in Fig. 5.17 and are in good agreement with the results of [23].

The second example is presented to calculate the input impedance of a circular patch antenna fed with a coaxial line, a schematic of which is shown in Fig. 5.18. The related parameters are as follows: the radius $R = 20$ mm, the center of the circular patch is assumed to be at the coordinate origin, so the feeding position is $x = 7$ mm and $y = 0$ mm, the radius of the coaxial line core is $r_0 = 0.45$ mm, the relative dielectric constant is $\varepsilon_r = 2.33$, the dielectric thickness is $d = 2.1844$ mm, and the working frequency is 2–3 GHz. The circular patch is divided into 128 triangles for the calculation. The input impedance of the antenna is as shown in Fig. 5.18 and is in good agreement with the experimental results of [24].

The third example calculates the scattering characteristics of a circular microstrip antenna fed with a coaxial line. The microstrip antenna and the input impedance

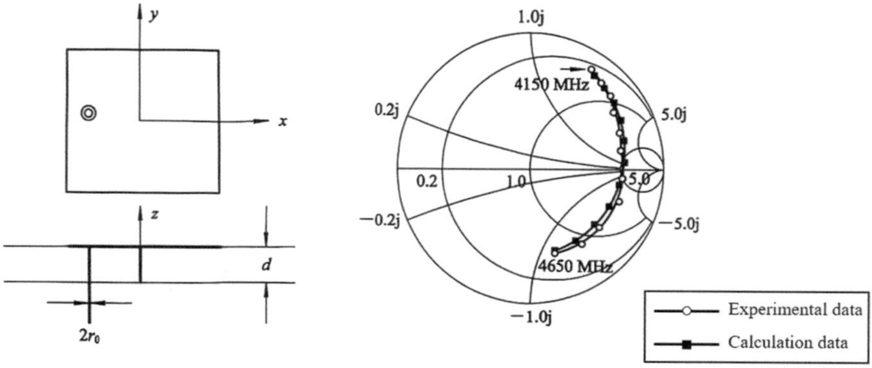

Fig. 5.17 Schematic of a rectangular patch antenna and the input impedance characteristics

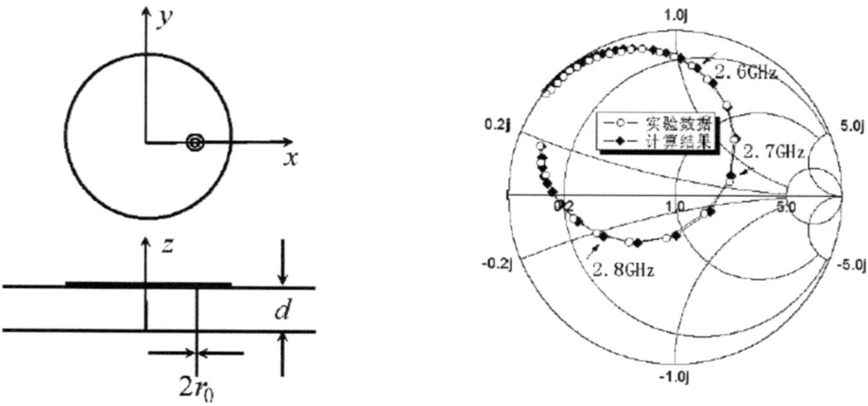

Fig. 5.18 Schematic of a circular microstrip antenna and the input impedance characteristics

characteristics are shown in Fig. 5.18. The relative dielectric constant is $\varepsilon_r = 2.2$, the loss tangent angle is $\tan\delta = 0.0009$, and the dielectric thickness is $h = 0.787$ mm. The radius of the coaxial line core is $a = 0.432$ mm, the outside diameter $b = 1.397$ mm, and the feeding position is $x = 2.5$ mm and $y = 0$ mm. The incident angle of the plane wave is $\theta_i = 60°$, $\varphi_i = 0°$, and the incident intensity is 1 V/m.

The RCS characteristics of the antenna with the feeding port loaded with $Z_l = 50\ \Omega$ from 6–15 GHz are shown in Fig. 5.19. Good consistency was achieved between the calculated results and those of [11]. The calculation method used in [11] was also based on the spatial domain moment method of the mixed-bit integral equation. However, the numerical integral method was used to solve the spatial domain Green's function, which was slower than using the secondary discrete complex image method. Compared with the results of [11], the calculated results at low frequencies agreed well with the experimental results, but at high frequencies the results were

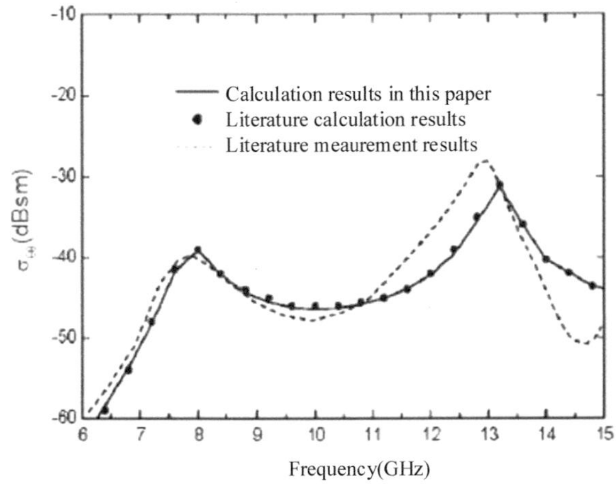

Fig. 5.19 RCS of a circular microstrip antenna with $Z_i = 50\,\Omega$

not consistent, owing primarily to measurement errors and the poor performance of the SMA connector at high frequencies [24].

5.2 Rapid Calculation Method of Microstrip Antenna Array RCS

5.2.1 Introduction

Since it was created in the 1960s, the MoM has been widely used to solve EM problems, including the calculation of the electrical properties of antennas and the scattering properties of complex scatterers. It is a low-frequency method that requires a primary function to simulate the electric surface distribution of the target. Owing to the general setting of a wavelength being divided into ten parts, an unknown larger number is required when the target size is too large. This can result in greater demand for computer memory and longer calculation times. Consequently, the traditional MoM can only be used to analyze objects with dimensions of several wavelengths.

To overcome these shortcomings, rapid calculation methods—such as the conjugate gradient fast Fourier transform (CGFFT), AIM, and fast multipole method (FMM)—have been developed in recent years. Based on the convolution property of the Green's functions, the CGFFT increases the speed of the matrix vector multiplication [25]. The current distribution can then be obtained by solving the matrix equation based on the iterative method. However, using the FFT to speed up the matrix vector multiplication requires the use of a rectangular or cuboid network (structure) to divide the target. A target that is not a uniform rectangle or cuboid can cause a step approximation error and result in the reduction of the calculation precision. The AIM [26] can be used to analyze the characteristics of irregular objects, first by using triangular facets or tetrahedra to divide the object and then by mapping the basis function and divergence or curl defined on the triangular facets or tetrahedra to the nearby rectangular network nodes using a uniform partition. Finally, the current distribution can be obtained by solving the impedance matrix equation using the CGFFT. The AIM can effectively reduce the required computer storage requirements as it does not need to store the impedance matrix elements in the far field (by replacing the basis function with multiple point resources).

The second part of this section presents the basic principles and operational process of the AIM. To reduce the computational complexity and memory requirements of the analysis of a large perfect electric conductor (PEC), the AIM method can be used to calculate the scattering properties of a large metal plate and metal ball. The analysis of the electrical properties of a large microstrip antenna array using the AIM method is discussed in the third part of this section, and the computational complexity and memory requirements of the analysis are compared to the traditional MoM method.

5.2.2 *Adaptive Integration Method (AIM)*

To address the large memory requirements and lengthy computational time of the impedance matrix equation for analysis of the electrical characteristics of large electrical targets using the traditional MoM, Bleszynski proposed [26] applying the AIM to analyze their radiation and scattering performance. Compared with the traditional MoM, the complexity and storage requirements of the impedance matrix equation calculations were reduced considerably. The basic principles and operational processing are discussed here, with the AIM being used to analyze the scattering performance of a large electrical conductor.

1. AIM: A brief description of the basic principles and operational processes of the AIM [26] is presented below using an analysis of the scattering characteristics of a PEC as an example. First, we can analyze the current distribution using the traditional MoM. Based on the continuous boundary condition of the tangential electric field on the conductor surface, the electric field integral equation (EFIE) can be expressed as follows:

$$-\vec{E}^{i}_{\tan}(\vec{r}) = [-j\omega\vec{A}(\vec{r}) - \nabla\phi(\vec{r})]_{\tan}, \vec{r} \text{ is on the S} \tag{5.35}$$

where E^{i}_{tan} denotes the incident field on the conductor surface, and $A(r)$ and $\phi(r)$ denote bit functions:

$$\vec{A}(\vec{r}) = \frac{\mu}{4\pi}\int\limits_{s}\vec{J}(\vec{r}')\frac{e^{-jkR}}{R}ds' \tag{5.36}$$

$$\phi(\vec{r}) = -\frac{1}{j4\pi\omega\varepsilon} \int_s \nabla' \cdot \vec{J}(\vec{r}') \frac{e^{-jkR}}{R} ds' \tag{5.37}$$

where $J(r')$ denotes the current distribution density on the conductor surface, k denotes the wavenumber in free space, and $R = |r - r'|$ denotes the distance from the field point to the source point.

When using the MoM to solve Eq. (5.35), the current distribution can be expanded to a linear combination of N basis functions $f_n(r')$, as follows:

$$\vec{J}(\vec{r}') = \sum_{n=1}^{N} I_n \vec{f}_n(\vec{r}') \tag{5.38}$$

We can obtain the impedance matrix equation by bringing Eq. (5.38) into the EFIE and taking the inner product of both sides of the EFIE using the Galerkin method, as follows:

$$[Z][I] = [V] \tag{5.39}$$

in which:

$$\begin{aligned} Z_{mn} &= -j\omega\mu \int_{S_m} \int_{S_n} \vec{f}_m(\vec{r}) \cdot \vec{f}_n(\vec{r}') G(\vec{r}, \vec{r}') ds' ds \\ &= -\frac{1}{j\omega\varepsilon} \int_{S_m} \int_{S_n} \nabla \cdot \vec{f}_m(\vec{r}) \nabla' \cdot \vec{f}_n(\vec{r}') G(\vec{r}, \vec{r}') ds' ds \end{aligned} \tag{5.40}$$

The impedance matrix $[Z]$ is a dense matrix and storage of $O(N^2)$ is required when using the traditional MoM. However, the computational complexity and memory requirements decrease when using the AIM.

Figure 5.20 shows the processing that brings the MoM basis function into the AIM grid points in the analysis of the scattering characteristics of a metal disk. As shown in Fig. 5.20a, a metal disk that is surrounded by a rectangle is divided into triangular patches. $N_x \times N_y$ grid points are evenly distributed on the rectangle, and the spacings in the x- and y-directions are Δx and Δy, respectively. The basis function $f_n(r')$ used in the MoM is the RWG function, which can be written as Eq. (5.3).

Next, the linear combination of the δ function that is located at the $(M+1)^2$ grid points around the center of the basis function approximately represents each basis function $f_n(r')$ and its divergence $\nabla \cdot f_n(r')$ from the original MoM [26] (Fig. 5.20b). If $\psi_n(r')$ X is used to represent $f_n(r')$, then:

$$\vec{\Psi}_n(\vec{r}\prime) = \sum_{q=1}^{(M+1)^2} \delta(x - x_{nq}\prime)\delta(y - y_{nq}\prime)\left[\Lambda_{nq}^{x}\hat{x} + \Lambda_{nq}^{y}\hat{y}\right] \tag{5.41}$$

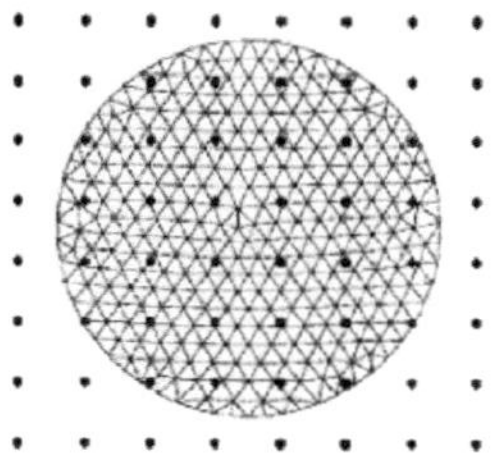

(a) Discretized metal patches placed on uniform grid points

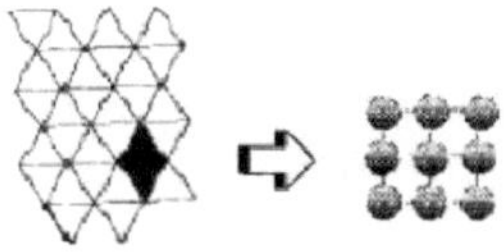

(b) δ function used to represent a basis function

Fig. 5.20 Process of transforming the moment basis function to AIM grid points

where r'_{nq} denotes the location vector of the $(M+1)^2$ grid point.

To solve the coefficient $\Lambda_{nq}^{x,y}$, the following equation should be solved:

$$\vec{M}^{n}_{q_1,q_2} = \vec{F}^{n}_{q_1,q_2} \tag{5.42}$$

in which:

$$\begin{aligned}\vec{M}^{n}_{q_1,q_2} &= \int_{-\infty}^{\infty}\int_{-\infty}^{\infty} \vec{\Psi}_n(\vec{r}\prime)(x'-x_0)^{q_1}(y'-y_0)^{q_2}dx'dy' \\ &= \sum_{q=1}^{(\mathrm{M}+1)^2}\left[\Lambda_{nq}^{x}\overset{\Lambda}{x}+\Lambda_{nq}^{y}\overset{\Lambda}{y}\right](x_{mq}-x_0)^{q_1}(y_{mq}-y_0)^{q_2}\end{aligned} \tag{5.43}$$

$$\vec{F}^{n}_{q_1,q_2} = \int_{-\infty}^{\infty}\int_{-\infty}^{\infty} \vec{\Psi}_n(\vec{r}\prime)(x'-x_0)^{q_1}(y'-y_0)^{q_2}dx'dy' \tag{5.44}$$

If $\Psi_n^d(r) = \nabla \cdot f_n(r')$, then:

$$\vec{\Psi}_n^d(\vec{r}\prime) = \sum_{q=1}^{(M+1)^2} \delta(x-x_{nq}\prime)\delta(y-y_{nq}\prime)\Lambda_{nq}^{d} \tag{5.45}$$

To solve the coefficient Λ_{nq}^{∂}, the following equation should be solved:

$$\int_{-\infty}^{\infty}\int_{-\infty}^{\infty} \vec{\Psi}_n^d(\vec{r}\prime)(x'-x_0)^{q_1}(y'-y_0)^{q_2}dx'dy' = \sum_{q=1}^{(M+1)^2} \Lambda_{nq}^{d}(x_{mq}-x_0)^{q_1}(y_{mq}-y_0)^{q_2} \tag{5.46}$$

where $0 \le q_1$, and $q_2 \le M$.

After the expansion coefficient (Λ) of every basis function is obtained, the AIM divides the impedance matrix elements of the MoM into near-field and far-field elements, which can then be solved using the traditional MoM and the approximation method, respectively. By bringing the basis and test functions that are expressed by Eqs. (5.41) and (5.42), respectively, into Eq. (5.40), the far-field element can be expressed as follows:

$$Z_{mn}^{far} = \sum_{u=1}^{(M+1)^2} \sum_{v=1}^{(M+1)^2} \left[-j\omega\mu \left(\Lambda_{mu}^{x} \Lambda_{mv}^{x} + \Lambda_{mu}^{y} \Lambda_{mv}^{y} \right) - \frac{1}{j\omega\varepsilon} \Lambda_{mu}^{d} \Lambda_{mv}^{d} \right] G(\vec{r}_u, \vec{r}_{v'}) \tag{5.47}$$

If the near-field element is defined as follows:

$$Z_{mn}^{near} = \begin{cases} Z_{mn} - Z_{mn}^{far}, & if\ |Z_{mn} - Z_{mn}^{far}| \ge \varepsilon_{near} |Z_{mn}|, \nexists d_{mn} \le d_{near} \\ 0, & \text{other} \end{cases} \tag{5.48}$$

then the impedance matrix can be written as the sum of $[Z]^{near}$ and $[Z]^{far}$, as follows:

$$[Z] = \left([Z] - [Z]^{far} \right) + [Z]^{far} = [Z]^{near} + [Z]^{far} \tag{5.49}$$

where d_{mn} denotes the distance between the mth testing function and the nth basis function center, d_{near} denotes the near-field threshold, and ε_{near} denotes the near-field tolerance error.

The process of multiplying the impedance matrix by the current matrix in solving the Eq. (5.39) based on the iterative method can be expressed as follows:

$$\begin{aligned} [V] &= [Z][I] = [Z]^{near}[I] + [Z]^{for}[I] \\ &= [Z]^{near}[I] + [\Lambda]\mathbb{F}^{-1}\left\{ \mathbb{F}([G])\mathbb{F}\left([\Lambda]^T [I] \right) \right\} \end{aligned} \tag{5.50}$$

where $F(\cdot)$ denotes the Fourier transform.

Here, $[Z]^{near}$ requires little memory because it is a sparse matrix; additionally, little computational complexity is involved in multiplying it with the current vector. Based on the convolution property of the Green's function, rapid multiplication of $[Z]^{far}$ and the current vector using the FFT can be achieved. Consequently, the AIM lays the foundation for solving electrically large objects using the MoM because the AIM has lower memory requirements than the traditional MoM and reduces the computation for each multiplication. The storage requirement to analyze a plate structure based on the AIM is $O\left(N^{3/2}\right)$, and the computation required is $O\left(N^{3/2} \lg N\right)$. The calculation processing is similar to Eqs. (5.41)–(5.50) with required storage of $O(N)$ and computation of $O(N \lg N)$.

Two processes are involved in the analysis of the electrical characteristics of the target using the AIM—that is, establishing the matrix and matrix vector multiplication.

1. Establishing the matrix
   (1) Establish the AIM grid points on the target.
   (2) Approximate the function $\Psi_n(r')$ using a linear combination of the δ function that is defined on the AIM $(M+1)^2$ grid points. The coefficients of the linear combination should be determined.
   (3) Establish the Green's function matrix $[G]$ by computing the Green's function on the AIM grid points.
   (4) Compute the near-field element using the traditional MoM.
2. Matrix vector multiplication
   (1) Compute $[\hat{I}] = [\Lambda]^T[I]$.
   (2) Use the FFT to compute $[\tilde{I}] = F[\hat{I}]$, $[\tilde{G}] = F[G]$.
   (3) Calculate the inverse Fourier transform $[\hat{V}] = F^{-1}[\tilde{V}] = F^{-1}([\tilde{G}][\tilde{I}])$.
   (4) Calculate $[V]^{far} = [\Lambda][\hat{V}]$, $[V]^{near} = [Z]^{near}[I]$.
   (5) The sum is $[V] = [V]^{near} + [V]^{far}$.

It should be noted that a (M + 1)2δ function is required to approximate the function $\Psi_n(r)$ for a 2D-surface structure, and a (M + 1)3δ function is required to approximate the function $\Psi_n(r)$ for a 3D-surface structure.

2. Comparative analysis of methods: The storage required by the AIM can be divided into three parts. The first part is required to store the expansion coefficients that represent each basis function using a (M + 1)3δ function. Additionally, the storage is $(\boldsymbol{M}+1)^3\boldsymbol{N}_{basis}$, in which $\boldsymbol{N}_{basis}$ denotes the number of basis functions. The second part is the storage $\boldsymbol{O}(\boldsymbol{N}_{near})$ required for the near-field elements, in which $\boldsymbol{N}_{near}$ denotes the number of near-field elements. The third part is the storage $\boldsymbol{O}(\boldsymbol{N}_{grid})$ required to store the Fourier transform in the iterative processing, in which $\boldsymbol{N}_{grid}$ denotes the number of grids. The following two equations are approximations for complex 3D electrically large objects:

$$n_{near} \leq \left(\frac{20a}{\lambda}\right)^2 (M + d_{near})^2, \text{ surface structure} \tag{5.51}$$

$$N_{\text{grid}} = \left(\frac{\lambda}{10a}\right)^3 N_{basis}^{3/2}, \text{ surface structure} \tag{5.52}$$

where a denotes the grid point spacing, and d_{near} denotes the near-field threshold. The storage required for the first and second parts is $O(N_{basis})$, and the storage for the third part is $O\left(N_{basis}^{3/2}\right)$. For the surface result, the storage required by the AIM is $O\left(N_{basis}^{3/2}\right)$. For the example described here, $N_{near}N_{basis} \gg N_{grid}$. Consequently, only the storage required for the near-field elements is considered in the comparison of the storage.

The computation required by the AIM includes three parts—that is, the computation $O(N_{basis})$ in solving the expansion coefficients $[\Lambda]$, $O(N_{basis})$ in solving the near-field elements, and $O(N_{grid} \lg N_{grid})$ required in each iterative process by solving the Fourier transform, the computation being $O\left(N_{basis}^{3/2} \lg N_{grid}\right)$ using Eq. (5.18). Therefore, the computation required by the AIM is $O\left(N_{basis}^{3/2} \lg N_{basis}\right)$ for the surface result. The number of iterations for analyzing the electrical characteristics of the target is nearly the same as in the traditional MoM and AIM. Thus, the comparison of the calculations shown below only includes a comparison of the time for each iteration.

The storage volume of the traditional MoM $O(N_{basis}^2)$ is required to store the impedance matrix and the calculation complexity is $O(N_{basis}^3)$. For the surface result, the storage and the calculation complexity using the AIM are $O\left(N_{basis}^{2/3}\right)$ and $O\left(N_{basis}^{2/3} \lg N_{basis}\right)$, respectively. It is evident that the AIM requires less storage and computation than the traditional MoM for analyzing electrically large objects. When the electrical size of the target is large, the traditional MoM is not suitable because of the computer memory limitations. Consequently, the AIM can be used to accurately analyze the electrical characteristics of the target.

3. **Numerical results**: Analysis of the scattering properties of a perfect conductor using the AIM is presented below. The first example presents an analysis of bistatic scattering of a conductor plate. The working frequency was 300 MHz. The size of the conductor plate was 3 × 3 m (electric size $3\lambda \times 3\lambda$), and the plate was located in the *xoy* plane. The normal incident wave had θ-polarization with an electric field intensity $E = 1$ V/m. The conductor plate was divided into 1,800 triangles with 2,640 common sides for the calculation, so there were 2,640 unknowns. The storage and computational times in calculating the scattering of the $3\lambda \times 3\lambda$ plate using the traditional MoM and AIM are presented in Tables 5.1 and 5.2, respectively. The parameters used in the AIM were as follows—the grid spacing was $\lambda/12$, expansion order $M = 2$. Single precision variables were used in the computer program, the program being run in BenTeng IV3.0 C, which was the same as the previous example.

As shown in Tables 5.1 and 5.2, the matrix filling time is 15.86 s when analyzing the scattering of the plate using the AIM, which is less than the 27.06 s required by the MoM. The time for the inverse operation of the AIM is 23.91 s, which is 55% of the 43.17 s required by the MoM. We can conclude that the speed of the AIM is faster than that of the MoM.

Table 5.1 Storage and computation time using the MoM (calculation of scattering from a $3\lambda \times 3\lambda$ square conductor plate)

Number of matrix elements	Matrix filling time (s)	Matrix inversion time (s)
6,969,600	27.06	43.17

Table 5.2 Storage and computation time using the AIM

Time for expansion coefficient Λ (s)	Number of elements in the near area	Element filling time in the near area (s)	Number of iterations	Total iteration time (s)
0.77	370,064	15.86	46	23.91

It should be noted that 370,064 near-field elements were required for the AIM analysis AIM, and $2,640 \times 2,640 = 696,9600$ matrix elements had to be stored. Consequently, the AIM, for which the sparsity of the matrix elements was 5.31, could save a substantial amount of memory. The results of the plate scattering characteristics calculated by the MoM and AIM are shown in Fig. 5.21. Good consistency is evident between the MoM and AIM. Moreover, the AIM can quickly and accurately analyze the scattering characteristics of the conductor plate.

The scale of the impedance matrix is large when the MoM is used to analyze electrically large objects. The computation for the direct matrix inversion is $O(N^3)$. Additionally, the computation for each matrix vector multiplication in solving the matrix using the iterative method is $O(N^2)$. Therefore, the iterative method is generally used to solve a matrix $[Z][I] = [V]$ with large sizes. The computation for each matrix vector multiplication using the FFT in the AIM operation is $O(N \lg N)$. Consequently, the iteration speed of the AIM is faster than that of the MoM.

To further demonstrate the advantages of the AIM compared with the MoM in calculations for electrically large objects, a second example of a $10\lambda \times 10\lambda$ plate is presented. The working frequency was 300 MHz. The size of the conductor plate was 10×10 m, and the plate was located in the xoy plane. The normal incident wave

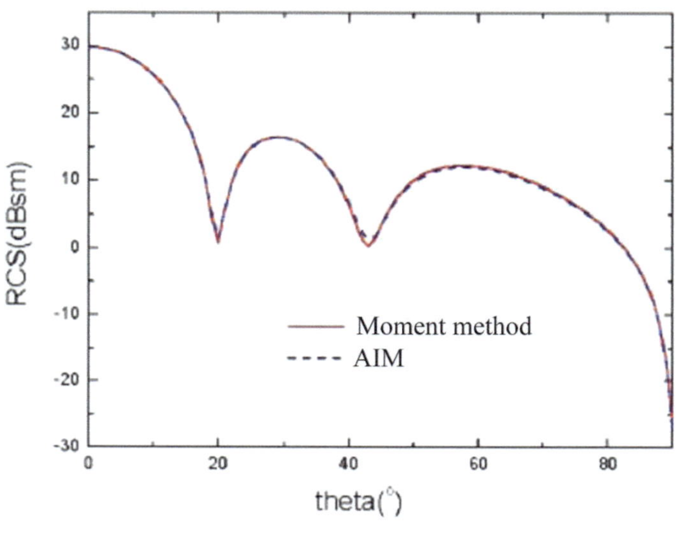

Fig. 5.21 Bistatic scattering of a $3\lambda \times 3\lambda$ square substrate

was θ-polarized with an electric field intensity $E = 1$ V/m. The plate was divided into 20,000 triangles, and there were 29,800 unknowns in the MoM analysis. A total of 29, 800 $\times$ 29, 800 matrix elements had to be stored when using the traditional MoM, the required storage being 6.62 Gb, which exceeded the maximum memory of the personal computer, thus limiting the use of the traditional MoM. If the AIM was used, 461,3053 near-field elements and 35.19 Gb of storage were required. By using the AIM, the time required to calculate the expansion coefficient was 8.78 s, and 186 iterations were used. Each iteration required 3.2 s to solve the current vector. The bistatic RCS of the $10\lambda \times 10\lambda$ plate is shown in Fig. 5.22.

The third example is the calculation of the scattering of a metal ball with a radius of 0.75 mm. The working frequency was 300 MHz. The normal incident wave was θ-polarized with an electric field intensity $E = 1$ V/m. The metal ball was divided into 2,304 triangles with 3,456 common edges. Consequently, there were 3,456 unknowns when using the MoM. A schematic diagram of the division of the metal ball is shown in Fig. 5.23.

The storage and computation time required to analyze the scattering of the metal ball using the MoM and AIM in BenTeng IV 3.0 G are presented in Tables 5.3 and 5.4, respectively. As shown, the storage required by the AIM is 7.66 MB, which is 8.4% of the 91.125 MB required by the MoM. Each iteration takes 1.8 s to solve the MoM impedance matrix using the iterative method. Using the AIM, the time taken is 1.0 s, which is 55.5% of that required by the MoM. The scattering results of the metal ball using the two methods are shown in Fig. 5.24, with good agreement between the MoM and the AIM being evident. The parameters used in the AIM were as follows—the grid spacing was half the wavelength, $d_{near} = 2$, and $M = 2$.

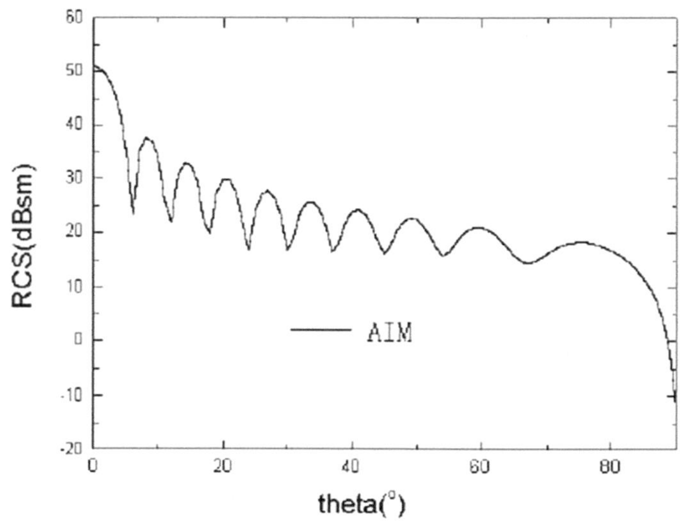

Fig. 5.22 Bistatic scattering of a $10\lambda \times 10\lambda$ square substrate

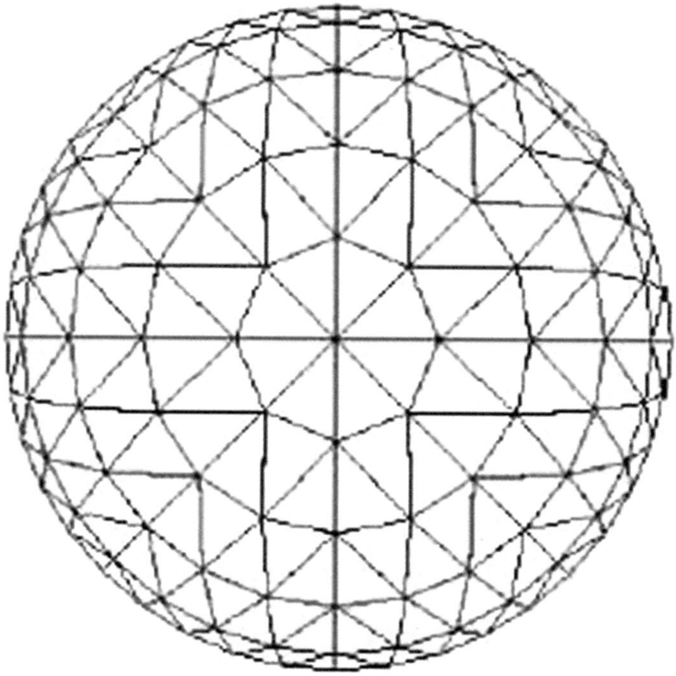

Fig. 5.23 Schematic diagram of the partitioning of a conductor sphere

Table 5.3 Storage and time required by the MoM to solve the scattering of a metal ball

Number of matrix elements	Required memory (Mb)	Iteration time (s)
11,943,936	91.125	1.8

Table 5.4 Storage and time required by the AIM to solve the scattering of a metal ball

Time for Λ (s)	Number of elements in the near area	Required memory (Mb)	Iteration time (s)
9.26	1,004,460	7.66	1.0

The three examples show that the AIM can rapidly analyze the scattering of electrically large conductors by using less storage. Additionally, it addresses the problems of the large storage volumes and long calculation times of the traditional MoM.

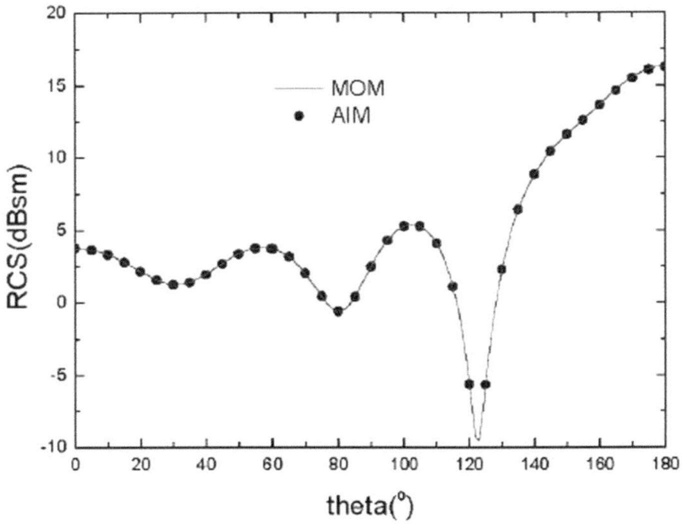

Fig. 5.24 Bistatic scattering of conducting spheres

5.2.3 *Rapid Calculation of the Electrical Characteristics of a Microstrip Patch Array Antenna Based on the AIM*

The limitations in analyzing and designing microstrip patch array antennas include the long times and large computer memory resources required. Thus, it is important to develop rapid calculation methods for the design of microstrip patch array antennas. Chao-Fu Wang analyzed [27] the characteristics of microstrip patch array antennas based on the BCG-FFT. The improved speed of analysis results from the convolutional properties of the Green's function allowing the use of the FFT to speed up the multiplication of the matrix vectors, failing which uniform grids are required when analyzing the target using the FFT. Moreover, the AIM is needed for an irregular target. Reference [28] calculated the properties of microstrip patch array antennas using the precorrected fast Fourier transform (PFFT). In [29], the AIM was used to analyze the *S* parameter of the microstrip current. In this section, the rapid calculation of the radiation and scattering characteristics of microstrip patch array antennas is realized using a combination of the adaptive integral method and the spatial domain MoM based on the mixed volume integral equation.

1. Theoretical analysis: The previous section introduced a method for analyzing microstrip antennas based on the spatial domain MoM. Here, we can obtain the elements of the impedance matrix produced by the MoM using Eqs. (5.1)–(5.7), such that:

$$Z_{mn} = -j\omega\mu_0 \int\limits_{S_m}\int\limits_{S_n} \left[\vec{f}_m(\vec{r})\cdot\vec{f}_n(\vec{r}')G_{xx}^A(\vec{r},\vec{r}') - \frac{1}{k_0^2}(\nabla\cdot\vec{f}_m(\vec{r}))(\nabla'\cdot\vec{f}_n(\vec{r}'))G^{\varphi}(\vec{r},\vec{r}')\right]ds'ds. \tag{5.53}$$

Using a similar process to Eqs. (5.41)–(5.46), the functions $f_n(r')$ and $\nabla'\cdot f_n(r')$ can be approximated by a linear combination of δ functions on the $(M+1)^2$ grids surrounding the center of the basis functions [26]. We can then calculate the corresponding expansion coefficient. The impedance element of Eq. (5.53) can be written as:

$$Z_{mn}^{far} = -j\omega\mu_0 \sum_{u=1}^{(M+1)^2}\sum_{v=1}^{(M+1)^2} [(\Lambda_{mu}^x\Lambda_{mv}^x + \Lambda_{mu}^y\Lambda_{mv}^y)G_{xx}^A(\vec{r}_u,\vec{r}'_v)] - \frac{1}{k_0^2}\Lambda_{mu}^d\Lambda_{mv}^d G^{\varphi}(\vec{r}_u,\vec{r}'_v) \tag{5.54}$$

The impedance element Z_{mm} in Eq. (5.53) can be expressed accurately by Z_{mn}^{far} in Eq. (5.54) when the test function is far from the basis function. The far-field matrix $[Z]^{far}$ can then be expressed as follows:

$$[Z]^{far} = -j\omega\mu_0\left\{[\Lambda_x][G_{xx}^A][\Lambda_x]^T + [\Lambda_y][G_{xx}^A]\left[\Lambda_y\right]^T - \frac{1}{k_0^2}\left[\Lambda_d][G\varphi][\Lambda_d]^T\right.\right\} \tag{5.55}$$

where $[\Lambda_x]$, $[\Lambda_y]$, and $[\Lambda_d]$ denote sparse matrixes in which each row contains only $(M+1)^2$ elements. The matrix element Z_{mm} can be represented accurately by Z_{mn}^{far} when the test function is close to the basis function. Similar to the method presented above, the impedance matrix can be divided into two parts by introducing the near-field matrix $[Z]^{near}$, as follows:

$$[V] = [Z][I] = [Z]^{near}[I] + [Z]^{far}[I] \tag{5.56}$$

To speed up the matrix vector multiplication of the second term on the right side of Eq. (5.56), we can use the iteration method, the convolution characteristics of the Green's function $[G_{xx}^A]$ and $[G_{\varphi}]$, and the FFT to solve the matrix equation, as follows:

$$\begin{aligned} {}^{far}[I] = &-j\omega\mu_0\{[\Lambda_x]\mathbb{F}^{-1}\{\mathbb{F}(G)\mathbb{F}([\Lambda_x]^T[I])\} \\ &+[\Lambda_y]\mathbb{F}^{-1}\{\mathbb{F}(G)\mathbb{F}([\Lambda_y]^T[I])\} - \frac{1}{k_0^2}[\Lambda_d]\mathbb{F}^{-1}\{\mathbb{F}(G)\mathbb{F}([\Lambda_d]^T[I])\}\} \end{aligned} \tag{5.57}$$

in which $F(\cdot)$ denotes the Fourier transform.

Little memory is required because $[Z]^{near}$ is a sparse matrix, and less computation is needed when it is multiplied by the current vector. The storage required for the FFT is $O(N_{grid})$, where N_{grid} denotes the number of AIM grids. The computation required for each FFT is $O(N_{grid}\lg N_{grid})$. Therefore, we can conclude that using the AIM can

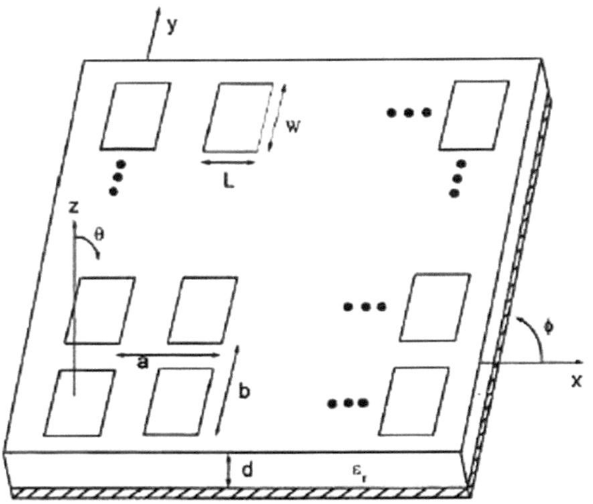

Fig. 5.25 Diagram of a microstrip patch array antenna with an infinite grounding substrate

substantially reduce the required storage for the traditional MoM impedance matrix and reduce the computation of the matrix vector multiplication. Moreover, we can calculate the characteristics of electrically large microstrip array antennas.

2. Numerical results: First, the application of the AIM to the analysis of microstrip patch array scattering characteristics was studied. Figure 5.25 shows a finite microstrip patch array placed on an infinite grounding substrate ($\varepsilon_r = 2.17$, operating at 3.7 GHz).

The AIM and MoM were used to calculate the monostatic RCS of a 3×3 microstrip patch array. θ was 0–90, and the electric field amplitude of incidence $E = 1$ V/m. In the calculation, the patch array was divided into 1,260 triangles with 1,737 common sides, resulting in 1,737 basis functions. The time and storage required in computations are shown in Tables 5.5 and 5.6.

Tables 5.5 and 5.6 show that the AIM requires 2.44 Mb of memory, which is less than the 23.02 Mb required by the MoM. The time required for each AIM iteration is

Table 5.5 Storage and computation time required by MoM (calculation of 3×3 microstrip patch array)

Number of matrix elements	Required memory (Mb)	Iteration time (s)
3,017,169	23.02	0.42

Table 5.6 Storage and computation time required by AIM (calculation of 3×3 microstrip patch array)

Time for Λ (s)	Number of elements in the near area	Required memory (Mb)	Iteration time (s)
1.656	319 773	2.44	0.22

0.22 s, which is 52.4% of the 0.42 s required for the MoM. The scattering characteristics of the 3 × 3 microstrip patch array is shown in Fig. 5.26 Good consistency is evident between the calculated results of the AIM and those of [30]. The grid spacing parameter used in analyzing the scattering of the microstrip patch array antennas by the AIM was λ/20 (Fig. 5.27).

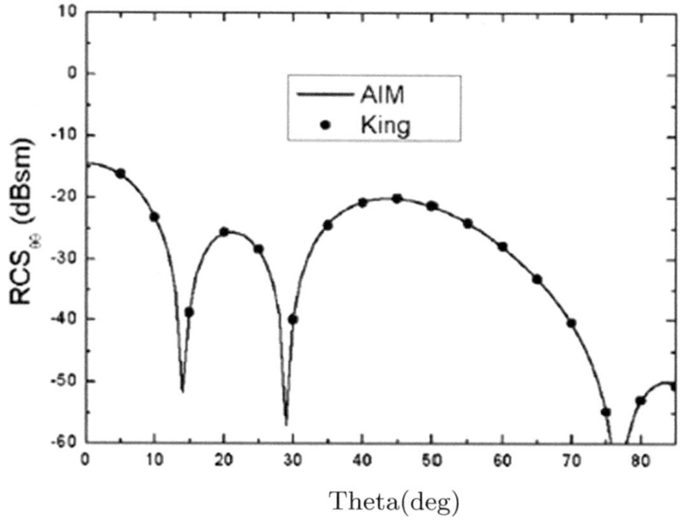

Fig. 5.26 3 × 3 single station scattering map of a microstrip patch array

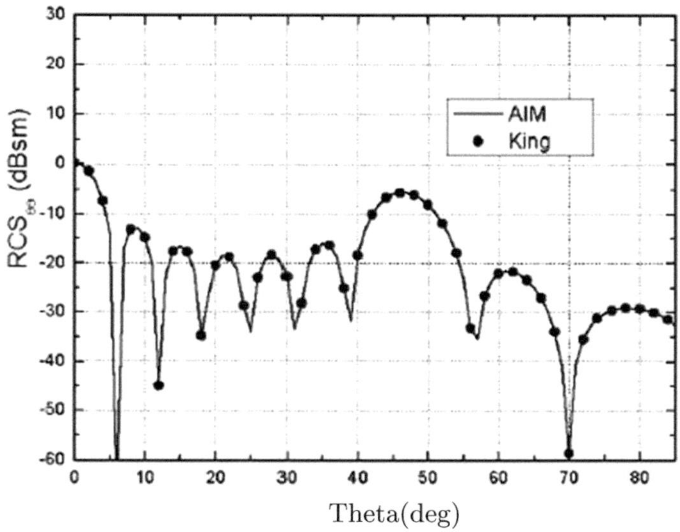

Fig. 5.27 7 × 7 single station scattering map of a microstrip patch array

The next analysis is of the scattering characteristics of a 7 × 7 microstrip patch array with similar parameters as the 3 × 3 array. Here, the patch array was divided into 3,430 triangles with 4,557 basis functions. The time and storage required for the computations of the MoM and AIM are presented in Tables 5.7 and 5.8. Good agreement is evident between the calculated results of the AIM and those reported in [30].

The results show that the AIM requires 5.72 Mb of memory, which is 3.6% of the 158.43 Mb required by the MoM. Additionally, the time consumed for each AIM iteration is 1.55 s, which is 47% of the 3.28 s required for the MoM. Consequently, the AIM can be used to save memory and calculation time. A grid spacing of $\lambda/16$ was used in analyzing the scattering of microstrip patch array antennas by the AIM.

The following analysis is presented for the scattering characteristics of an 11 × 11 microstrip patch array, in which the size of the patch, array element spacing, substrate, and working frequency were similar to those in the two previous examples. For the calculation, the patch array was divided into 8,470 triangles with 11,253 basis functions. The MoM required 966.11 Mb of memory, whereas 1,010,013 near-field elements had to be stored in the AIM, which required 7.7 Mb of memory, 0.79% of that required for the MoM. The calculation results are shown in Fig. 5.28. Good consistency is evident between the calculated results of the AIM and those of [30]. Tables 5.9 and 5.10 show that the time required for each AIM iteration is 4.78 s, which is 14.6% of the 32.68 s required for the MoM.

The fourth example is the calculation of the radiation performance of a four element microstrip patch array antenna fed with a microstrip line based on the AIM. The antenna structure is shown in Fig. 5.29, in the measurement units of which are in mm. A substrate with a relative permittivity of 2.1 and thickness of 1.5748 mm was used. The array antenna operated at 9.42 GHz. The microstrip antenna was divided into 1,252 triangles with 1,730 basis functions. This example is the same as the example of the microstrip array antenna fed with a microstrip line in the previous section, and the traditional MoM was used. Tables 5.11 and 5.12 show the time and storage required for the MoM and AIM, respectively. The radiation pattern of the

Table 5.7 Storage and computation time required by MoM (calculation of 7 × 7 microstrip patch array)

Number of matrix elements	Required memory (Mb)	Iteration time (s)
20,766,249	158.43	3.28

Table 5.8 Storage and computation time required by AIM (calculation of 7 × 7 microstrip patch array)

Time for Λ (s)	Number of elements in the near area	Required memory (Mb)	Iteration time (s)
1.343 375	749,119	5.72	1.55

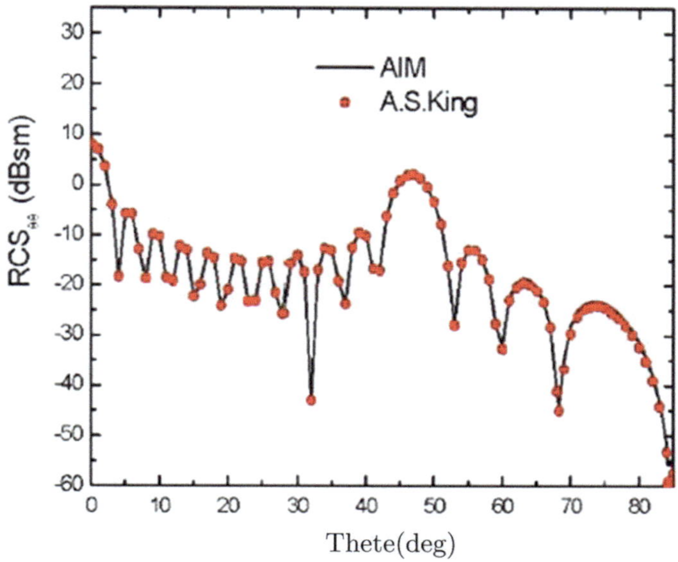

Fig. 5.28 11 × 11 single station scattering map of a microstrip patch array

Table 5.9 Storage and computation time required by MoM (calculation of 11 × 11 microstrip patch array)

Number of matrix elements	Required memory (Mb)	Iteration time (s)
126,630,009	966.11	32.68

Table 5.10 Storage and computation time required by AIM (calculation of 11 × 11 microstrip patch array)

Time for Λ (s)	Number of elements in the near area	Required memory (Mb)	Iteration time (s)
14.515 38	1,010013	7.7	4.78

antenna in the *E*-plane is shown in Fig. 5.30. Good consistency is evident between the calculated results of the AIM and those reported in [19].

Tables 5.11 and 5.12 show that the AIM requires 4.02 Mb of memory, which is 17.6% of the 22.83 Mb required by the MoM. The time required for the AIM iteration is 0.23 s, whereas the time for the MoM is 0.41 s. Compared with the MoM, the AIM reduces the storage and computation required for the analysis. Consequently, the AIM can be used to rapidly analyze the characteristics of microstrip patch antennas.

The analysis of the radiation performance of an 8 × 4 microstrip patch array using the AIM is presented below, the antenna structure being as shown in Fig. 5.31. A substrate with a relative permittivity of 2.2 and a thickness of 1.59 mm was used. The array antenna operated at 9.42 GHz. The microstrip patch array antenna was divided

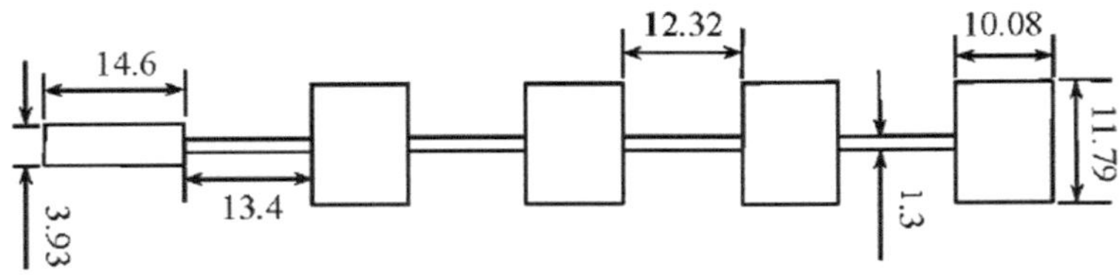

Fig. 5.29 Schematic diagram of a microstrip patch array

Table 5.11 Storage and computation time required by MoM (calculation of four elements microstrip patch antenna)

Number of matrix elements	Required memory (Mb)	Iteration time (s)
2,992,900	22.83	0.41

Table 5.12 Storage and computation time required by AIM (calculation of four elements microstrip patch antenna)

Time for Λ (s)	Number of elements in the near area	Required memory (Mb)	Iteration time (s)
1.671 985	526,298	4.02	0.23

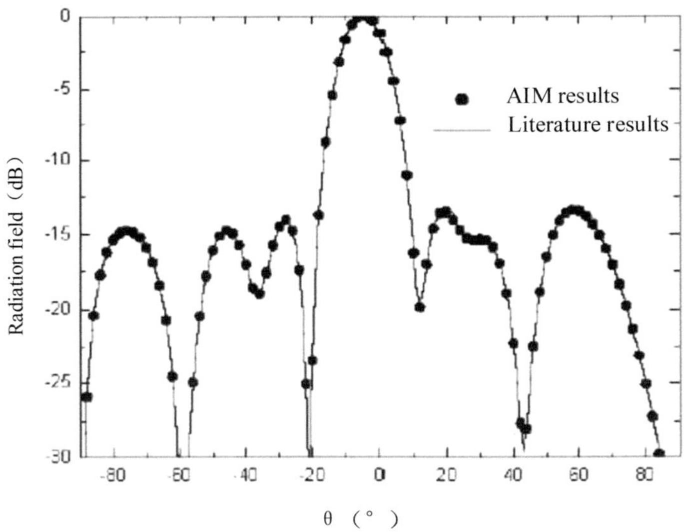

Fig. 5.30 *E*-face pattern of a microstrip-fed microstrip patch array

into 6,607 triangles, which produced 8,724 unknowns. During the calculation by the AIM, there were 1,455,095 near-field elements, which required 11.1 Mb of memory. The MoM required 580.7 Mb of memory. The time taken was 13.87 s for the MoM

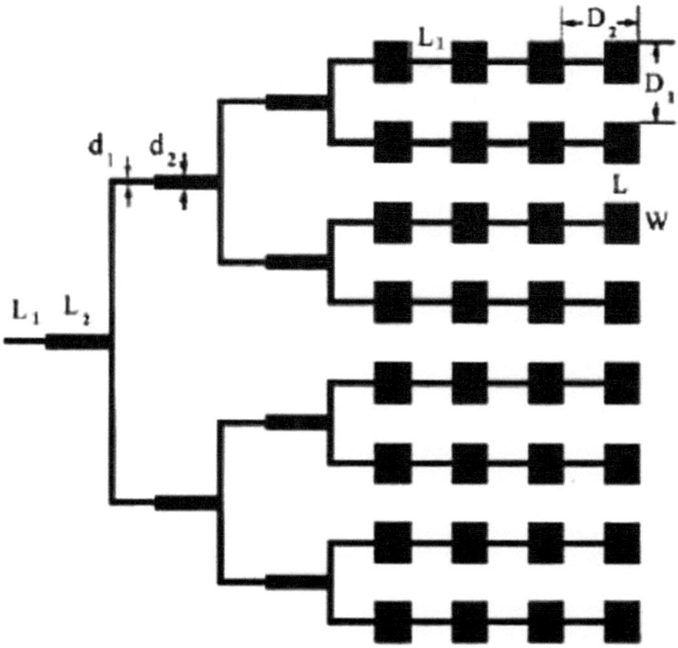

Fig. 5.31 Schematic diagram of an 8 × 4 microstrip patch array

and 2.68 s for the AIM. The radiation patterns at $\varphi = 0°$ and $\varphi = 90°$ are shown in Fig. 5.32. Good consistency is evident between the calculated results and those reported in [29].

5.3 Summary

The spatial domain MoM based on the mixed-potential integral equation was examined in this chapter. The MoM was used to analyze the electrical properties of various microstrip patch antennas with different excitations and feeding structures, including the scattering properties of a microstrip patch, the radiation and scattering characteristics of a microstrip antenna fed with a microstrip line, and the radiation and scattering characteristics of a microstrip antenna fed with a coaxial line. The application of the AIM in calculating the electrical characteristics of electrically large objects was studied. The scattering characteristics of electrically large conductors and the scattering and radiation characteristics of large microstrip array antennas were calculated using the AIM. The storage and calculations required by the MoM analysis for electrically large objects were reduced, and the scope of the MoM was effectively expanded.

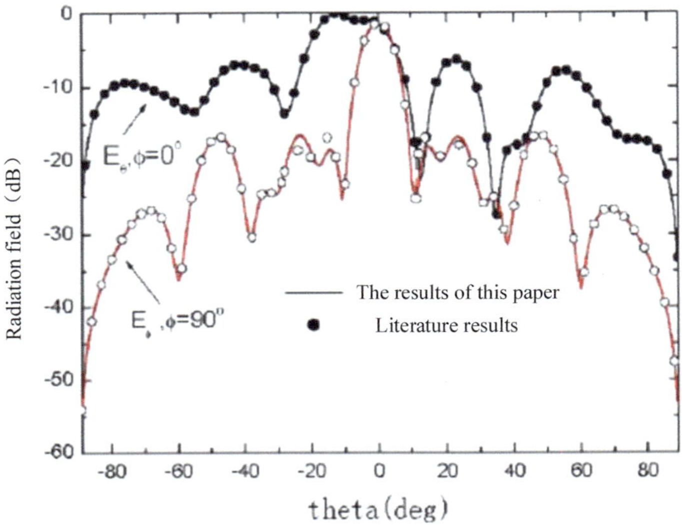

Fig. 5.32 Radiation pattern of an 8 × 4 microstrip patch array

References

1. Y.T. Lo, W.F. Richards, Theory and Experiment on Microstrip Antennas, IEEE Trans. Antennas Propagation, Mar, Vol. 27, No. 2, pp. 137–145, 1979.
2. V. Palanisamy, R. Garg, Analysis of Arbitrary Shaped Microstrip Patch Antennas Using Segmentation Technique and Cavity Model, IEEE Trans. Antennas and Propagation, Vol. 34, No. 10, pp. 1208–1213, 1986.
3. S.O. Park, C.A. Balanis, Analysis Evaluation of the Asymptotic Impedance Matrix of a Grounded Dielectric Slab with Roof-top Functions, IEEE Transaction on Antennas and Propagation, Vol. 47, No. 2, pp. 251–259, 1998.
4. S.M. Rao, D.R. Wilton, A.W. Glisson, Electromagnetic Scattering by Surface of Arbitrary Shape, IEEE Transactions on Antennas and Propagation, Vol. 30, No. 3, pp. 409–418, 1982.
5. Mouping Jin, Research on Generalized Resonance in Multi-conductor Scattering, Xidian University, Doctoral Dissertation, 2000.
6. D.R. Wilton, S.M. Rao, A.W. Glisson, et.al, Potential Integrals for Uniform and Linear Source Distributions on Polygonal and Polyhedral domains, IEEE Transactions on Antennas and Propagation, Vol. 32, No. 3, pp. 276–281, 1984.
7. Jinglin Guo, Research on the Scattering of Rivets and Screws on Aircraft, Xidian University, Master Thesis, 2002.
8. E.H. Mewman, Generation of Wideband Data from the Method of Moments by Interpolating the Impedance Matrix, IEEE Transactions on Antennas and Propagation, Vol. 36, No. 4, pp. 1820–1824, 1991.
9. K.L. Vriga, Y. Rahamat–Samii, Generation of Wideband Antenna Performance by [Z] and [Y] Matrix Interpolation in the Method of Moments, IEEE Transactions on Antennas and Propagation, Vol. 47, No. 1, pp. 65–76, 1999.
10. Xiulian He, Shuxi Gong, Yicai Ji, Qizhong Liu, Fast Calculation of the Electrical Characteristics in the Wideband of Communication Antennas, Chinese Journal of Radio Science, Vol. 17, No. 2, pp. 187–191, 2002.
11. Krzysztof A. Michalski, RCS Computation of coax-loaded microstrip patch antennas of arbitrary shape, Electromagnetics, Vol. 14, No. 1, pp. 33–62, 1994.

12. Davidovitz, M. Lo, Y.T. Rigorous analysis of a Circular Patch Antenna excited by a Microstrip Transmission Line, IEEE Transactions on Antennas and Propagation, Vol. 37, No. 8, pp. 949–958, 1989.
13. Katehi, P.B., Alxopoulos, N.G., Frequency-Dependent Characteristics of Microstrip Discontinuities in Millimeter-Wave Integrated Circuits, IEEE Transactions on Microwave Theory and Techniques, Vol. 33, No. 10, pp. 1029–1035, 1985.
14. Edwin K.L. Yeung, John C. Beal, Yahia M.M. Antar, Multilayer Microstrip Structure Analysis with Matched Load Simulation, IEEE Transactions on Antennas and Propagation, Vol. 43, No. 1, pp. 143–149, 1995.
15. Jian-Ying Li, Zaw Zaw Oo, Le-Wei Li, Pang-Shyan Kooi, Mook-Seng Leong, Improvement of Characteristics of Microstrip Antennas Using Unbalanced Structures, IEEE Antennas and Wireless Propagation Letters, Vol. 1, No. 1, pp. 71–73, 2002.
16. D.C. Chang, J.X. Zhang, Electromagnetic Modeling of Passive Circuit Elements in MMIC, IEEE Transaction on Microwave Theory Technology, Vol. 40, No. 9, pp. 1741–1747, 1992.
17. R.W. Jackson, D.M. Pozar, Full-wave Analysis of Microstrip Open-ended and Gap Discontinuities, IEEE Transaction on Microwave Theory Technology, Vol. 33, No. 10, pp. 1038–1042, 1985.
18. Ming-Ju Tsai, De Flaviis, F. Fordham, O. Alexopoulos, N.G., Modeling Planar Arbitrarily Shaped Microstrip Elements in Multilayered Media, IEEE Transaction on Microwave Theory Technology, Vol. 45, No. 3, pp. 330–337, 1997.
19. F. Ling, J.M. Jin, Scattering and Radiation Analysis of Microstrip Antennas using discrete Complex Image Method and Reciprocity Theorem, Microwave Optical Technology Letters, Vol. 16, No. 4, pp. 212–216, 1997.
20. D. Zheng, K.A. Michalski, Analysis of Arbitrarily Shaped Coax-fed Microstrip Antennas with Thick Substrates, IEE Electronics Letters, Vol. 26, No. 12, pp. 794–795, 1990.
21. K.A. Michalski, Discrete Complex Image Mixed-potential Integral Equation Analysis of Microstrip Patch Antennas with Vertical Probe Feeds, Electromagnetics, Vol. 15, No. 4, pp. 377–392, 1995.
22. S.U. Hwu, D.R. Wilton, S.M. Rao, Electromagnetic Scattering and Radiation by Arbitrary Conducting Wire/Surface Configurations, IEEE International Symposium on AP, pp. 890–893, 1988.
23. M.D. Deshpande, M.C. Bailey, Input Impedance of Microstrip Antennas, IEEE Transaction on Antennas and Propagation, Vol. 30, No. 7, pp. 645–650, 1982.
24. J.T. Aberle, D.M. Pozar, Evaluation of Input Impedance and Radar Cross Section of Probe-fed Microstrip Patch Element Using an Accurate Feed Model, IEEE Transactions on Antennas and Propagation, Vol. 39, No. 12, pp. 1691–1696, 1991.
25. Manuel F. Catedra, Emilio Gago, Luis Nuno, A Numerical Scheme to Obtain the RCS of Three-dimensional Bodies of Resonant Size Using the Conjugate Gradient Method and the Fast Fourier Transform, IEEE Trans. Antennas and Propagation, Vol. 37, No. 5, pp. 528–537, 1989.
26. E. Bleszynski, M. Bleszynski, T. Jaroszewicz, AIM: Adaptive Integral Method for Solving Large-scale Electromagnetic Scattering and Radiation Problems, Radio Science, Vol. 31, No. 5, pp. 1225–1251, 1996.
27. Chao-Fu Wang, Feng Ling, Jian-Ming Jin, A Fast Full-wave Analysis of Scattering and Radiation from Large Finite Arrays of Microstrip Antennas, IEEE Transactions on Antennas and Propagation, Vol. 46, No. 10, pp. 1467–1474, 1998.
28. Ning Yuan, Tat Soon Yeo, Xiao-Chun Nie, Le Wei Li, A Fast Analysis of Scattering and Radiation of Large Microstrip Antenna Arrays, IEEE Transactions on Antennas and Propagation, Vol. 51, No. 9, pp. 2218–2226, 2003.

29. Feng Li, Chao-Fu Wang, Jian-Ming Jin, An Efficient Algorithm for Analyzing Large-scale Microstrip Structures using Adaptive Integral Method Combined with Discrete Complex-imaged Method, IEEE Transactions on Microwave Theory and Techniques, Vol. 48, No. 5, pp. 832–839, 2000.
30. Adrian S. King, Wallace J. Bow, Scattering from a Finite Array of Microstrip Patches, IEEE Transactions on Antennas and Propagation, Vol. 40, No. 7, pp. 770–774, 1992.

Chapter 6
Design of a Microstrip Antenna with a Low RCS

6.1 Scattering Mode Analysis of a Patch Antenna

The RCS reduction problem of a rectangular patch antenna is studied in this section. First, the relationship between the resonant mode of the rectangular patch antenna and the scattering and radiation properties of the patch antenna is analyzed, the relationship between the resonant mode and the scattering properties being the main focus. Next, a method of reducing the patch antenna's RCS by means of slot cutting is proposed based on the surface current distribution of the antenna patch. Based on the proposed method, a low-RCS microstrip patch antenna with good radiation performance is designed. Finally, several related problems are discussed.

Figure 6.1 shows a diagram of a general rectangular patch in which L denotes the long side along the x-axis, and W denotes the short side along the y-axis. Based on the cavity model theory, the area between the microstrip patch and the ground plate can be considered to be a resonant cavity whose upper and lower surfaces are electric walls, and the other surfaces are magnetic walls.

The resonant frequency of each resonant mode of the microstrip antenna can be determined as follows:

$$f_{pq} = \frac{0.15}{\sqrt{\varepsilon_r}}\sqrt{\left(\frac{p}{L}\right)^2 + \left(\frac{q}{W}\right)^2} \tag{6.1}$$

where p and q denote the number of sinusoidal variations with a half period along the long and short sides, respectively. The unit of f_{pq} is GHz.

The patch antenna resonates at its eigenmode when it is excited by the feed or a plane wave. It should be noted that the radiation characteristics of the antenna can be calculated using the cavity mode theory [1], and the peak value of the antenna's RCS frequency response can be labeled using the antenna's resonance point obtained by the cavity mode theory [2]. The patch antenna irradiated by a plane wave operates at

Y. Liu et al., *Antenna Radar Cross Section: Theory and Design*,
https://doi.org/10.1007/978-981-96-1226-0_6

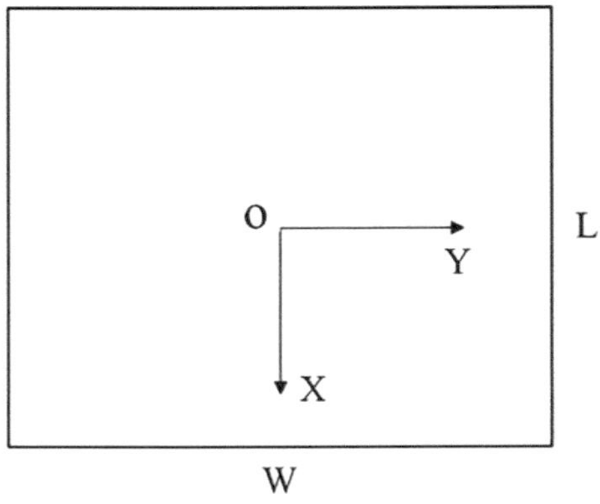

Fig. 6.1 Diagram of the rectangular patch antenna

numerous resonant modes, including the main mode—that is, the TM_{10} mode—as well as higher-order modes.

To illustrate this concept in detail, the RCS response of a rectangular aperture coupled patch antenna irradiated by plane waves in different directions can be calculated. The thickness of the substrates of the antenna and feed is 2 mm, and the relative permittivity is 2.65. The size of the antenna patch is 28 × 37 mm, the size of the ground is 40 × 40 mm, and the size of the coupled aperture located in the center of the ground is 1.5 × 14.1 mm. The incident angles of the θ-polarized plane wave is $\theta = 60°$ and $\varphi = 0°$, $\theta = 60°$ and $\varphi = 45°$, and $\theta = 60°$ and $\varphi = 90°$. The operating frequency of the antenna is 2.85 GHz. The calculated RCS response to being irradiated by different incident angles is shown in Fig. 6.2.

The results show that the RCS response of the patch antenna has numerous peaks under plane waves with different incident angles. Considering the precision, the obtained frequency points of each mode calculated by Eq. (6.1) generally agree with the positions of the RCS peak points. The TM_{m0} modes on the long side are excited when the incident angles are $\theta = 60°$, $\varphi = 0°$, the TM_{0n} modes on the broad side are excited when the incident angles are $\theta = 60°$, $\varphi = 90°$, and all of the TM_{mn} modes are excited when the incident angles are $\theta = 60°$, $\varphi = 45°$.

To illustrate the relationship between the RCS response and the resonant mode of the rectangular patch antenna differently, the RCS response of the rectangular patch antenna operating at different frequencies irradiated by a plane wave with the same incident angle can be calculated. Figure 6.3 shows the RCS response of the rectangular patch antenna irradiated by plane waves at 3 and 4 GHz when the incident angles are $\theta = 60°$, $\varphi = 45°$. The results show that although the different patch sizes and substrate parameters result in different operating frequencies of the rectangular patch antenna, they exhibit the same resonant mode, as reflected in the similar peak distribution of their RCS response curves. The difference between them is that when the patch antenna operates at a lower frequency, the corresponding resonant frequency of each mode is lower, and the distance between the resonant frequency of each mode is small; by contrast, when the patch antenna operates at a higher frequency, the corresponding resonant frequency of each mode is higher, and

Fig. 6.2 RCS response of an antenna irradiated by different plane waves

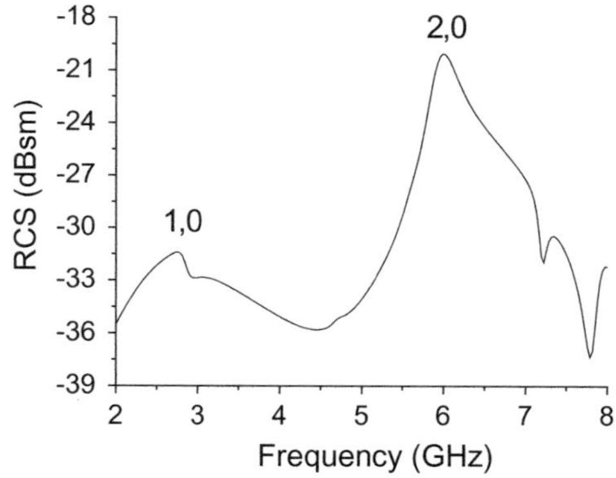

(a) Incident angle: $\theta = 60°$, $\varphi = 0°\theta = 60°$, $\varphi = 0°$

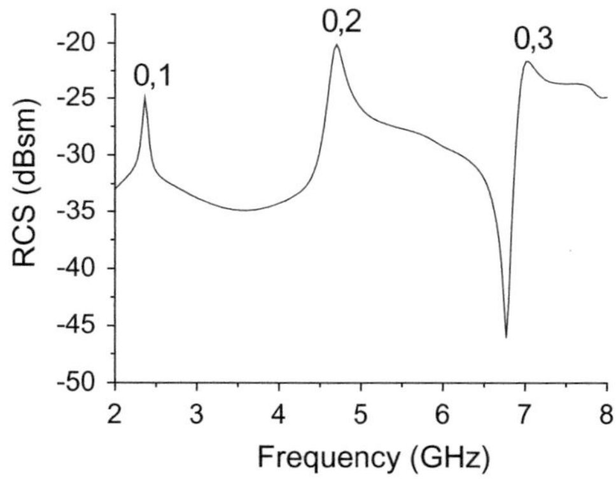

(b) Incident angle: $\theta = 60°$, $\varphi = 90°$

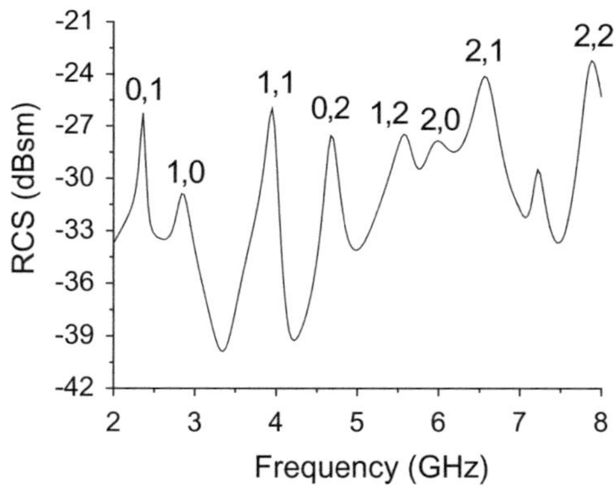

(c) Incident angle: $\theta = 60°$, $\varphi = 45°$

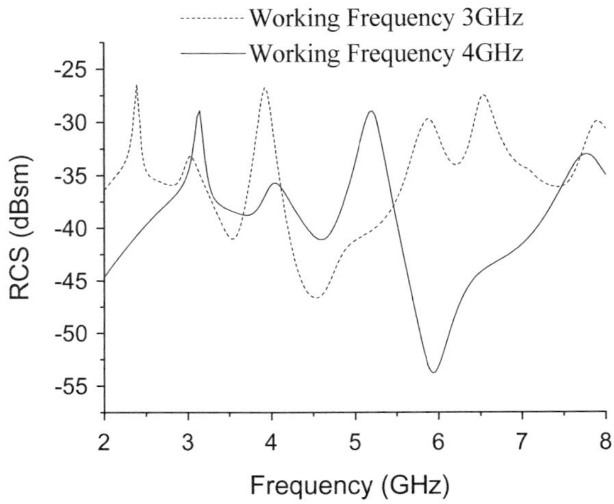

Fig. 6.3 RCS response of antennas operating at different frequencies irradiated by a plane wave with incident angles of $\theta = 60°, \varphi = 45°$

the distance between the resonant frequency of each mode is large (according to Eq. (6.1)).

From the above analysis, the following conclusions can be drawn:

(1) The working frequency of the patch antenna should be higher only from the point of view of the RCSR. If the patch antenna operates at a higher frequency, the corresponding resonant frequency of each mode is higher, and the distance between the resonant frequencies of each mode is large; thus, for a fixed observation frequency, the number of RCS peak points is relatively small. Moreover, the physical size of an antenna with a higher working frequency is smaller, and its corresponding structural mode scattering is lower, which also contributes to the RCSR.

(2) The working frequency of the patch antenna should be selected to be lower from a reduction method point of view. RCSR is generally studied over a range of frequencies similar to those in this work (2–8 GHz), and lower antenna operating frequencies allow for more RCS peaks over the same frequency range, which increases the complexity of the RCSR study, although the results are more valuable.

The radiation and scattering fields of the microstrip antenna are both caused by the induced current on the patch surface. For the sake of analysis, we can examine the resonant mode characteristics of the microstrip patch antenna from the surface current distribution of the patch.

Based on the surface current distribution corresponding to the different resonant modes shown above, the following two points can be determined:

(1) The current distribution on the surface of the patch is based on the half-wavelength distribution. The TM_{mn} mode indicates a current distribution of m half-wavelengths along the x-axis on the surface of the patch and n half-wavelengths along the y-axis. When the length and width of the patch antenna

are equal to an integral multiple of half of the wavelength of the excitation EM wave, the patch antenna resonates.

(2) Based on the surface current distribution on the patch, these resonant modes can be divided into three categories. Here, we classify only the *y*-directional currents in the patch as the first type, which can be excited by the incident plane waves at $\theta = 60°$, $\varphi = 45°$ and $\theta = 60°$, $\varphi = 90°$, with all of the TM_{0n} ($n = 1$, 2, 3, …) modes falling into this category. The second category includes only the *x*-directional surface currents on the patch, which can be excited by the incident plane waves at $\theta = 60°$, $\varphi = 45°$and $\theta = 60°$, $\varphi = 0°$, with all the TM_{m0} ($m = 1$, 2, 3, …) modes falling into this category. It should be noted that the TM_{10} mode—which is the operating mode—also falls into this category when the patch antenna is radiating. The third type of mode is a mode with both *x*- and *y*-direction currents on the patch surface. Only plane waves with an incident angle of $\theta = 60°$, $\varphi = 45°$ excite these modes, and all of the TM_{mn} ($m = 1$, 2, 3, …, $n = 1$, 2, 3, …) modes fall into this category.

The resonant characteristics make it possible for the patch antenna to radiate, but they also cause RCS peaks. After analyzing the relationship between the resonant mode and the radiation and scattering characteristics of the rectangular microstrip patch antenna, we can examine how to use the relationship to reduce the RCS of the patch antenna.

6.2 Aperture-Coupled Microstrip Antenna with a Low RCS

6.2.1 *Study of the Slot-Cutting Method*

The purpose of reducing the RCS of a microstrip patch antenna is to reduce its peak RCS while maintaining its radiation properties. Patch antennas usually radiate in the TM_{10} mode. However, the incident plane wave can excite not only the TM_{10} mode but also many higher-order modes. Consequently, for out-of-band RCSR, the resonant modes that generate RCS peaks should be suppressed or eliminated; however, the TM_{10} mode cannot be affected. In-band RCSR can be difficult as the radiation and scattering properties are both related to the TM_{10} mode. Therefore, the gain and efficiency of the antenna is sacrificed to obtain better scattering properties. These factors indicate that the RCSR of a microstrip antenna requires a compromise between radiation and scattering.

This chapter focuses on out-of-band RCSR. We discuss how to suppress or eliminate the resonant modes that generate the RCS peaks without affecting the TM_{10} mode. Analysis of the surface current of a patch shows that this goal can be achieved by controlling the distribution of the surface current. Specifically, the surface current distribution of the TM_{10} mode must be maintained, while the surface current distribution of the other resonant modes are suppressed. Cutting slots on the ground and radiation patch to control the surface current distribution has been widely used to realize the preferred radiation properties. Similarly, these methods can be used in RCSR. This will be discussed in detail below.

In general, cutting slots affects the currents that run perpendicular to the slot direction; a slot along the *x*-axis affects the currents that flow along the *y*-axis, and a slot along the *y*-axis affects the currents that flow along the *x*-axis. Because the current direction is only along the *y*-axis, slots can be cut along the *x*-axis (with a slot length *L*). This divides the patch into several narrow strips (printed dipoles). The currents that flow along the *y*-axis are completely cut off. Considering the half-wavelength periodicity of the surface currents of high resonant modes, multiple slots are needed, as shown in Fig. 6.5a. Thus, the TM_{0n} modes can be eliminated, and the RCS peaks generated by these modes reduced. However, the surface current of the TM_{10} mode is not affected. Therefore, the radiation properties can be maintained. A similar structure has been used to broaden the antenna bandwidth [3].

For modes that only have surface currents in the *x*-direction, a slot can be cut in the *y*-direction (with a length *W*), which cuts off all of the currents that flow in the *x*-direction. Moreover, the RCS peak is removed as the TM_{m0} mode is eliminated. However, this method can also eliminate the main mode—that is, the TM_{10} mode—so the antenna cannot work properly. Consequently, considering the current path of the TM_{10} mode, the slot length (*L*) should be shorter than the patch width (*W*). Because the currents in the *x*-direction are not completely cut off, this retains some of the TM_{m0} mode. In fact, the TM_{10} mode is affected even if the slot in the *y*-direction is short. Additionally, the antenna gain and bandwidth deteriorate considerably [4]. An alternative technique is to cut a slot on the ground. However, the slot should not be too long. A long slot can be replaced by several short slots, with a certain distance being maintained between adjacent slots, as demonstrated in Chiou and Wong [5]. Similarly, the ground needs several groups of slots because of the high mode, as shown in Fig. 6.5b. Slotting on the ground also plays the role of aperture coupling. In this way, the TM_{m0} mode can be restrained by the *y*-direction slots on the ground and the RCS peak can be reduced without dramatically affecting the radiation performance.

Finally, all of the TM_{mn} modes can be restrained by combining the *x*-direction slots on the patch and the *y*-direction slots on the ground. The corresponding RCS peaks can be reduced without destroying the radiation properties of the antenna.

6.2.2 Antenna Characteristics

An aperture coupled microstrip antenna can be designed to check the effectiveness of the previous analysis. Here, both the patch and the ground have slots in them with the slots on the ground having the effect of aperture coupling. The antenna structure and the slot sizes on the patch and ground are shown in Fig. 6.6. The thicknesses of the antenna substrate and the feeding substrate are both 2 mm with a permittivity of 2.65. The microstrip feeding line has a width of 4.5 mm and a length of 31 mm. An aperture coupled microstrip antenna can then be designed, which has the same operating band as a reference antenna, to compare the antennas' radiation and scattering properties. The patch size of the antenna is 27.6 × 37 mm, and the ground size is 40 × 49 mm. The thickness and permittivity of the antenna substrate and the feeding substrate remain unchanged. The rectangular coupling aperture is at the center of the ground with a size of 1.4 × 14.3 mm. The width of the microstrip feeding line is 14.5 mm, and the length is 33.5 mm.

The designed antenna operates at 2.85 GHz with a bandwidth of 85.5 MHz. The antenna gain at the center frequency is 5.9 dBi. The reference antenna has a bandwidth of 57 MHz, and the antenna gain at the center frequency is 6.4 dBi. The bandwidth of the proposed antenna is slightly wider, but the antenna gain is 0.5 dB lower. Comparisons of the S_{11} and radiation patterns of these two antennas are shown in Fig. 6.7. These two antennas were fabricated and measured. Owing to the material properties and machining errors, the operating bandwidth exhibited slight shifts; however, the bandwidth and gain were identical to the simulation results.

The monostatic RCSs of these two antennas under a θ-polarized plane wave at typical incident angles can be calculated. Figure 6.8a shows that every scattering peak of the microstrip patch antenna is reduced under an incidence of $\theta = 60°$, $\varphi = 45°$. As shown by the previous analysis, the RCS peaks of the TM_{01} mode and the TM_{02} mode are completely eliminated. This is evident in the RCS response at an incidence of $\theta = 60°$, $\varphi = 90°$, as shown in Fig. 6.8b. The RCS peaks of the TM_{10} and TM_{20} modes, which change along the long edge, are effectively reduced. This is evident in Fig. 6.8c at an incidence of $\theta = 60°$, $\varphi = 0°$.

Owing the symmetry of the slots on the patch and ground, the RCS values are the same under symmetric incident angles. Figure 6.9 shows comparisons of the RCS values at four angles of incidence—that is, $\theta = 60°$, $\varphi = 0°$, $\theta = 60°$, $\varphi = 135°$, $\theta = 60°$, $\varphi = 225°$, and $\theta = 60°$, $\varphi = 315°$. The antenna has low-RCS properties in all directions, although the RCS values have small differences under different incident angles.

6.2.3 *Discussion of the Problems*

Combined with the calculation results, we discuss the previous analysis below.

1. **The resonance of the TM_{on} mode is eliminated**: The previous analysis shows that slots in the x-direction with a patch length L can restrain the resonance of the TM_{on} mode. We can prove this by analyzing the distribution of the surface currents. The surface current distribution of the second reduced peaks in Fig. 6.8b is shown in Fig. 6.10. Only the direction of the surface currents differ based on the observations of the antennas' surface currents. However, the magnitude of the surface current of the slotted antenna is smaller than that of the reference antenna. Consequently, the RCS peaks are reduced.

Additionally, the RCS peaks increase at several frequencies, as shown in Fig. 6.8b. Although the resonance property of the patch is eliminated, the scattering property of the antenna ground as a common scatter remains. Reference [6] presented the calculation formula for the monostatic RCS of a rectangular conductor plate of size $a \times b$, as follows:

$$\sigma \cong \frac{1}{\pi}(kab\cos\theta_0 \frac{\sin\alpha}{\alpha}\frac{\sin\beta}{\beta})^2 \tag{6.2}$$

where $k = 2\pi/\lambda$, $a = ka \ \sin\theta_0 \ \cos\varphi_0$, $\beta = kb \ \sin\theta_0 \ \sin\varphi_0$, θ_0, φ_0 denote the angle of incidence, and λ denotes the incident wavelength. As shown in Eq. (6.2), the RCS increases by 6 dB if the incident wavelength decreases by half. Additionally, the RCS response has a ladder-like distribution. The difference between the adjacent ladders is almost 6 dB, which is why the RCS of some frequency points increase.

2. **The resonance of the TM_{m0} modes and TM_{mn} modes is restrained**: The resonance of the TM_{m0} modes and TM_{mn} modes cannot be completely eliminated as we must ensure the operation of the TM_{10} mode, so we can only partially restrain them. Consequently, the RCS peaks corresponding to these modes cannot be suppressed completely. The RCS response of the slotted antenna shows two peaks at 3.14 and 5.97 GHz. Figure 6.11 shows the surface current distribution of these two frequency points. It is evident that the current distribution of the resonance modes are similar to the current distribution of the TM_{11} and the TM_{21} modes (note the directions of the currents on both sides of the patch) compared with the surface current distribution shown in Fig. 6.4. Although the peaks are not completely eliminated, their magnitudes are considerably smaller than those of the reference antenna.
3. **Reduction of in-band RCS peaks**: In-band RCSR can be difficult. The radiation properties are based on the TM_{10} mode, which also lead to the RCS peaks in the operating band. Thus, we have to control the resonance of the TM_{10} mode. The scattering property can only be realized at the cost of radiation efficiency and antenna gain. In the calculations shown above, the slots on the antenna ground affect the resonance of the TM_{10} mode, leading to gain loss and in-band RCSR. Moreover, the in-band RCS peaks are reduced, as shown in Figs. 6.8a and c.

6.3 Design of a Low-RCS Antenna Using a Comprehensive Method

6.3.1 Introduction

For a rectangular microstrip antenna, the length of the nonradiation edge is approximately half of the wavelength, as shown in Fig. 6.21a. The electric field at the center of the nonradiation edge is zero [7], as shown in Fig. 6.12a. In theory, loading the resistor on the centerline of the nonradiation edge has little effect on the radiation performance of its operating frequency (TM_{10} mode). However, for other modes

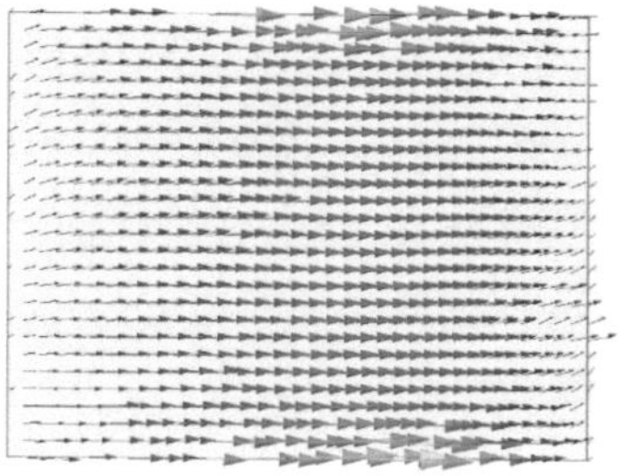

(a) TM_{01} mode

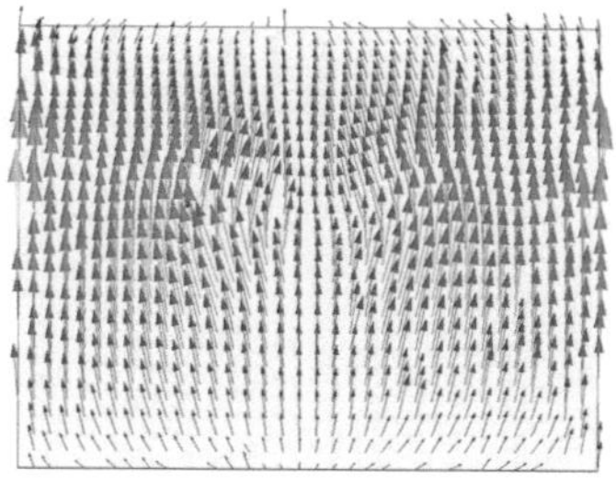

(b) TM_{10} mode

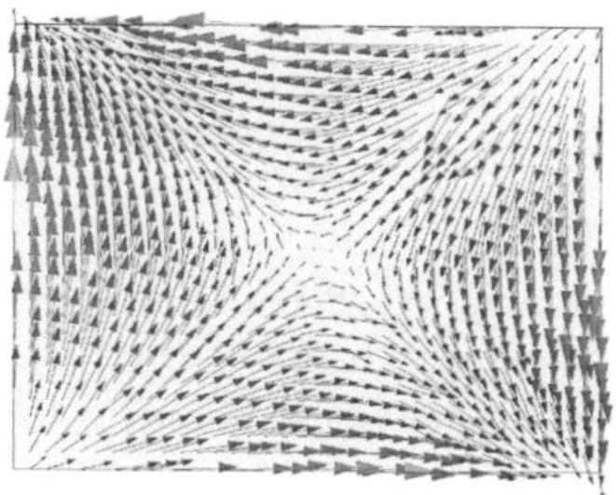

(c) TM_{11} mode

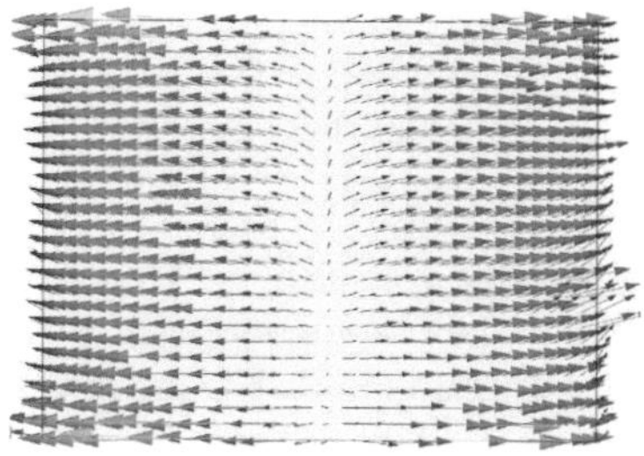

(d) TM_{02} mode

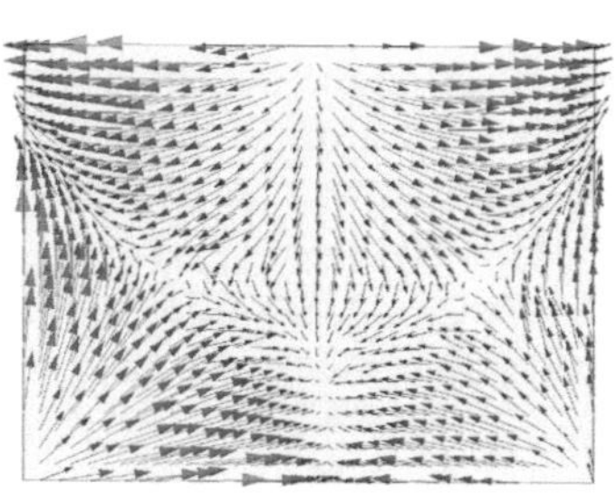

(e) TM_{12} mode

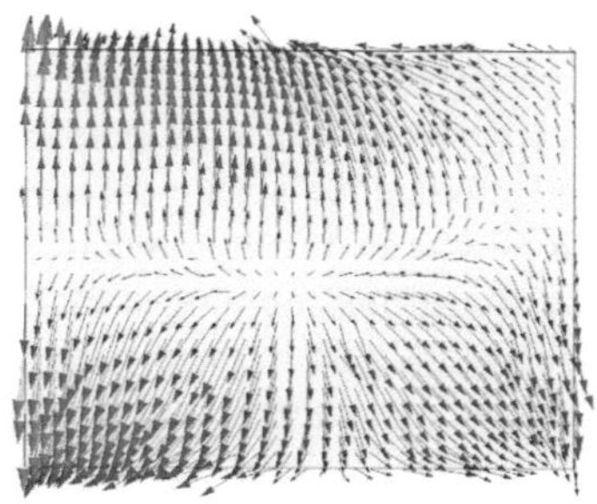

(f) TM_{20} mode

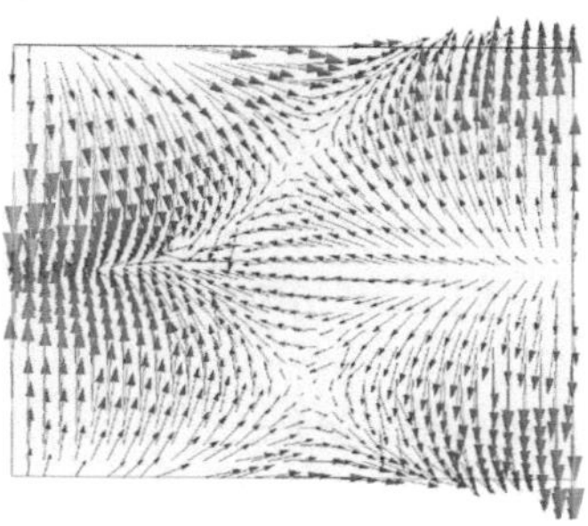

(g) TM_{21} mode

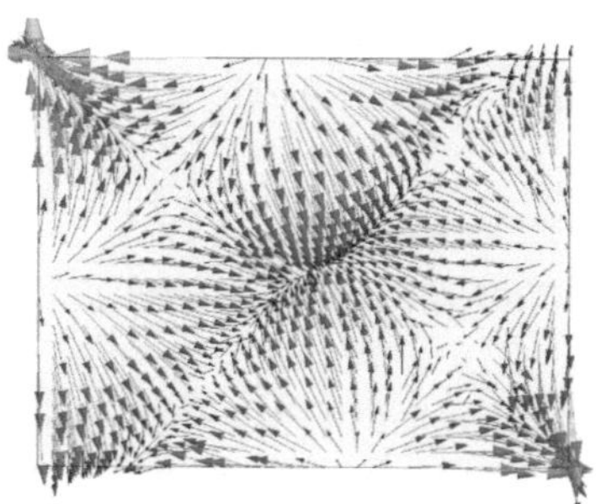

(h) TM_{22} mode

Fig. 6.4 Current distribution on the patch surface corresponding to different resonant modes

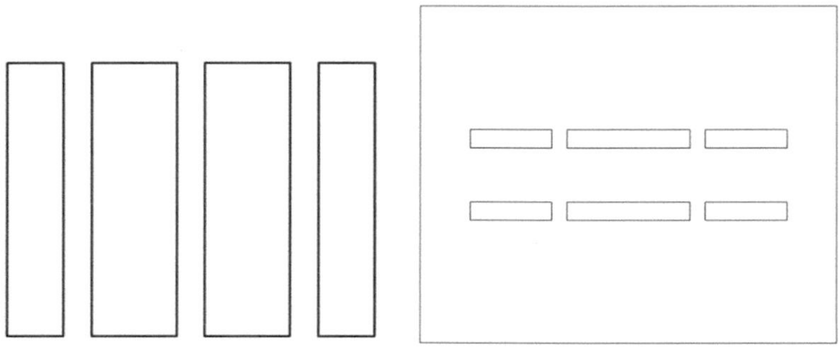
(a) Slots along the x-direction (b) Slots along the y-direction

Fig. 6.5 Diagrams of the slots

(frequencies), the voltage at the center of the nonradiation edge is not zero. Owing to the loading of the resistor, there is no influence on the radiation performance of the antenna's dominant mode. For the remaining modes, owing to the nonzero electric field in the center of the nonradiation side, the energy is absorbed due to the resistance loading to achieve the RCSR.

However, for the working mode of the microstrip antenna (in the working frequency band), the loading resistance has no effect on the radiation (scattering) performance because its electric field is zero at the center of the nonradiation side (the detailed simulation data are described in the next section). To achieve good in-band RCSR, slots can be cut in the ground based on the impedance loading, which achieves multi-aperture coupling of the aperture coupled antenna. The grooves on the floor can change the path of the induced current, which can then affect the scattering characteristics of the antenna to achieve RCSR.

Through the integrated use of loading the impedance at the center of the nonradiation edge and the multi-aperture coupling technology, the RCSR of the aperture coupled microstrip antenna can be realized across the entire frequency band. To further reduce the RCS of the antenna, we can adjust the antenna structure by cutting two rectangular slots on the floor [8–14].

6.3.2 *Structure and Characteristics of the Antenna*

The structure and size of the designed antenna are shown in Fig. 6.13.

As shown in Fig. 6.13, the center coordinates of the two coupling slots are (0, 7) and (0, -7), the center coordinates of the rectangular slot on the floor are (12.25, 0) and (-12.25, 0), and the width of the slot is 0.5 mm. To illustrate the characteristics of the antenna, we selected and constructed a reference antenna. The structure and size of the reference antenna are shown in Fig. 6.14.

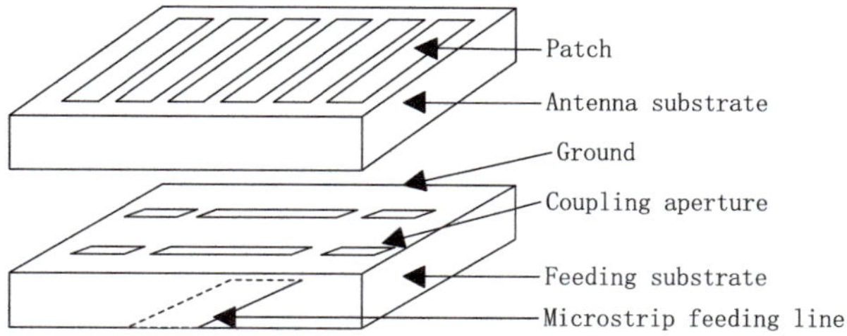

(a) Diagram of the antenna structure

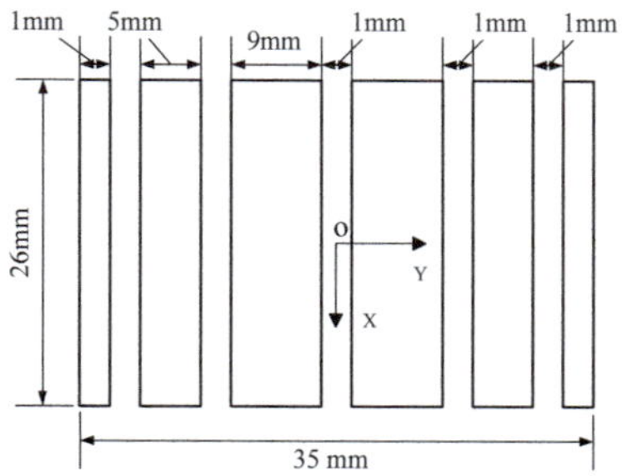

(b) Patch structure and size

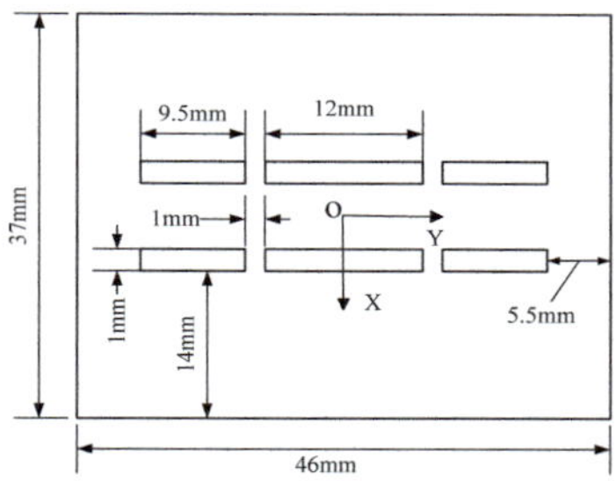

(c) Ground structure and size

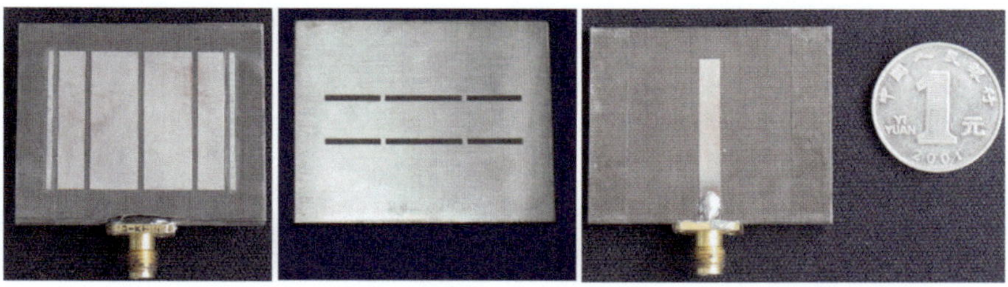

Upper patch Medium ground Bottom microstrip feeding line

(d) Photographs of the proposed antenna

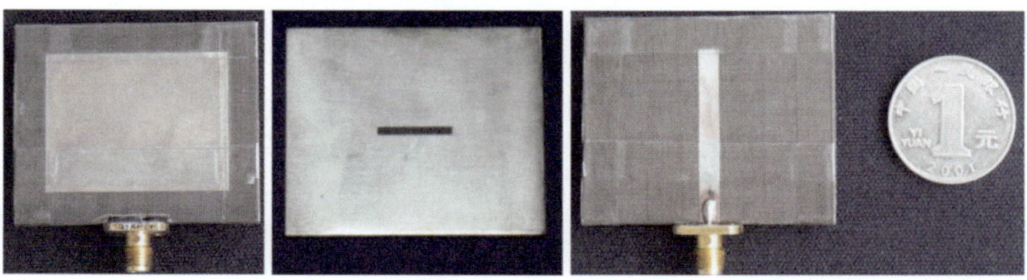

Upper patch Medium ground Bottom microstrip feeding line

(e) Photographs of the reference antenna

Fig. 6.6 Diagram of the antenna structure and photographs of the patch and ground

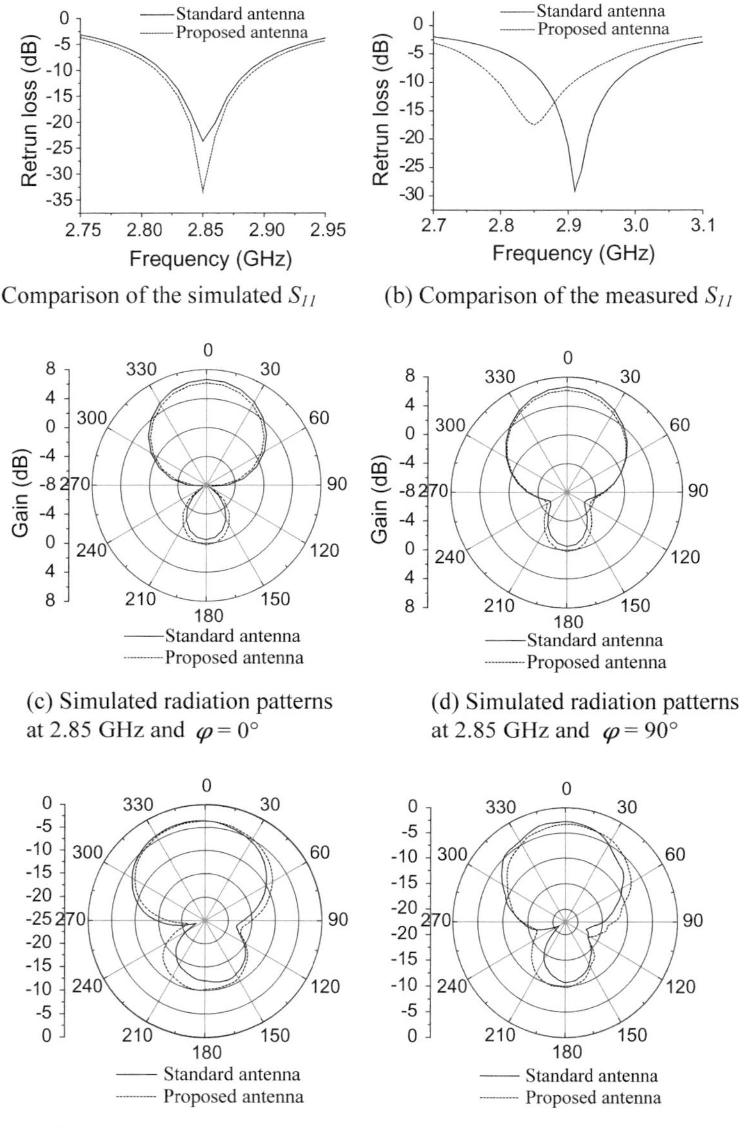

(a) Comparison of the simulated S_{11}

(b) Comparison of the measured S_{11}

(c) Simulated radiation patterns at 2.85 GHz and $\varphi = 0°$

(d) Simulated radiation patterns at 2.85 GHz and $\varphi = 90°$

(e) Measured radiation patterns at $\varphi = 0°$

(f) Measured radiation patterns at $\varphi = 90°$

Fig. 6.7 Comparisons of the S_{11} and radiation patterns of the two antennas

The radiation and scattering properties of the two antennas could then be compared. Figure 6.15 shows a comparison of the radiation patterns of the two antennas. Figure 6.16 shows a comparison of the S_{11} values of the two antennas, and Fig. 6.17 shows a comparison of the scattering performance of the two antennas. The radiation performance of the proposed antenna exhibits little loss compared

Fig. 6.8 Comparison of the RCSs of the two antennas under different angles of incidence

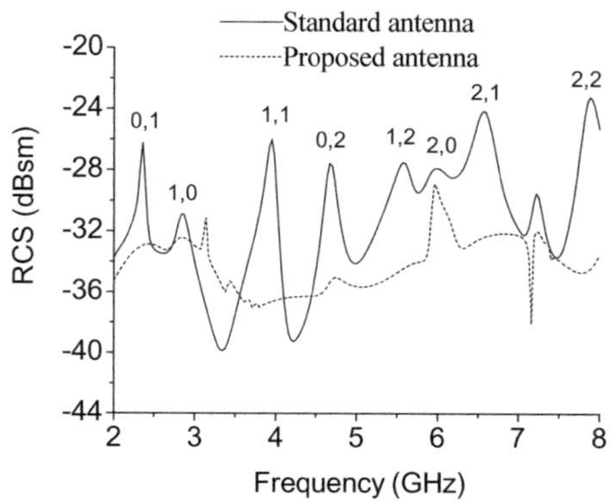

(a) Angle of incidence: $\theta = 60°$, $\varphi = 45°$

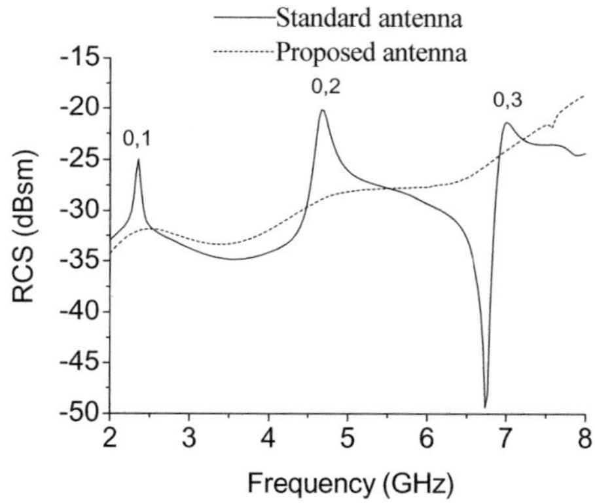

(b) Angle of incidence: $\theta = 60°$, $\varphi = 90°$

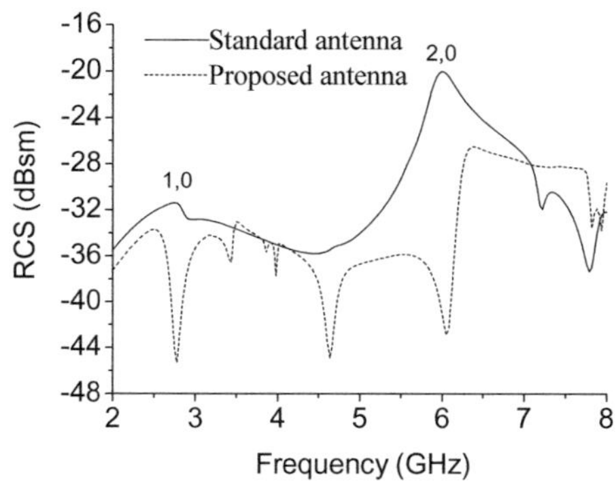

(c) Angle of incidence: $\theta = 60°$, $\varphi = 0°$

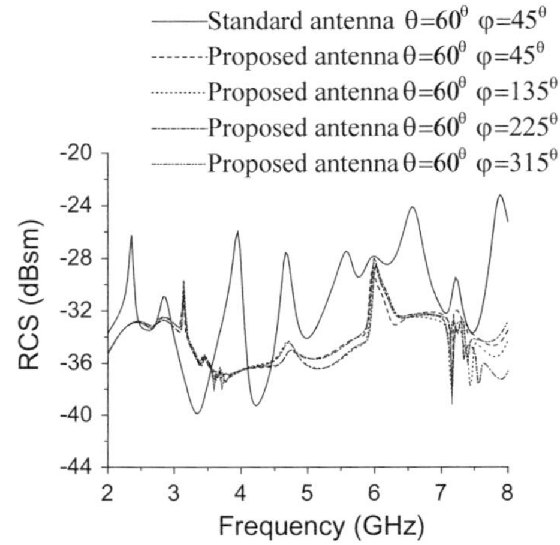

Fig. 6.9 Antenna RCS values under symmetric angles of incidence

with the reference antenna. Additionally, the gain loss is less than 1 dB. However, the scattering characteristics result in a better RCSR over a wide frequency range.

The antenna was fabricated based on the design, as shown in Fig. 6.18. The measured results of the two antennas are shown in Figs. 6.19 and 6.20.

This section discusses the RCSR of the aperture coupled microstrip antenna by combining impedance loading technology, floor slotting, and miniaturization technology. The simulation design was achieved using HFSS 9.0. Electromagnetic simulation software, and the experimental results of the fabricated antenna were consistent with the simulation results. It is relatively easy to reduce the RCS of the antenna through the combination of impedance loading, floor slotting, and miniaturization achieving the desired result at low cost.

The results in the figures show that with incident waves with the directions of $\theta = 60°, \varphi = 0°$ and $\theta = 60°, \varphi = 45°$, large reductions in the range of 2–8 GHz can be achieved. The in-band RCS of the proposed antenna at three angles was reduced to approximately −35 dB—that is, by more than an order of magnitude. Thus, by comprehensive use of impedance loading, miniaturization, and floor slotting, the designed antenna could maintain good radiation performance as well as achieve effective RCSR. Additionally, the antenna structure achieved using this method was relatively simple, and the cost was relatively low. We also found the combination of impedance loading, and floor-slotting technology to be suitable for probe and caliber coupling.

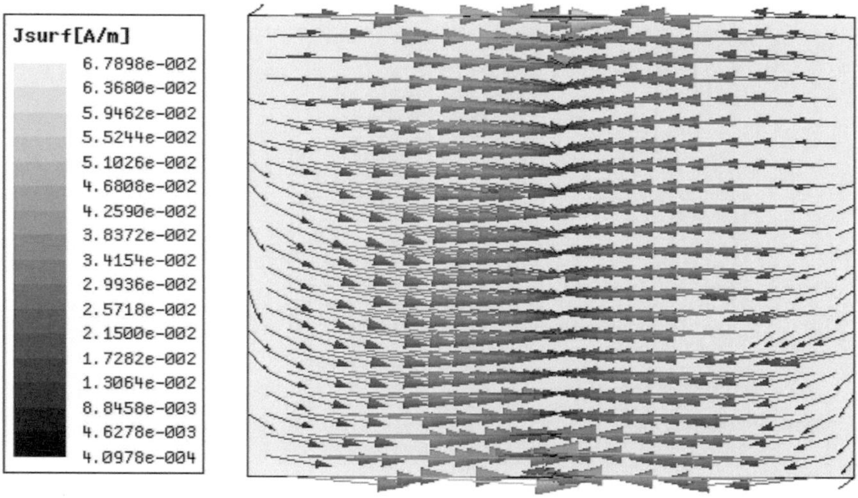

(a) Surface current distribution of the reference antenna at 4.67 GHz

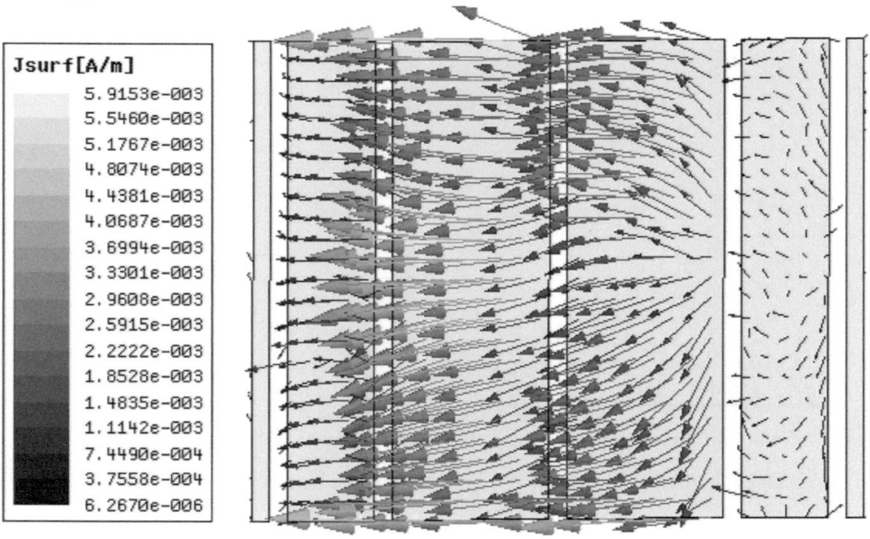

(b) Surface current distribution of the proposed antenna at 4.67 GHz

Fig. 6.10 Comparison of the surface current distributions of the two antennas

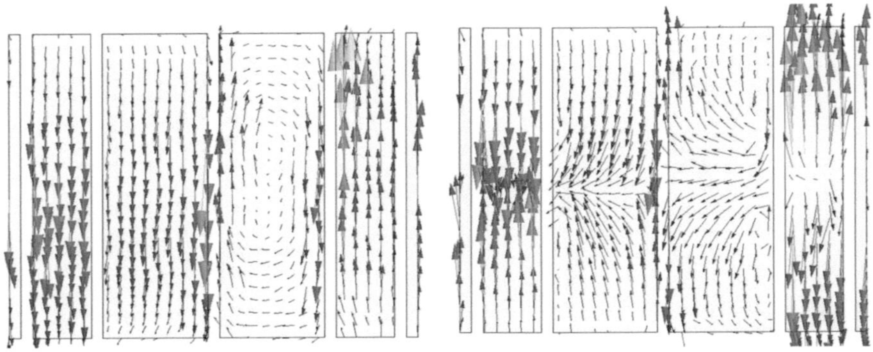

(a) Current distribution similar to the TM_{11} mode

(b) Current distribution similar to the TM_{21} mode

Fig. 6.11 Surface current distribution of the slotted antenna

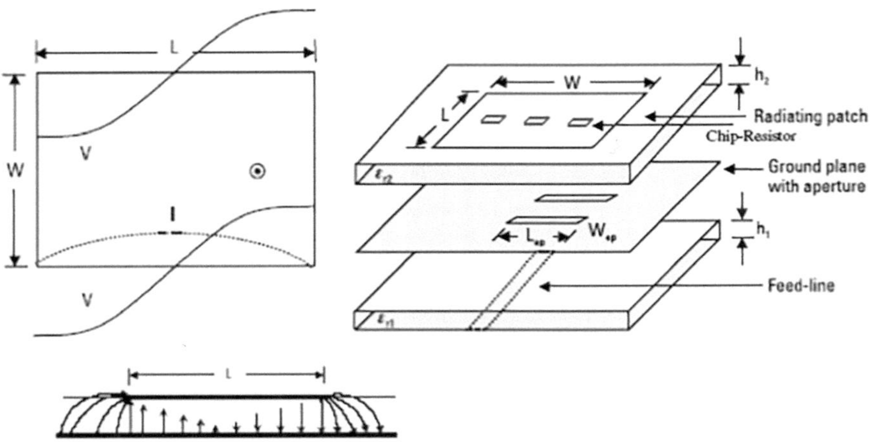

(a) Distribution of the voltage and electric field of the rectangular microstrip antenna for the TM_{10} mode

(b) Diagram of the multi-aperture coupled antenna loaded with resistances

Fig. 6.12 Rectangular microstrip antenna

6.4 Design of a Fractal Antenna with a Low RCS

The Koch fractal curve has unique spatial filling characteristics. The resonant frequency of the antenna can be effectively reduced if the straight line is replaced by the Koch fractal curve, reducing the antenna's size. Previous studies have shown that the resonant frequency of a monopole antenna using the Koch fractal curve

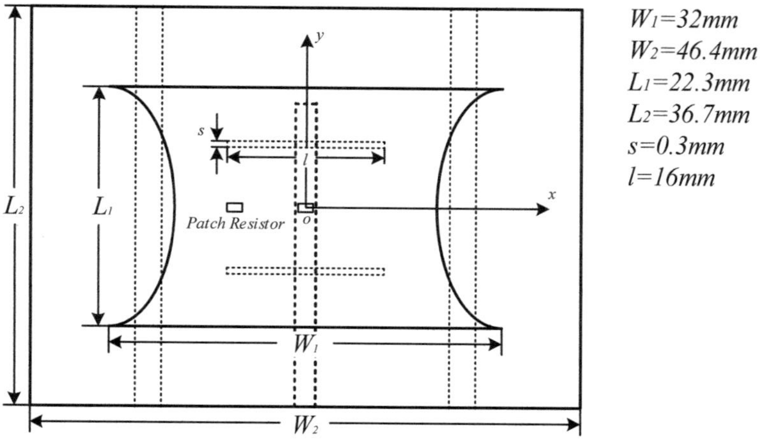

Fig. 6.13 Dimensions of the antenna with slots on the floor

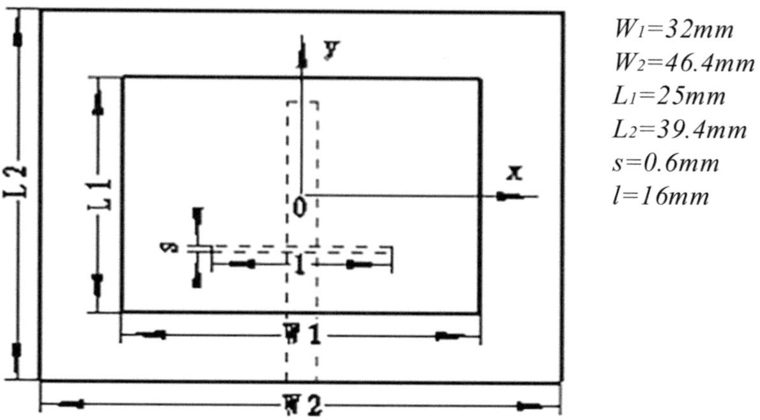

Fig. 6.14 Structure of the reference antenna

is 1.6 times lower than that of a traditional linear monopole antenna. Moreover, the height of the Koch monopole antenna is 40% smaller than that of a traditional monopole antenna. References [15, 16] conducted research on the Koch monopole, a photograph of which is shown in Fig. 6.21.

The Koch fractal curve can be used in the design of the resonant edges of a microstrip antenna, as shown in Fig. 6.22. The irregular fractal property is beneficial for maintaining the radiation characteristics of the antenna. The antenna operates at the TM_{10} mode. Owing to this kind of fractal boundary, the current does not flow strictly along the boundary; it flows along the curved fractal boundary similar to the way vehicles bypass an obstacle. This guarantees the operational performance of the antenna in the original working mode. Moreover, the antenna exhibits lower

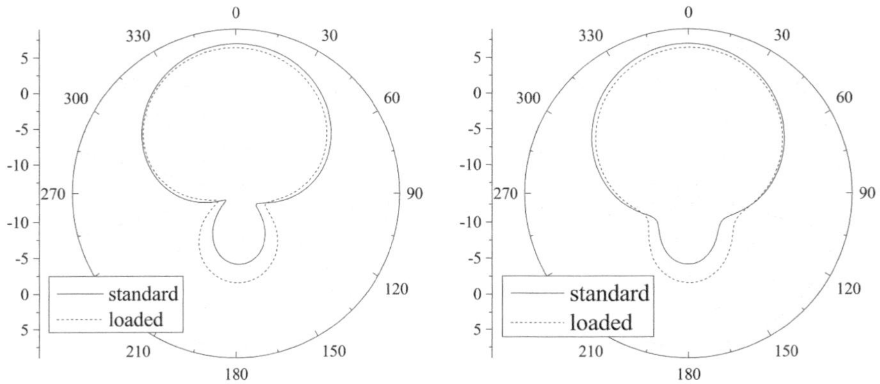

Fig. 6.15 Comparison of the radiation patterns

Fig. 6.16 Comparison of S_{11}

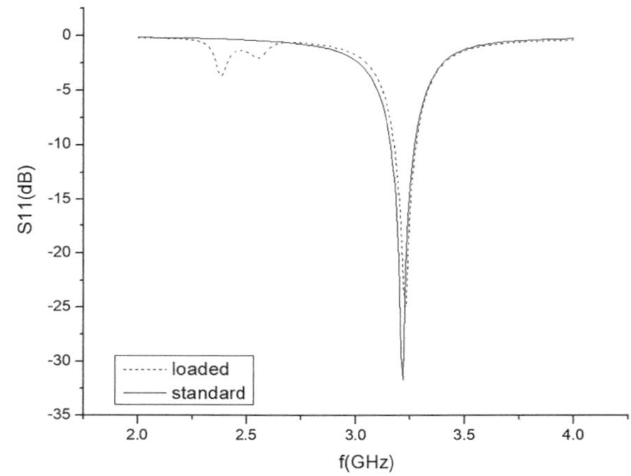

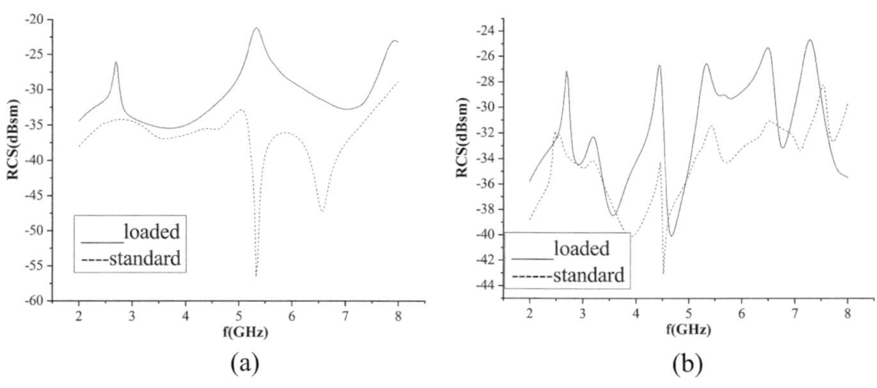

Fig. 6.17 Comparison of the scattering properties of the reference and multi-aperture coupled microstrip antennas

(a) Top view of the reference multi-aperture coupled antenna

(b) Bottom view of the reference multi-aperture coupled antenna

(c) Top view of the loading aperture coupledantenna

(d) Bottom view of the loading aperture coupled antenna

Fig. 6.18 Photographs of the two antennas

Fig. 6.19 Comparison of the measured S_{11} values between the reference rectangular patch antenna and the loaded antenna

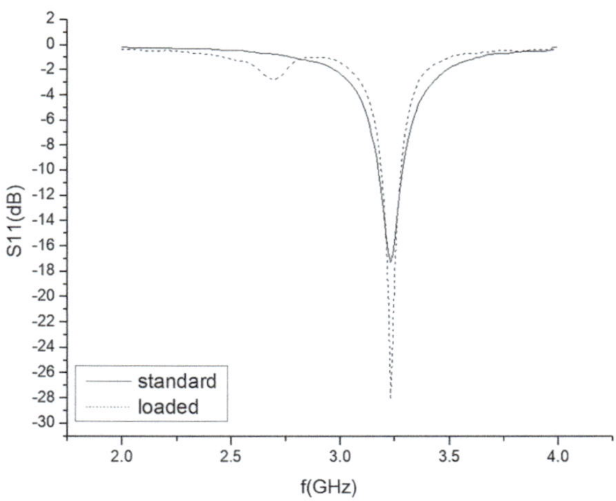

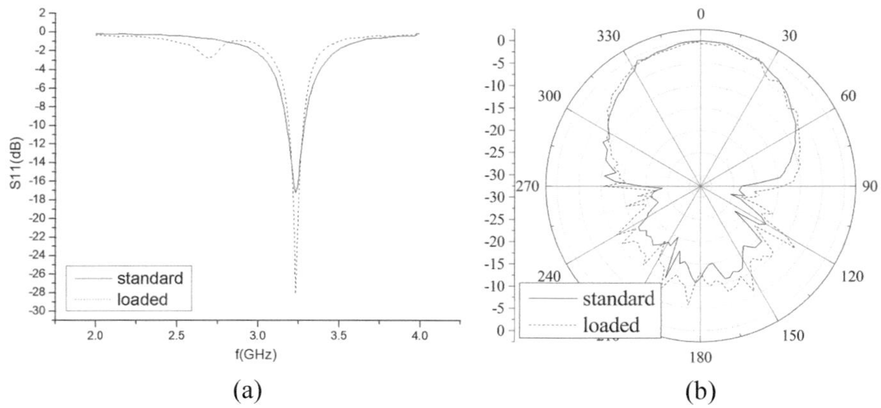

Fig. 6.20 Comparison of the measured radiation patterns

Fig. 6.21 The Koch monopole

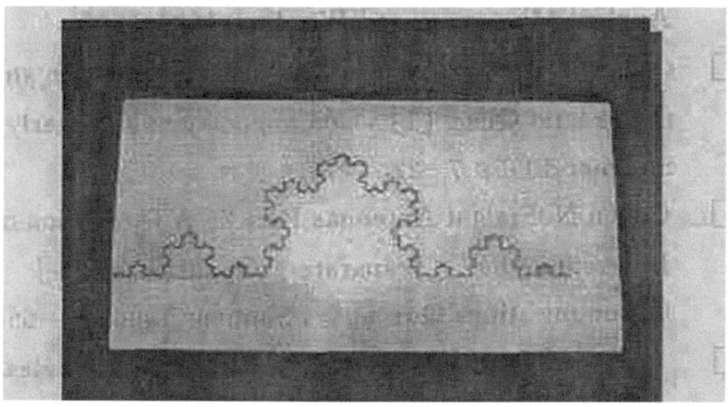

RCS properties owing to the low loss in radiation characteristics. Comparisons of the radiation and scattering performance of the fractal and the reference antennas are shown in Figs. 6.23, 6.24, 6.25.

Fig. 6.22 Structure of the fractal antenna

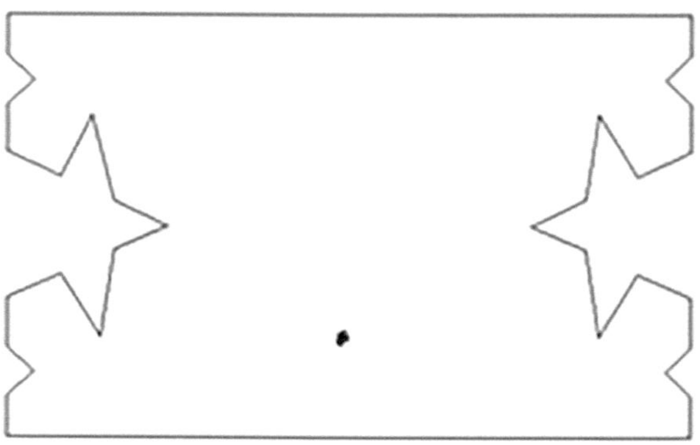

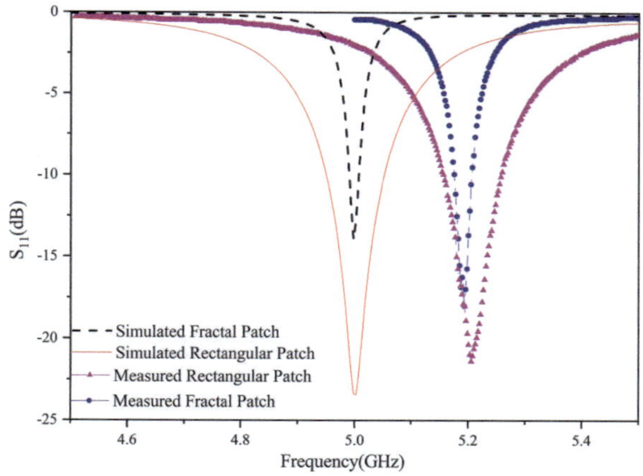

Fig. 6.23 S_{11} values of the reference and fractal antennas

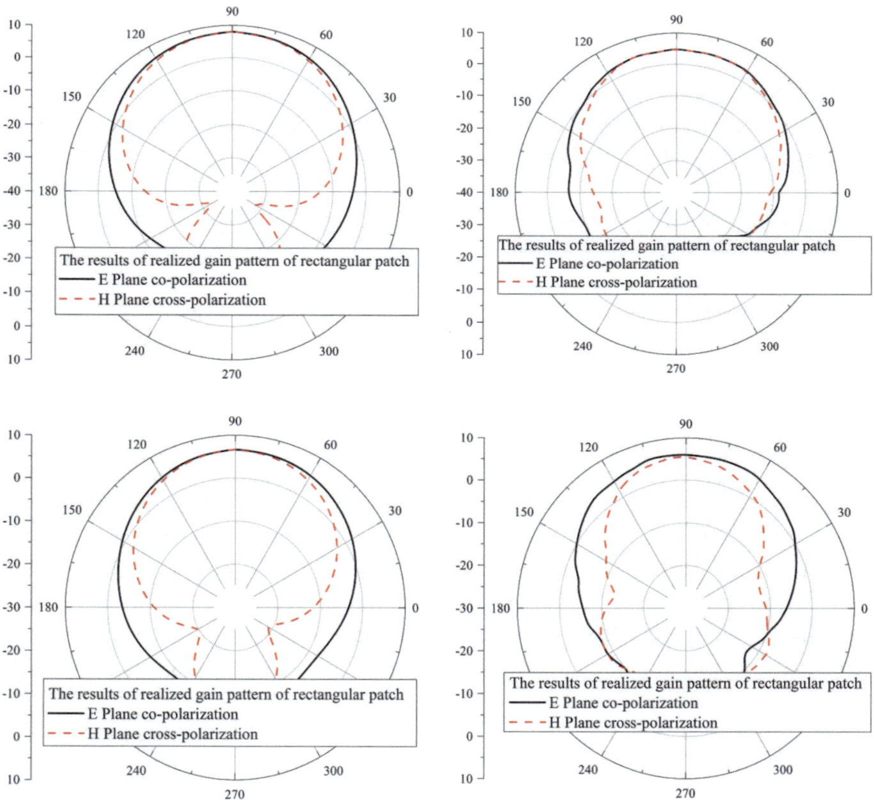

Fig. 6.24 Radiation performance of the reference and fractal antennas

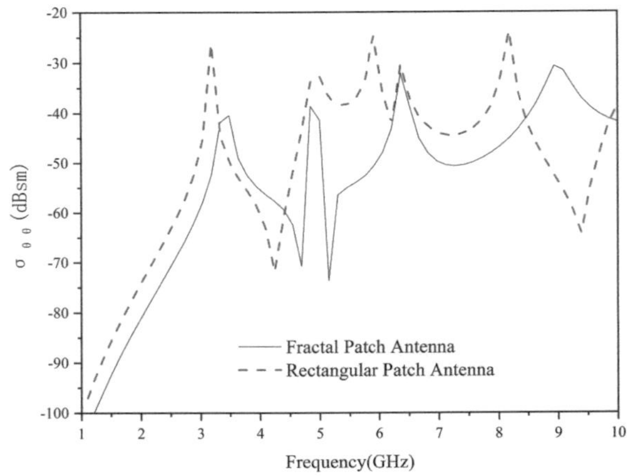

Fig. 6.25 Scattering performance of the reference and fractal antennas

Cutting rectangular slots parallel or perpendicular to the edges of the rectangular patch or adding short pins can also be used to reduce the antenna's RCS [17]. Here, we attempted to further reduce the RCS of the antenna based on a microstrip antenna with three-order fractal radiation edges. The distance between any two points on the radiation edges should be equal based on the radiation principles of the microstrip. Consequently, when using the Koch fractal curves as the two radiation edges, the current path could be changed by cutting two one-order Koch fractal slots parallel to the Koch fractal edges on the patch.

This could be designed in accordance with the following ideas—that is, keeping the resonant edges unchanged, reducing the length of the fractal edges, or cutting two one-order Koch fractal slots parallel to the Koch fractal edges on the patch. The results are shown in Fig. 6.26.

The antenna size can be reduced because the Koch fractal curve effectively increases the electric length of the radiation edges. The fractal antenna was modified to resonate at 5 GHz to compare it with the reference antenna. The parameters were $W = 18.1$ mm, $L = 10.12$ mm, and $h = 2.5$ mm. The distance of the one-order fractal slots was optimized to $d = 8.14$ mm. The two antennas were both fed by 50-Ω coaxial lines operating at the TM_{01} mode. Because the fractal microstrip antenna was much smaller than the reference antenna, a substrate with a lower dielectric constant was chosen to improve the radiation performance. The electrical properties of the substrate were: $\varepsilon_r = 2.2$, $\tan\delta = 0.0009$.

The surface current distribution of the patch is shown in Fig. 6.27, and a comparison of the simulated reflection losses is shown in Fig. 6.28.

As is evident from the results, the two first-order Koch fractal slots change the current patch. Starting from one of the radiation edges, the currents have to bypass two fractal slots to reach the other radiation edge. Consequently, the paths of the currents are extended in a limited space and the electric length increases. The combination of the first-order fractal slots and the third-order fractal radiation edges decreases the resonant frequency of the antenna. This results in the proposed antenna's size being

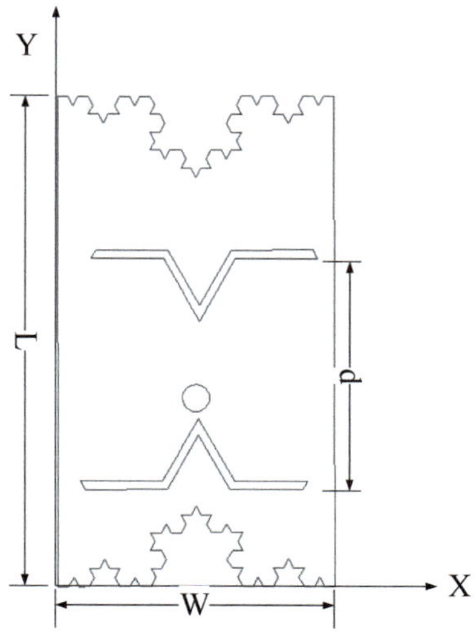

Fig. 6.26 Diagram of the Koch fractal microstrip patch antenna

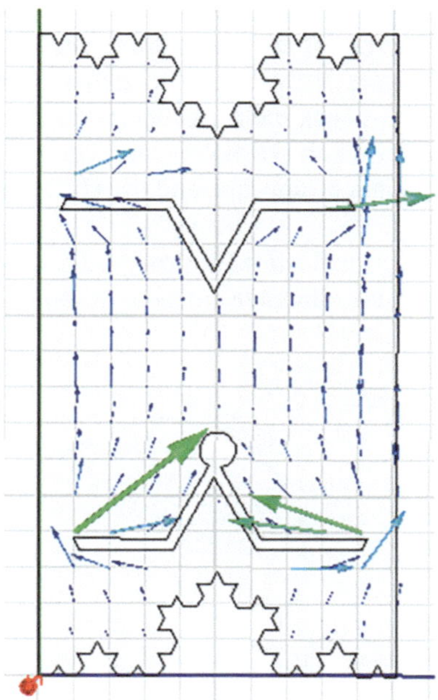

Fig. 6.27 Surface current distribution of the Koch fractal patch antenna

much smaller than that of the traditional antenna—that is, the fractal antenna area is only 40% that of the traditional antenna.

Next, the radiation performances and scattering performances were compared. The cost of RCSR is the loss of bandwidth. For example, the fractal antenna's operational bandwidth was 4.6%, while the traditional antenna's bandwidth was 5.8%.

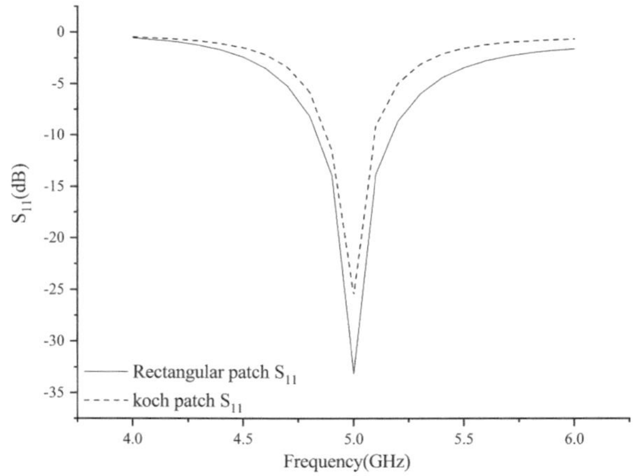

Fig. 6.28 Comparison of the simulated reflection losses

Although the bandwidth was affected, the cost was acceptable for these narrow-band antennas.

The far-field radiation patterns are shown in Fig. 6.29. The far-field radiation patterns exhibit little difference. The gain of the Koch fractal antenna is 6.2 dB, and the gain of the reference antenna is 6.9 dB, the gain decreasing by just 0.7 dB. For the *E*-plane, the back lobe of the Koch fractal antenna is smaller, but the half power beam width (HPBW) is not changed. For the *H*-plane, the back lobe of the fractal antenna disappears, and the HPBW is larger. This causes a reduction in the antenna gain.

In conclusion, the proposed fractal antenna exhibited lower radiation performance than the reference antenna. In some applications, the proposed antenna may be better

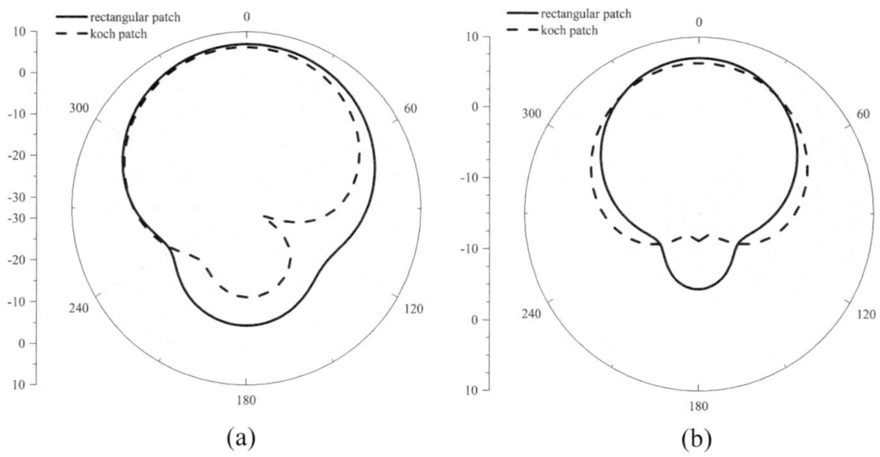

Fig. 6.29 Simulated radiation patterns

because of the reduction of the back lobe. Moreover, the miniaturization of the antenna based on the fractal has little influence on the radiation performance.

The results of the radiation performance show that this fractal antenna exhibits radiation performance similar to that of the reference antenna.

Next, the scattering performance of these two antennas was analyzed.

Figure 6.30a shows the scattering performances under an incident plane wave incident at an angle of $\theta = 60°$, $\varphi = 60°$. The RCS is reduced throughout the entire band, particularly above 7 GHz. A reduction of more than 10 dB is evident from 7–11 GHz. The RCS peaks are completely eliminated from 9–10 GHz, the RCS being reduced by more than 20 dB.

Figure 6.30b shows the RCS results of these two antennas under a θ-polarized plane wave incident from $\theta = 60°$, $\varphi = 45°$. The RCS of this incident angle is the most crucial because the plane wave excites the most modes from this incident angle. The RCS of the fractal antenna exhibits a visible reduction compared with that of the reference antenna from 2–12 GHz, the in-band RCS being reduced by approximately 3 dB. The RCS peaks at approximately 4 and 6 GHz are completely eliminated, the reduction being more than 10 dB. The RCS is also reduced by more than 10 dB above 9 GHz. Owing to the shift in frequency, a new RCS peak emerges at 7 GHz. This makes the RCS of the fractal antenna approximately 7 GHz greater than that of the reference antenna.

Figure 6.30c shows the RCS results of these two antennas under a θ-polarized plane wave incident from $\theta = 60°$, $\varphi = 90°$. Reductions of more than 10 dB are evident from 2–5 GHz and above 7.5 GHz. The RCS increases from 5–7.5 GHz because of the shift in frequency. The Koch fractal antenna has a lower RCS both within and outside the operating band. It therefore exhibits better stealth performance than does the traditional patch antenna.

The rectangular antenna and the Koch fractal antenna were fabricated and measured to verify the feasibility of the theory. A substrate with a dielectric constant of 2.2 is rare, and the thickness does not satisfy the design requirements. The substrates were changed to Teflon substrates with $h = 1.5$ mm and $\varepsilon_r = 2.65$. Photographs of the two antennas are shown in Fig. 6.31.

A comparison of the simulated and measured reflection losses are shown in Fig. 6.32.

These two antennas were measured in the same environment, the measured results of which are shown in Fig. 6.33. The normalized gain of the fractal antenna is 1 dB lower than that of the rectangular patch antenna—that is, the gain loss is acceptable for antenna RCSR.

As shown in the comparison of the simulated and measured results, the resonant frequency shifts slightly. This could be caused by fabrication errors. The resonant frequency can be tuned to 5 GHz by modifying the fabricated rectangular antenna and fractal antenna; the measured results should then be consistent with the simulated results. The measured radiation patterns fit the simulated results well, although small differences exists. Consequently, the results demonstrated the feasibility of the design method. Owing to measurement limitations, only the radiation performance (radiation patterns and reflection loss) were measured; the scattering properties were

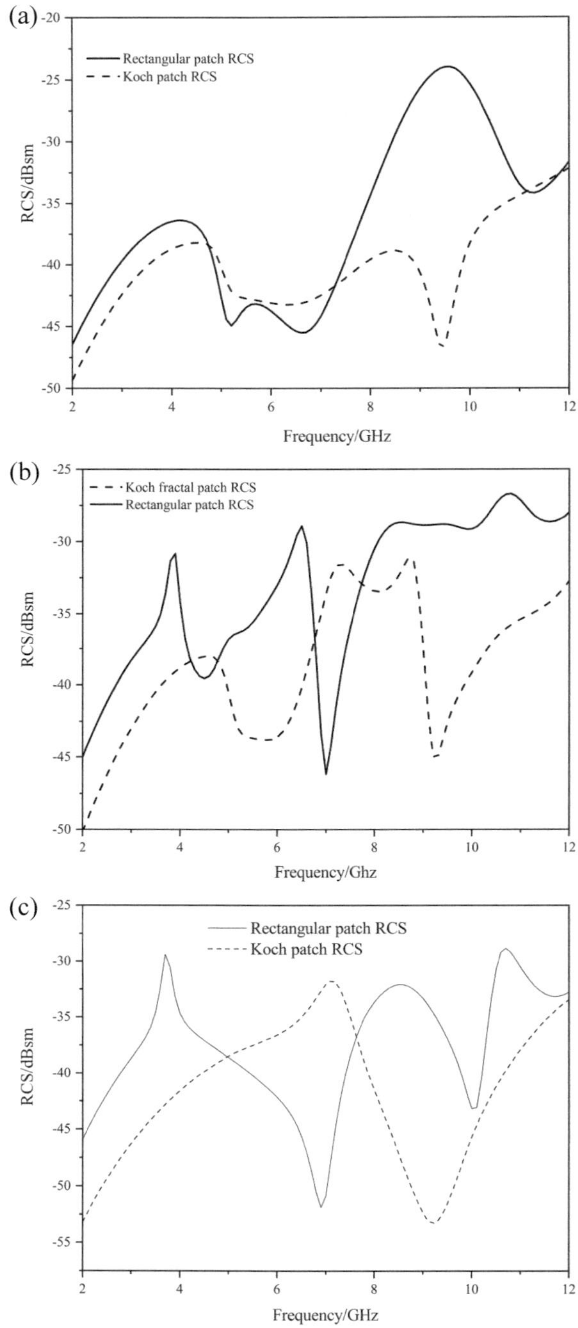

Fig. 6.30 Comparison of the scattering performance

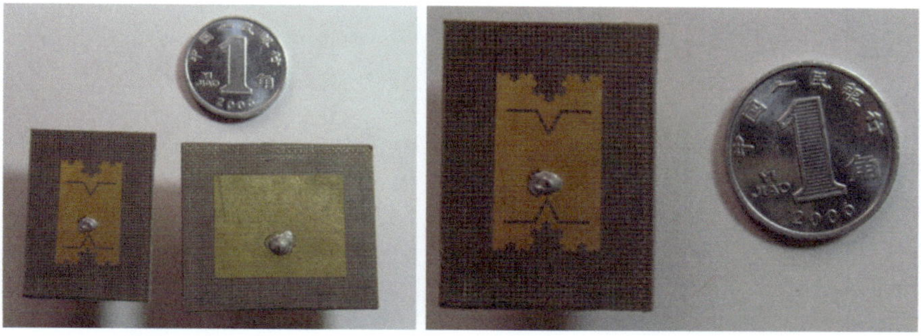

(a) The rectangular antenna (b) The fractal antenna

Fig. 6.31 Photographs of the fabricated antennas

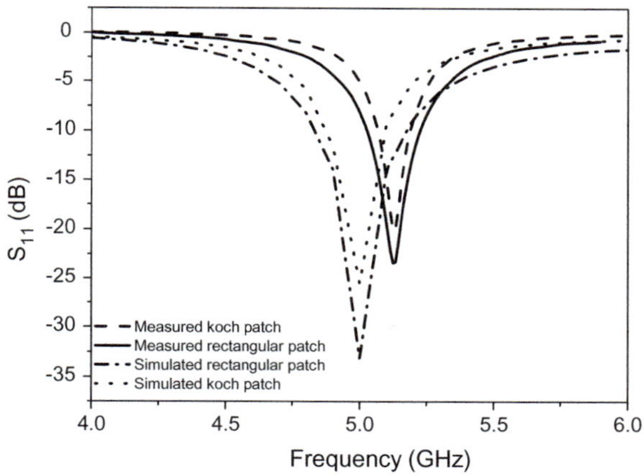

Fig. 6.32 Comparison of the reflection losses

not measured. However, as shown by the comparisons, the results simulated using Ansoft HFSS were similar to the measured results. Consequently, we believe that the Koch fractal antenna exhibited remarkable in-band and out-of-band RCSR compared with that of the reference antenna.

It can be difficult to reduce the antenna RCS while maintaining its radiation performance. In practice, the results are usually a compromise between these two factors. The method used above is the first time that a microstrip antenna's radiation edges have been designed as a Koch fractal structure. Additionally, two first-order fractal slots parallel to the radiation edges were cut on the antenna patch. The antenna RCS exhibited remarkable reductions both within and outside the operating band, and the radiation performance was maintained. It is evident that these results have reference value for antenna RCS research.

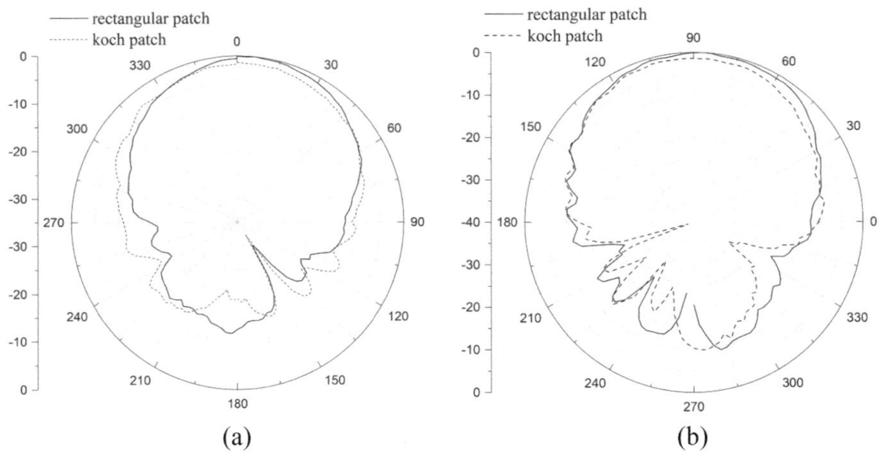

Fig. 6.33 Measured radiation patterns

6.5 Summary

This chapter proposed several designs for low-RCS antennas based on analysis of the distribution of the surface currents and operating properties of the operation modes, including an aperture coupled microstrip antenna using slot cutting and loading techniques and a low-RCS antenna based on fractal structures. These schemes better integrated the radiation and scattering characteristics of the antenna.

References

1. Y.T. Lo, D. Solomon, W.F. Richards, Theory and Experiment on Microstrip Antennas, IEEE Trans. Antennas Propagation, Mar, Vol. 27, No. 2, pp. 137–145, 1979.
2. R.H. Newman, Scattering from a Microstrip Patch, IEEE Trans. Antennas Propagation, Vol. 35, pp. 245–251, 1987.
3. C.K. Aanandan, P. Mohahan, Broad-band Gap Coupled Microstrip Antenna, IEEE Trans. Antennas Propagation, Vol. 38, No. 10, pp. 1581–1586, 1990.
4. Palaniswamy, V. R. Garg, Rectangular Ring and H Shape Microstrip Antenna-Alternative Approach to Rectangular Microstrip Antenna, Electronic Letters, Vol. 21, pp. 874–876, 1985.
5. T.W. Chiou, K.L. Wong, Designs of Compact Microstrip Antennas with a Slotted Ground Plane, Antennas Propagat Soc. Int. Symp. Dig., pp. 732–735, 2001.
6. Y.T. Lo, S.W. Lee, Antenna Handbook: Theory, Application and Design, Van Nostrand Reinhold Company, New York, pp. 2–23, 1988.
7. Girish Kumar, K.P. Ray, K. P. Ray, 'Broadband Microstrip Antennas' Artech House antennas and propagation library, 2003.
8. Yazidi, M. E., M. Himdi, and J. P. Daniel, Aperture Coupled Microstrip Antenna for Dual Frequency Operation, Electronics Letters, Vol. 29, pp. 1506–1508, 1993.
9. Croq, F., and D. M. Pozar, Multifrequency Operation of Microstrip Antennas Using Aperture Coupled Parallel Resonators, IEEE Trans. Antennas Propagation, Vol. AP-40, pp. 1367–1374, 1992.

10. Peter L. Sullivan and Daviel H. Schaubert, Analysis of an Aperture Coupled Microstrip Antenna, IEEE Transaction on Antenna and Propagation, Vol. 34, No. 8, pp. 977–984, 1986.
11. David M. Pozer, A Reciprocity Method of Analysis for Printed Slot and Slot- coupled Microstrip Antennas, IEEE Transaction on Antenna and Propagation, Vol. 34, No. 12, pp. 1439–1446, 1986.
12. Yang Xian Hua, Lotfollach Shafai, Characteristics of Aperture Coupled Microstrip Microstrip Antenna with Various Radiating Patches and Coupling Aperture, Microstrip Antennas, IEEE Transaction on Antenna and Propagation, Vol. 37, No. 4, pp. 72–78, 1985.
13. Apisak Ittipiboon, Ron Oostlander, Yhia M.M, *et al.*, A Modal Expansion Method of Analysis and Measurement on Aperture-Coupled Microstrip Antenna, IEEE Transaction on Antenna and Propagation, Vol. 39, No. 11, pp. 1567–1574, 1991.
14. I. Park and R. Mittra, Aperture-coupled Small Microstrip Antenna, Electronics letters, Vol. 32, No. 19, pp. 1741–1742, 1996.
15. C. Puente, J. Romeu, R. Pous, Small but Long Koch Fractal Monopole, Electronics Letters, Vol. 34, No. 1, pp. 9–10, 1998.
16. C. Puente, J. Romeu, R. Pous. The Koch Monopole a Small Fractal Antenna. IEEE Transactions on Antennas and Propagation, Vol. 48, No. 11, 2000.
17. Qingsong Shi, Shuxi Gong, Xiulian He, Two Methods for Reducing the Scattering Cross Section of Microstrip Antennas and Their Summarization, Radar Science and Technology, Vol. 8, No. 4, 2005.

Chapter 7
Design of a Low-RCS Antenna Based on an FSS

A frequency selective surface (FSS) [1–5] is a two-dimensional periodic array designed to reflect or transmit EM waves based on the frequency of the field. In this sense, an FSS is a type of filter. An FSS commonly comprises a periodic array of metal patches or periodic apertures arranged on a metal screen, which behave as a total reflection (for patches) or transmission (using apertures) in the neighborhood of the element resonant frequency.

An FSS is usually applied to out-of-band RCSR when the enemy and friendly radars operate at different frequencies. There are two ways to achieve this goal. The first is to cover the radar antenna and its accessory parts with a radome made of a bandpass FSS. Because the bandpass FSS exhibits good transmission properties—that is, the transmission coefficient is close to 1—within the working frequency band of our radar, the radar antenna can operate well. Additionally, a bandpass FSS—which has a streamlined outer surface—can reflect the out-of-band incident wave in nonthreatening directions to reduce the monostatic RCS. Another method is to use a band-stop FSS as the reflector or the ground plane of the antenna. The band-stop FSS behaves as a PEC—that is, the reflection coefficient is close to 1—within the operating frequency band of our radar, ensuring that the antenna can operate well. For an out-of-band incident wave, there is little reflected energy, so the monostatic RCS is reduced.

7.1 Research on the Bandpass FSS

7.1.1 General Research on the Bandpass FSS

The frequency characteristics of an FSS depend primarily on the shape and size of the unit cell and the grid arrangement of the periodic surface. Moreover, the dielectric loading and the incidence angle and polarization of the incident wave can also greatly

Y. Liu et al., *Antenna Radar Cross Section: Theory and Design*,
https://doi.org/10.1007/978-981-96-1226-0_7

influence the FSS properties. Consequently, the influence of the parameters of a single-ring-type FSS on its characteristics is discussed below.

Assume that the single-ring-type FSS is arranged in a square grid and that its parameters are the same as those shown in Fig. 7.1. Here, we can examine the FSS in terms of the shape and size of the unit cell, the dielectric loading, grid arrangement, element spacing, and incident angle and polarization of the incident wave.

1. **Shape and size of the unit cell**: The frequency response of an FSS depends primarily on the shape and size of the unit cell of the periodic surface. Figure 7.2 shows some of the most commonly used unit cells. Several of their characteristics without a dielectric substrate are listed in Table 7.1. Different unit shapes lead to different characteristics. Consequently, the first thing to do when designing an FSS is to select an appropriate unit cell geometry. Its size is another important parameter that influences the frequency response of the FSS (Table 7.2).

Based on the theory proposed by Munk [5], a ring-shaped unit resonates and reradiates energy when its perimeter is approximately equal to an integer multiple of the wavelength. To obtain an FSS with a compact structure and strong resonance,

Fig. 7.1 Geometry of the single-ring-type FSS

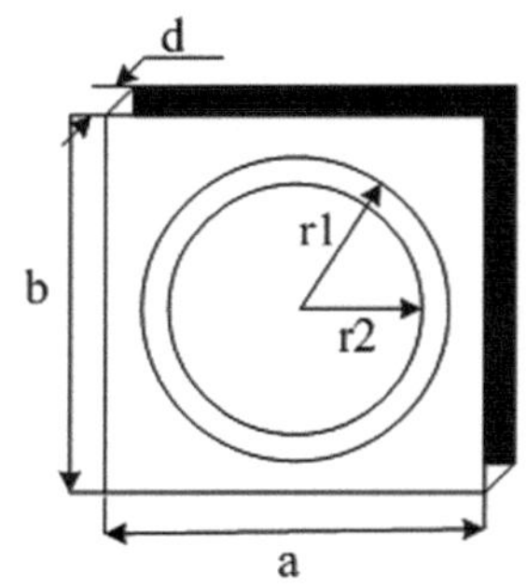

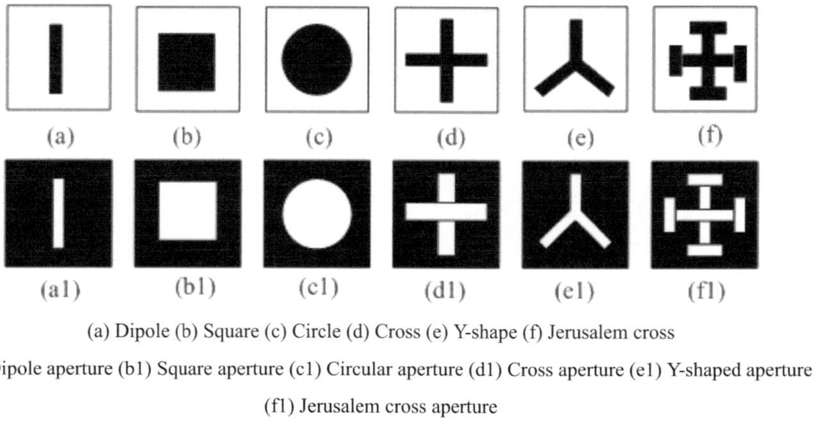

Fig. 7.2 Geometries of common unit cells of FSSs

Table 7.1 Parameters of the single-ring-type FSS

a	16 mm	b	16 mm
$r1$	7 mm	$r2$	5.5 mm
D	1 mm	ε_r	2.65
Tang(δ)	0.001		

Table 7.2 Comparison among different unit cells of FSSs (single layer without a dielectric substrate)

Unit type	Stability of resonant frequency	Ability to suppress cross-polarization	Bandwidth	
Dipole	Worst	Best	Worst	Best
Square	Best	Best	Best	Best
Circle	Best	Second best	Best	Best
Cross	Bad	Bad	Bad	Bad
Y-shape	Bad	Bad	Bad	Second best
Jerusalem cross	Second best	Bad	Second best	Second best

the perimeter of the ring should be selected to be one wavelength. As shown in Fig. 7.3, with increasing outer radius, the resonant frequency decreases, and the bandwidth increases. When r_1= 6.0 mm, the resonant frequency (f_0) is 7.50 GHz, which corresponds to the wavelength λ_0= 40 mm. The perimeter of the ring $c_1 \approx \pi(r_1 + r_2) = 36.11$mm.

2. **Dielectric loading**: The dielectric plate not only supports the metal structure of the FSS but also plays an important role in improving its performance. Figure 7.4 shows that the resonant frequency of the FSS can be greatly affected by the relative permittivity of the dielectric [6, 7]. When the relative permittivity increases, the

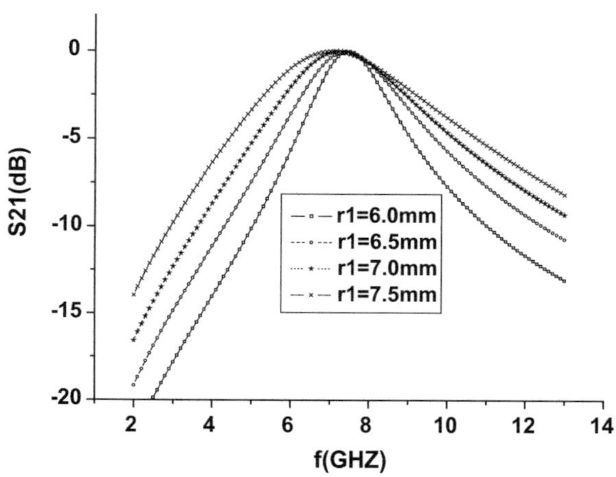

Fig. 7.3 Influence of the outer radius of the single ring on the FSS transmission coefficient

resonant frequency of the FSS shifts to a lower frequency band, and the bandwidth decreases, equivalent to increasing the size of the unit.

The thickness of the dielectric also influences the FSS characteristics. Figure 7.5 shows the frequency response of the FSS to different dielectric thicknesses. The resonant frequency decreases with increasing thickness, and the transmission loss also increases. However, when the thickness is increased to 10 mm, the transmission decreases, and the resonant frequency increases. This phenomenon illustrates that the resonant frequency of the FSS approaches a fixed value when the thickness of the dielectric reaches a certain value.

3. **Grid arrangement and element spacing**: The unit cells of the FSS can be arranged in different grids—such as squares, equilateral triangles, and isosceles triangles. Different grid arrangements also have an important effect on the FSS performance [8, 9]. Generally, the characteristics of the equilateral arrangement

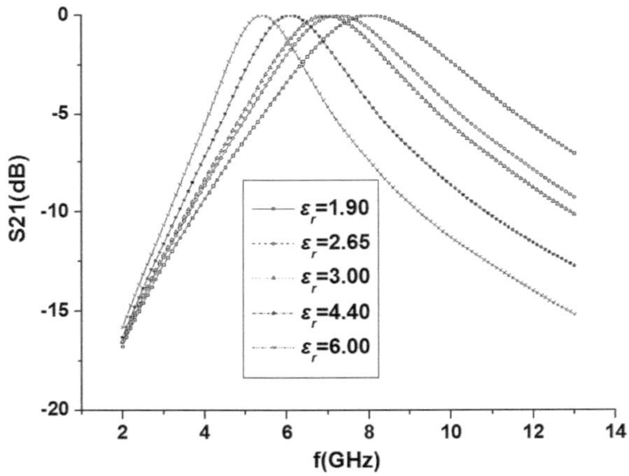

Fig. 7.4 Influence of the relative permittivity on the FSS

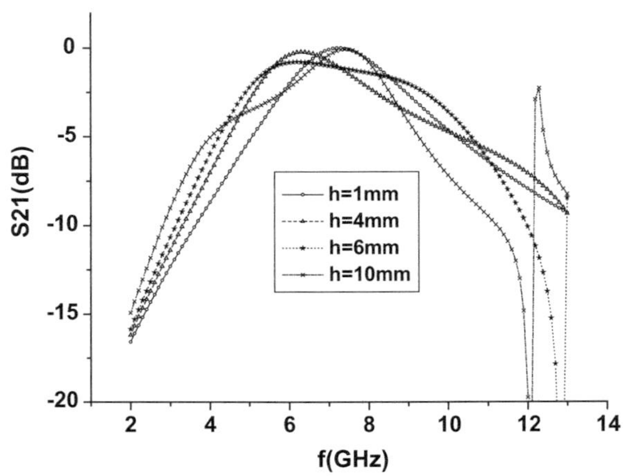

Fig. 7.5 Influence of the dielectric thickness on the FSS

are more stable than those of the square arrangement. Regardless of which arrangement is used, the unit spacing should meet the requirements shown in Table 7.3.

There are two special phenomena associated with an FSS—that is, the Wood's singularity and the grating lobe. For a patch-type FSS, when the phase difference of the current or voltage on adjacent units is $2n\pi$, a zero point appears in the reflection coefficient curve, which is called total transmission. Conversely, for an aperture FSS, total reflection occurs. This phenomenon is called the Wood's singularity. When the incident wave frequency is lower than the first Wood's singularity frequency, all of the FSS scattering energy either passes through the surface of the FSS in the incident direction or reflects along the FSS. However, when the incident wave frequency is higher than the first Wood's singularity frequency, the phase relationship between the current and voltage on the FSS results in some of the reflected energy propagating in the direction called the grating lobe direction. It is impossible to achieve total transmission or total reflection, so the performance of the FSS will be affected. Consequently, the unit spacing should be as small as possible to avoid a grating lobe in the actual design. When a transmission grating lobe does not appear or the FSS units are not in contact with one another, the unit spacing influences the FSS resonant bandwidth, which becomes wider as the unit spacing decreases. Generally, the FSS resonant bandwidth approximately doubles if the unit spacing *dx* or *dy* decreases by half. Additionally, the FSS resonant bandwidth approximately quadruples if the unit spacings *dx* and *dy* both decrease by half.

Figure 7.6 shows the frequency response curves of the ring aperture FSS at different unit spacings, the results of which are similar to the relationships between the unit spacing and bandwidth.

4. **Incident angle and polarization state**: Different incident angles and polarizations also have considerable effects on the frequency characteristics of an FSS. When a plane wave is incident on the *E*-plane, the transmission coefficient

Table 7.3 Arrangements of the FSS and the conditions required to avoid grating lobes

Arrangement form	Grid shapes	Maximum spacing	θ = 0°	θ = 45°
Square	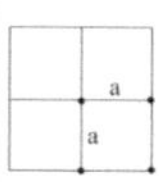	$\frac{d}{\lambda_0} < \frac{1}{1+sin\theta_0}$	$\frac{d}{\lambda_0} < 1$	$\frac{d}{\lambda_0} < 0.59$
Equilateral triangle	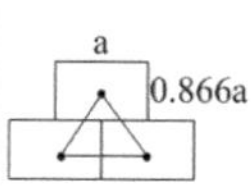	$\frac{d}{\lambda_0} < \frac{1.15}{1+sin\theta_0}$	$\frac{d}{\lambda_0} < 1.15$	$\frac{d}{\lambda_0} < 0.67$
Isosceles triangle	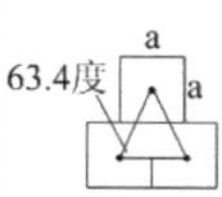	$\frac{d}{\lambda_0} < \frac{1.12}{1+sin\theta_0}$	$\frac{d}{\lambda_0} < 1.12$	$\frac{d}{\lambda_0} < 0.65$

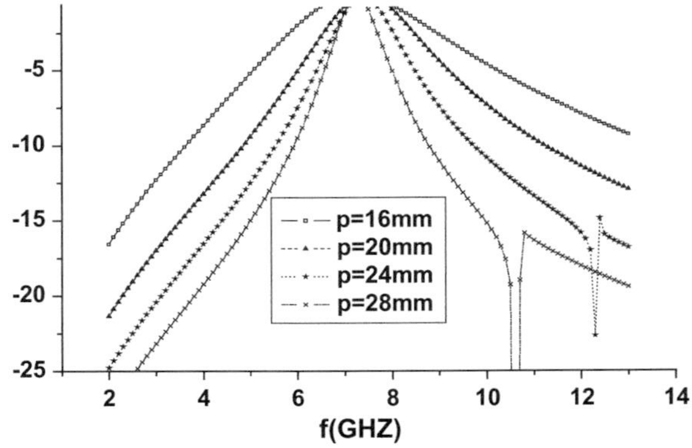

Fig. 7.6 Frequency response curves of the ring aperture FSS at different unit spacings

increases by 1/cos(θ) as the incident angle increases, in which $\theta = 0°$ corresponds to vertical incidence. When the plane wave is incident on the *H*-plane, the transmission coefficient decreases by cos(θ) as the incident angle increases. However, the opposite occurs for patch-type FSSs.

The incident angle affects the bandwidth but also influences the FSS resonant frequency, although the impact is much less on the FSS resonant frequency than on the bandwidth. Additionally, for FSSs of different shapes, the resonant frequency varies rapidly as the incident angle changes [10]. The effects of different incident angles and polarizations on the transmission coefficient are shown in Fig. 7.7.

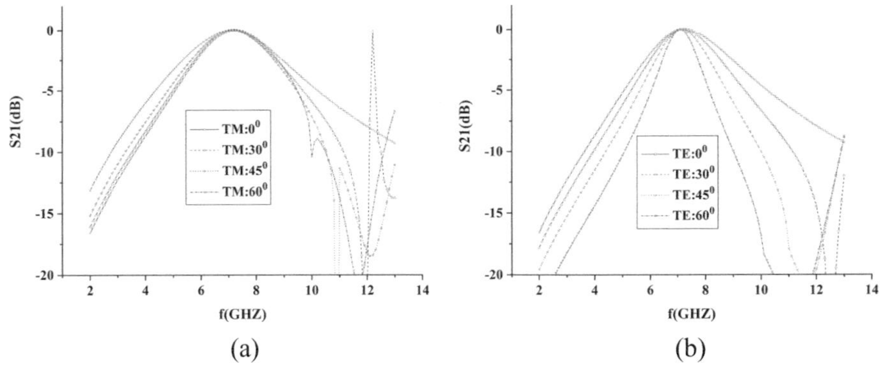

Fig. 7.7 Effects of different incident angles and polarizations on the transmission coefficient

7.1.2 *Research on Substrate-Integrated Waveguide Frequency Selective Surfaces (FSSs)*

Substrate-integrated waveguide (SIW) technology—which has been proposed in recent years—is a new waveguide structure that can be integrated into a dielectric substrate with low insertion loss and low radiation characteristics. This technology uses a metal through-hole arrangement on a low-loss dielectric substrate with metal layers on the upper and lower surfaces. In this way, the function of a traditional metal waveguide can be achieved on the media substrate. It has the advantages of being high-quality, easy to design, lightweight, and exhibits good selectivity [11, 12]. Here, we introduce SIW into the design of a Y-shaped FSS and a single-circular-ring FSS to examine the specific performance of improved FSSs.

1. **General FSS and improved substrate-integrated FSS**: CST simulation software was used during the design process, the simulation model being as shown in Fig. 7.8. A square-shaped grid was used, as shown in Fig. 7.9; thus, $\alpha = 90$. A 1-mm-thick F4B dielectric substrate was used, with a dielectric constant of 2.65, giving it little influence on the resonance frequency.

The Y-shaped FSS is simple [13, 14], and exhibits good stability for different modes of incident plane waves. Many studies have been performed on Y-shaped units owing to their consistency with conical radome surfaces. For a Y-shaped FSS,

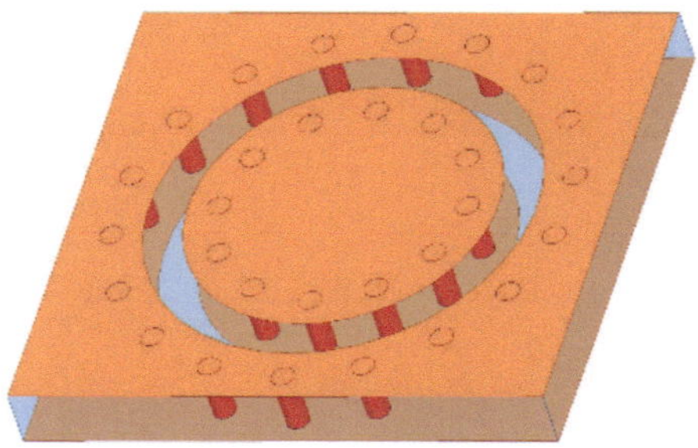

Fig. 7.8 CST simulation model

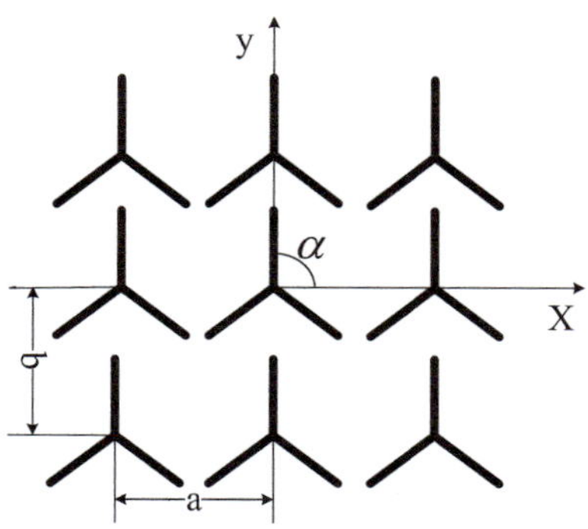

Fig. 7.9 Cell arrangement

the bandwidth is the primary problem that needs to be solved. For the simplest single-circular-ring FSS, a wide bandwidth problem exists, restricting the application of an ordinary bandpass FSS in engineering applications requiring a narrow bandwidth and good frequency selectivity.

Metal through-holes can be added without changing the ordinary Y-shaped FSS and single-circular-ring FSS units, and a good reduction in bandwidth can be achieved without changing the original cell spacing. The frequency selectivity of the improved FSS is better, as are other FSS features.

Figure 7.10 and Table 7.4 show the Y-shaped ring-hole unit, single-ring unit, and modified models and their parameters, respectively. A metal plate was etched into a unit shape and attached to a side of the dielectric substrate of the ordinary FSS. The upper and lower metal plates of the ordinary FSS were connected by metal through-holes, making up the SIW structure.

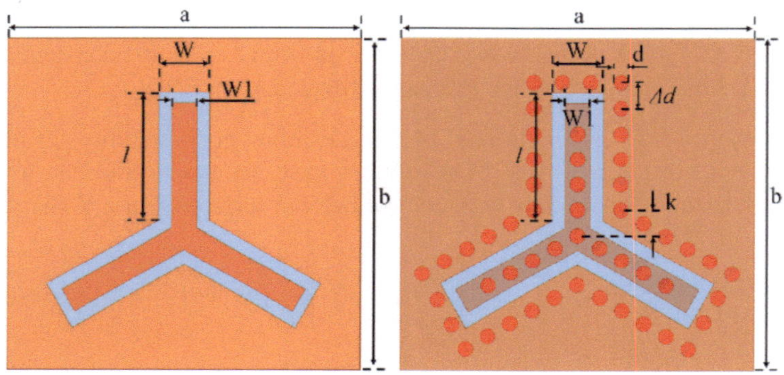

(a) Y-shaped ring-hole FSS

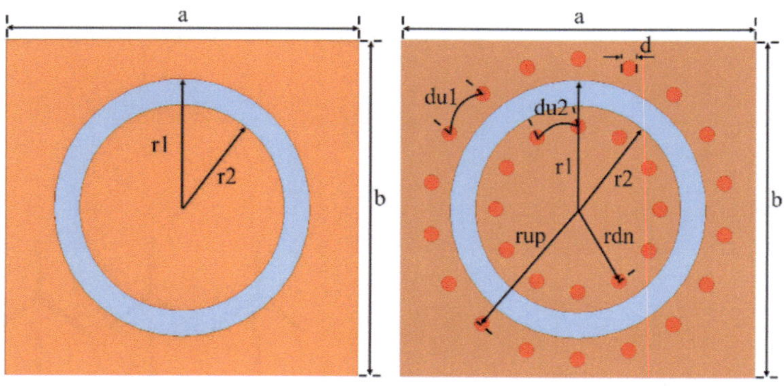

(b) Single-ring FSS

Fig. 7.10 Common Y-shaped ring-hole FSS, single-ring FSS and their modified models

Table 7.4 Parameters of the Y-shaped ring-hole FSS and the single-ring FSS

a	13 mm	b	13.8 mm	a	12 mm	b	12 mm
w	2 mm	$w1$	1 mm	r_{up}	5.5 mm	r_{dn}	3.5 mm
l	5 mm	k	1 mm	du_1	20°	du_2	30°
Δd	1 mm	d	0.6 mm	r_1	5 mm	r_2	4 mm
tang(δ)	0.001	ε_r	2.65	d	0.6 mm	ε_r	2.65
				tang(δ)	0.001		
(a) Y-shaped ring-hole FSS				(b) single-ring FSS			

Figure 7.11a and b show the frequency responses of the transmission coefficients of the common and improved FSSs of the Y-shaped ring-hole FSS and the single-ring FSS at vertical incidence, respectively. The −3 dB bandwidth is considerably reduced in both the Y-shaped ring-hole FSS and the single-ring FSS improved using the SIW structure. The frequency selectivity is better, and the resonance frequency remains nearly constant.

The impact of the FSS parameters show that the FSS resonance frequency is related primarily to the unit shape and unit size. Similarly, the frequency response of a Y-shaped ring-hole FSS and single-ring FSS depends primarily on the size of the unit on the periodic surface, the size of the cell determining the resonance frequency. The SIW structure is introduced based on the original FSS without changing its shape and size. Consequently, the modified frequency-selective resonance frequency shown in Fig. 7.11 is practically unchanged.

Considering this problem from the point of view of an equivalent circuit, using SIW technology in the design of an FSS is equivalent to using a parallel inductor. This means that the total inductance of the equivalent circuit increases. According to $Q = w_0 L/R$, as the inductance (L) increases, the quality factor (Q) increases, so

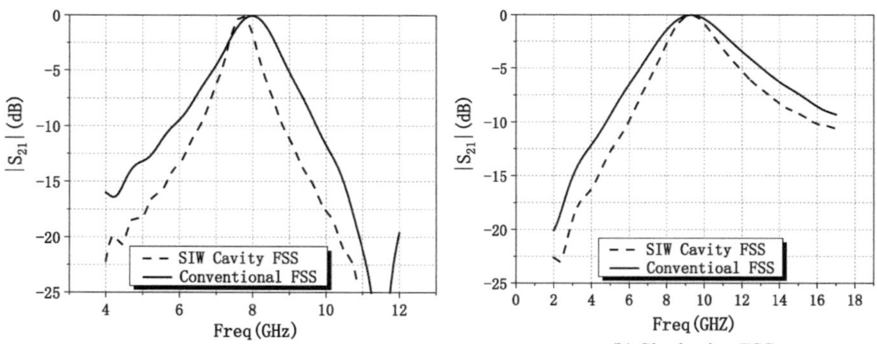

(a) Y-shaped ring-hole FSS (b) Single-ring FSS

Fig. 7.11 Frequency response curves of the Y-shaped ring-hole FSS, the single-ring FSS, and their improved models at vertical incidence the bandwidth decreases. Similarly, the resonant frequency decreases slightly based on the formula $w_0 = 1/\sqrt{LC}$.

2. **Parameter analysis of a substrate-integrated waveguide FSS**: To design an FSS with good performance, the appropriate unit must be chosen and several parameters considered, such as the unit size, unit spacing, and media. The impact of these parameters on the frequency response characteristics of the SIW FSS will be discussed below.

(1) The arm length l of the Y-shaped ring-hole FSS and the inner radius $r2$ of the single-ring FSS: The frequency characteristics of an FSS are mainly determined by the shape of the unit surface. Figure 7.12a and b show the impacts of the arm length (l) of a Y-shaped ring-hole FSS and the inner radius ($r2$) of a single-ring FSS on the frequency response of the modified FSSs—that is, the SIW FSSs. The resonance frequency varies with the arm length (l) and inner radius ($r2$). As the arm length (l) and inner radius ($r2$) increase, the resonance frequency decreases.

Additionally, the arm length (l) primarily affects the resonance frequency of the Y-shaped ring-hole FSS but has little effect on the bandwidth. The inner radius ($r2$) of the single-ring FSS not only influences the resonance frequency but also greatly influences the bandwidth. As $r2$ increases, the bandwidth decreases. The results show that the arm length (l) of the Y-shaped ring-hole FSS and the inner radius ($r2$) of the single-ring FSS have a similar influence on both the ordinary FSS and the modified FSS. Consequently, the SIW in the modified FSS has little effect on the new FSS when the cell size is changed.

(2) Diameter of the SIW: Fig. 7.13 shows the effect of the diameter of the SIW on the transmission coefficient of the modified FSS. The results show that the diameter of the SIW has little effect on its performance. To ensure the properties of the FSS, the FSS boards generally used in practice are relatively large and contain at least 20 × 20 units. Considering the practical applications of FSSs, the design

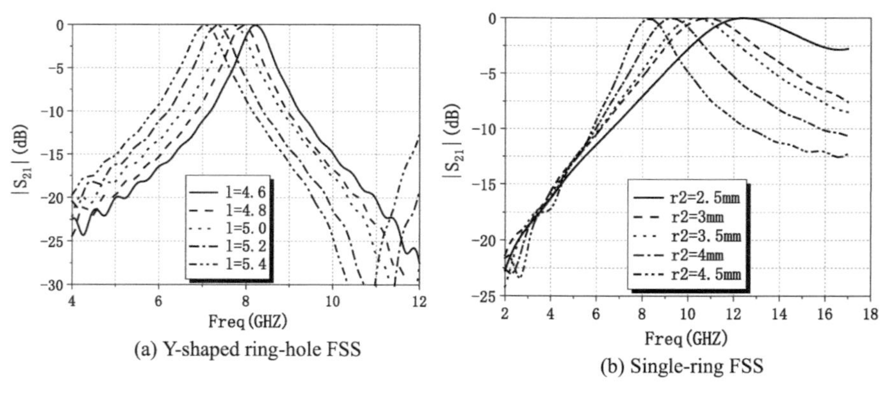

Fig. 7.12 Frequency responses of the arm length (l) of the Y-shaped ring-hole FSS and the inner radius ($r2$) of the single-ring FSS to the SIW FSS

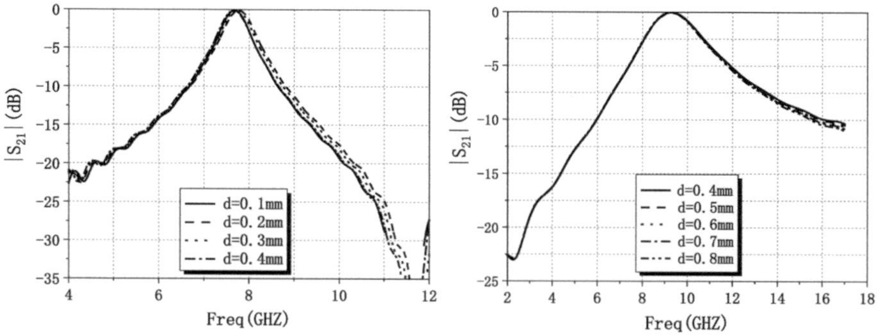

(a) Y-shaped ring-hole FSS (b) Single-ring FSS

Fig. 7.13 Effect of the SIW diameter on the transmission coefficient

of an FSS is expected to be light. Due to the presence of the metallized through-holes, modified SIW-based FSSs are lighter than ordinary FSSs. However, the aperture size has little effect on its performance. Consequently, the proper hole size should be designed based on the actual needs, with the weight of the selected surface being reduced to a minimum.

(3) Thickness of the dielectric board: Owing to the existence of substrate-integrated waveguides, the thickness of the dielectric plate can have a greater impact on the improved FSS. This differs from the common FSS. Figure 7.14 shows that the resonant frequencies and bandwidths of both the Y-shaped ring-hole FSS and the single-ring FSS decrease with increasing thickness of the dielectric board. The increase in the dielectric board thickness has little effect on the in-band insertion loss. However, when the dielectric board thickness increases slightly, the grating lobe appears more easily.

With the ordinary FSS, as the thickness of the dielectric board increases, the resonance frequency decreases, and the transmission loss in the band increases. When the board reaches a certain thickness, the transmission loss in the band

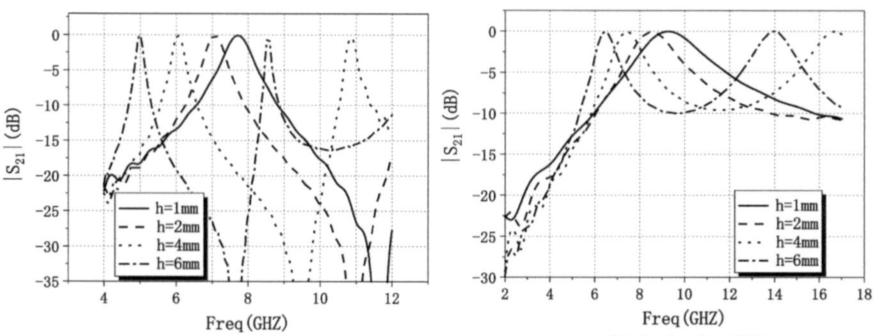

(a) Y-shaped ring-hole FSS (b) single-ring FSS

Fig. 7.14 Frequency responses of S_{21} with the dielectric plate thickness of the modified FSS decreases, and the resonance frequency increases. With increasing thickness of the substrate, the ordinary FSS also produces grating lobes. However, it is not as sensitive to the thickness of the substrate as is the improved FSS. This analysis shows that for an SIW-modified FSS, the SIW also plays an important role as the dielectric plate thickness changes.

(4) Unit spacing: The previous analysis shows that the unit spacing primarily affects the bandwidth—that is, the bandwidth decreases with increasing unit spacing. Figure 7.15 shows the effect of variations in the unit spacing on the transmission coefficient of the improved FSS. The improved FSS does not change the impact of the unit spacing on the FSS—that is, the bandwidth decreases with increasing unit spacing.

(5) Dielectric constant: The dielectric boards not only support the FSS but also improve the FSS properties. Owing to the presence of the dielectric board, the FSS resonant frequency decreases. Additionally, for different polarizations and different incident angles of the incoming wave, the resonant frequency and bandwidth are more stable than without the dielectric board. Figure 7.16 shows the effect of different dielectric constants of the dielectric board on the improved FSS transmission coefficient. Similar to the common FSS trends, the resonance frequency and bandwidth decrease with increasing dielectric constant.

3. **Measurement and study of the substrate-integrated waveguide FSS**: The size of the SIW Y-ring-type FSS [20–22] used as an example was 260 × 276 mm, as shown in Fig. 7.17. The area of the FSS is small in the physical model, but is infinite in the simulation. The background cancellation method was used to prevent the effect of diffraction during the test.

Figure 7.18 shows the actual measurement of the FSS. A waveguide probe was used as the transmitter, and a horn antenna was used as the receiver. To ensure that most of the EM waves emitted by the probe were irradiated on the FSS, the distance between the FSS and the probe was adjusted to approximately 10 cm. At such a small distance, the EM waves that reached the FSS were spherical waves, which

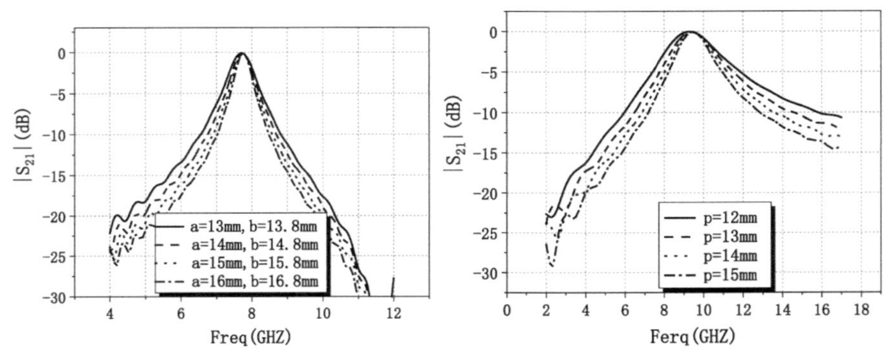

(a) Y-shaped ring-hole FSS

(b) Single-ring FSS

Fig. 7.15 Effect of the modified FSS unit spacing on S_{21}

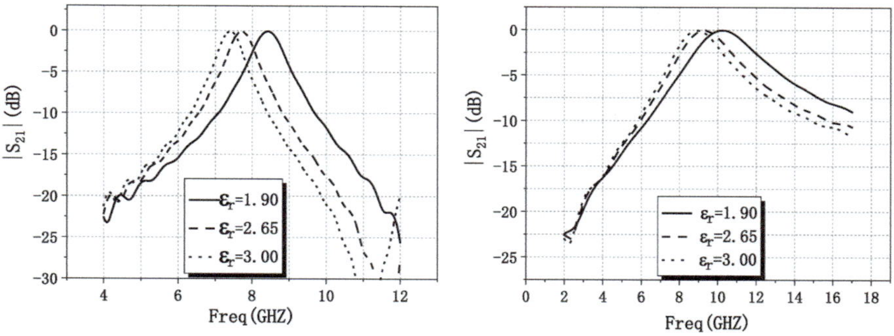

(a) Y-shaped ring-hole FSS (b) Single-ring FSS

Fig. 7.16 Effect of the dielectric constant on S_{21} of the modified FSS

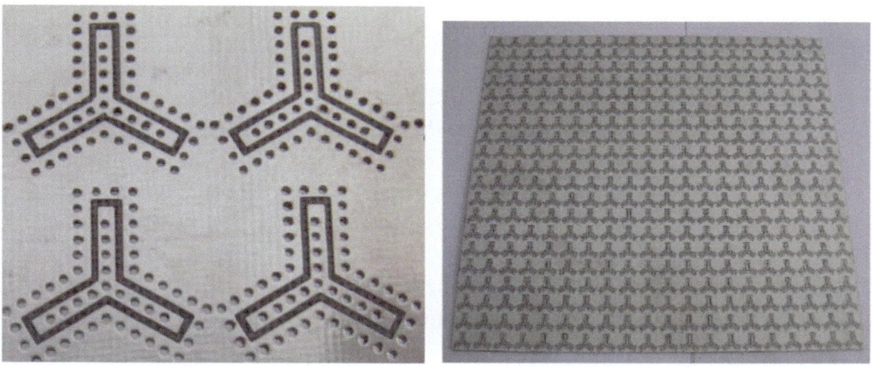

(a) Local structure diagram (b) Overall structure diagram

Fig. 7.17 Improved Y-ring-type FSS used in the test

is quite different from the assumed plane wave irradiation in the simulations but is similar to practical applications of an FSS in antenna engineering. The entire test was conducted in a microwave anechoic chamber and measurements were obtained using a vector network analyzer.

Fig. 7.18 FSS test schematic

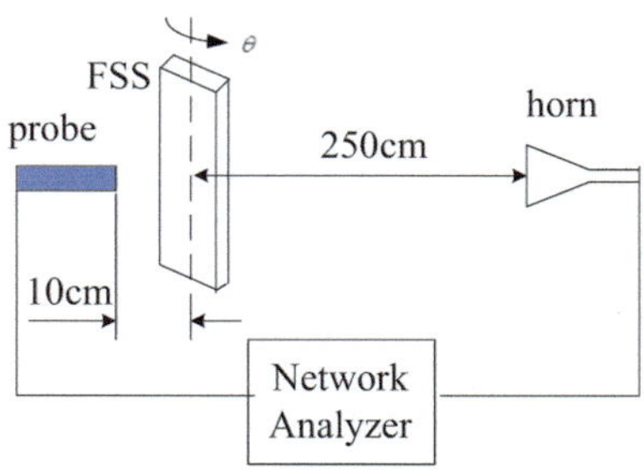

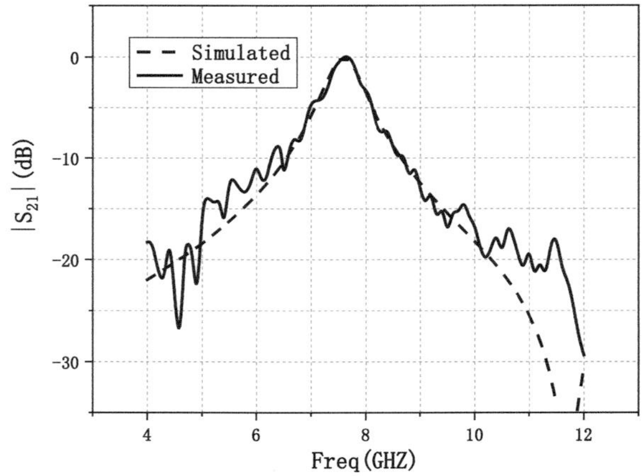

Fig. 7.19 Comparison of the simulation and test results

The specific test steps are as follows:

(1) Remove the FSS and measure S'_{21} (dB) and $E0$ (direct field).
(2) Place a 1-mm-thick aluminum plate that is as large as the machined frequency selection between the probe and the horn antenna. The aluminum plate is an ideal conductance plate, but the projectile is not good. Consequently, the energy received by the horn is caused by the diffraction of the EM wave. Measure the diffraction field of S''_{21} (dB).
(3) Put the FSS plate in the position of the aluminum plate, measure the S''_{21} (dB), and obtain $E2$ (the sum of the transmission field of the diffraction situation). We can then obtain:

The transmission coefficient:

$$T = S_{21}(dB) = 10\lg\frac{E_2 - E_1}{E_0} \tag{7.1}$$

The reflection loss:

$$\Gamma(dB) = S_{11}\left(dB\right) = 10\lg(1 - 10^{\frac{S_{21}(dB)}{10}}) \tag{7.2}$$

The test results are shown in Fig. 7.19. The measured results are generally consistent with the simulation results.

7.1.3 A New Composite Frequency Selective Surface (FSS)

As is well known, the ordinary FSS is composed of the most basic units. However, combining these common elements forms a composite FSS [15, 16], which can achieve performance that is not possible using ordinary frequency selection units.

The most commonly used two-way FSS can be used as an example [17–19] and combined with a cross-groove FSS to form a composite FSS. Here, the characteristics of the composite FSS can be studied in detail using the Designer simulation software.

1. **CDSLS-FSS model and frequency response**: For convenience, the new composite FSS is called the composite double square loop slot FSS (CDSLS-FSS), and the original common two-ring ring-type is called the double square loop slot FSS (DSLS-FSS). Figure 7.20a shows a three-dimensional model of the CDSLS-FSS. It clearly differs from the DSLS-FSS in that a cross-groove unit is added to the lower surface of the dielectric plate. An ordinary FSS adds a periodic unit metal plate layer on the upper side of the dielectric plate; this composite-type frequency selection can be used to add an FSS with different shape units on the upper and lower sides of the dielectric plate. The specific cell size and geometry of the CDSLS-FSS are shown in Fig. 7.20 and Table 7.5.

Figure 7.21 shows the frequency response characteristics of the CDSLS-FSS and DSLS-FSS transmission coefficients. The bandwidth of the CDSLS-FSS is considerably lower than that of the DSLS-FSS, and both resonance frequencies are shifted to lower frequencies.

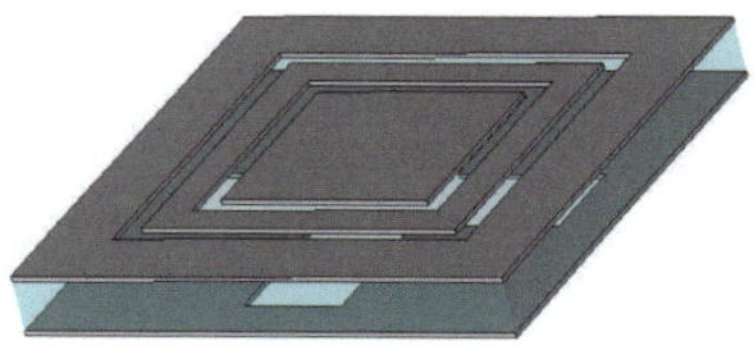

(a) 3D map

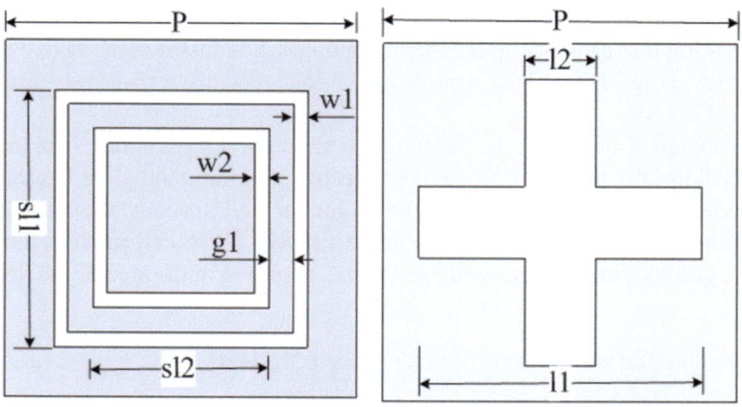

(b) Above (c) Below

Fig. 7.20 Geometric models of the CDSLS-FSS and DSLS-FSS

Table 7.5 Geometric parameters of the CDSLS-FSS and DSLS-FSS

P	10 mm	h	1 mm
$sl1$	7.2 mm	$sl2$	5.0 mm
$sw1$	0.4 mm	$sw2$	0.4 mm
$l1$	8 mm	$l2$	2 mm
ε_r	2.65	tang(δ)	0.001
$g1$	$(sl1 - 2 \times sw2 - sl1)/2$		

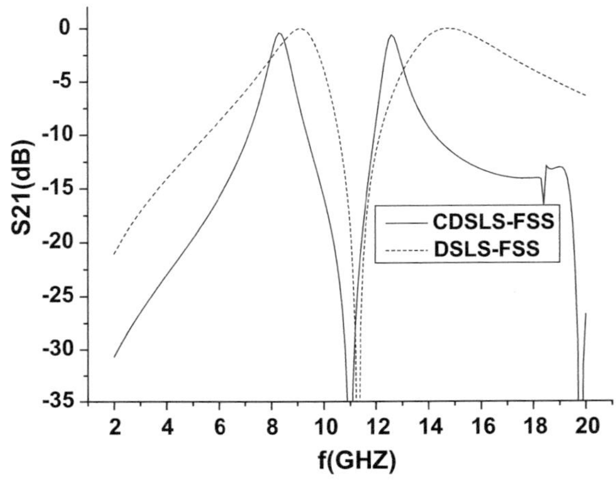

Fig. 7.21 Comparison of the transmission response of the CDSLS-FSS and DSLS-FSS with frequency

Table 7.6 shows a detailed comparison of the values of the CDSLS-FSS and DSLS-FSS. The first and second resonance frequencies of the CDSLS-FSS are reduced by 780 and 2,220 MHz, respectively, and the −3-dB bandwidth corresponding to the two resonance frequencies is reduced by 1,220 and 3,400 MHz, respectively. The in-band insertion loss also increases. The in-band insertion losses at the two resonance frequencies of the CDSLS-FSS increase by 0.37 and 0.60 dB, respectively, over those of the DSLS-FSS.

For a bandpass-type FSS, the bandwidth is an important parameter. The narrower the bandwidth, the better the frequency selectivity. Additionally, the bandwidth is related to the cell spacing—that is, the larger the cell spacing, the narrower the bandwidth. Moreover, the cell pitch can be limited. If the cell pitch is too large, however, grating lobes can be easily generated, resulting in the area of the final FSS

Table 7.6 Numerical analysis of the CDSLS-FSS and DSLS-FSS

	Resonant frequency		−3 dB bandwidth		Transmission loss	
	f_{01} (GHz)	f_{02} (GHz)	f_{01} (MHz)	f_{02} (MHz)	f_{01} (dB)	f_{02} (dB)
CDSLS-FSS	8.33	12.52	750	620	−0.39	−0.63
DSLS-FSS	9.11	14.74	1.970	4.020	0.02	−0.03

design being large. Here, when the other geometric parameters of the DSLS-FSS remain unchanged, the bandwidth can be reduced by adding a cross-shaped unit on the opposite side of the normal DSLS-FSS. Additionally, the bandwidth decreases by different amounts at both resonance points, the bandwidth being substantially reduced at the high-frequency resonance point.

The resonant frequency is also an important bandpass FSS parameter. Based on the Munk theory, for this type of ring-shaped FSS, resonance occurs when the circumference of the ring is an integral multiple of the resonance wavelength—that is, the larger the cell size, the longer the resonant wavelength, and the lower the corresponding resonant frequency. The ideal FSS is an infinite plane. However, in practical applications, to ensure the FSS performance, at least 20×20 units are processed. Consequently, in FSS research, the low FSS frequency and miniaturization of the cell size are contradictory goals. Many studies have investigated this contradiction. The CDSLS-FSS examined here has a different degree of resonance frequency reduction at the two resonance points than the DSLS-FSS, so a low FSS frequency at the double resonance point can be realized.

Of course, the cost of the CDSLS-FSS to achieve these advantages is that the insertion losses at its two resonance points increase, although the insertion losses of -0.39 and -0.63 dB at the two resonance points are acceptable. Consequently, this composite type of FSS has research and application value.

2. **Frequency response characteristics of the CDSLS-FSS parameters**: To understand the differences between the CDSLS-FSS and DSLS-FSS properties, the nature of the CDSLS-FSS can be considered from a parameter perspective.

(1) Groove lengths $L1$ and $L2$ of the cross slot: Fig. 7.22 shows the frequency response of the CDSLS-FSS with different slot lengths. The slot length l_1 of the cross slot has a greater effect on the CDSLS-FSS than that of l_2. As l_1 increases, both resonance points move to lower frequencies. The bandwidth of the first resonance point (f_{01}) increases, and that of the second resonance point (f_{02}) hardly changes. Moreover, the transmission loss at f_{01} is reduced, while the opposite occurs at f_{02}, showing a decreasing trend. As the slot length l_2 increases, the bandwidths at the resonance points f_{01} and f_{02} widen. Additionally, when the resonance frequency f_{02} decreases, f_{01} remains practically unchanged, and the transmission losses of f_{01} and f_{02} decrease. Consequently, to obtain better CDSLS-FSS performance, the appropriate slot length l_1 of the cross slot and l_2 should be chosen.

(2) Cell spacing: Fig. 7.23 shows the frequency responses of the CDSLS-FSS and DSLS-FSS with different cell spacings. Similar to the conventional DSLS-FSS, as the cell spacing increases, the bandwidths at the two resonance points decrease, and the two resonance frequencies remain practically constant. Moreover, the insertion loss of the transmission also increases. However, the CDSLS-FSS is more sensitive to the cell spacing than the DSLS-FSS. Consequently, in the design of the CDSLS-FSS, the cell spacing should not be too large—that is, if the cell spacing is too large, it is easy to generate grating lobes at the high-frequency resonance point.

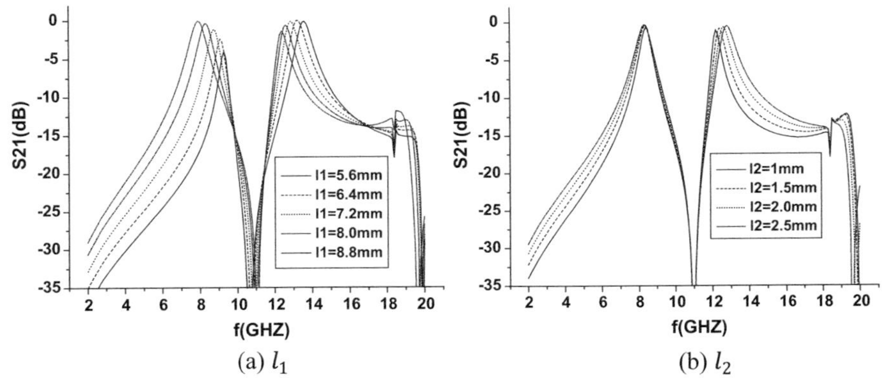

Fig. 7.22 Frequency response of the CDSLS-FSS corresponding to different slot lengths

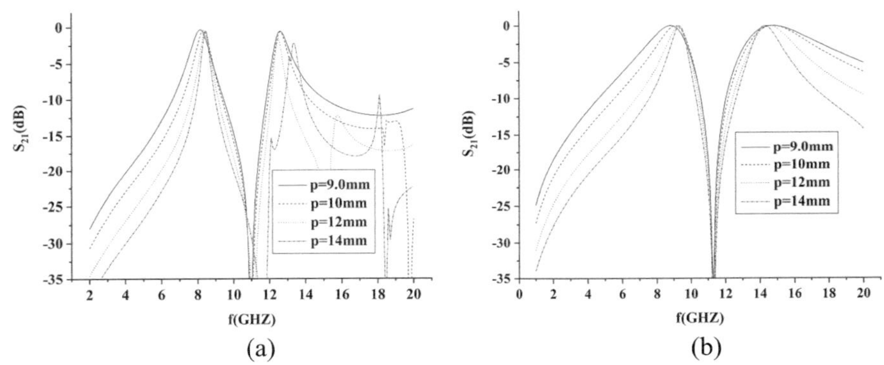

Fig. 7.23 Frequency responses of different cell spacings for the CDSLS-FSS and DSLS-FSS

(3) Dielectric plate thicknard thicknesses: Fig. 7.24 shows the S_{21} responses of the CDSLS-FSS and DSLS-FSS to different dielectric board thicknesses. The frequency response of the CDSLS-FSS is more sensitive than that of the DSLS-FSS. As the thickness of the dielectric plate increases, the two resonant frequencies f_{01} and f_{02} move toward higher frequencies, the amplitude of the side-lobe increases, and the transmission loss in the band increases. Consequently, to design a CDSLS-FSS with good performance, the substrate should not be too thick.
(4) Dielectric constant: The dielectric constant has important properties for the FSS as it both supports and improves its performance. Figure 7.25 shows the frequency responses of the CDSLS-FSS and DSLS-FSS with different dielectric constants. The response trends are the same as that of the normal DSLS-FSS, with the dielectric constant of the dielectric plate only affecting the resonance frequency. As the dielectric constant increases, the resonance frequency

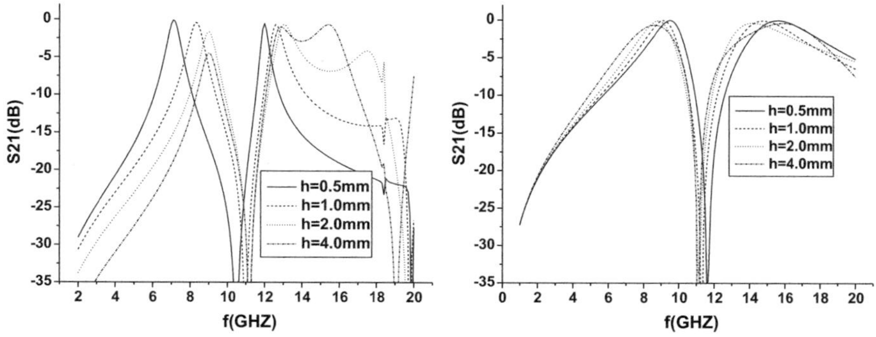

(a) CDSLS-FSS (b) DSLS-FSS

Fig. 7.24 S_{21} responses of the CDSLS-FSS and DSLS-FSS to the dielectric plate thickness

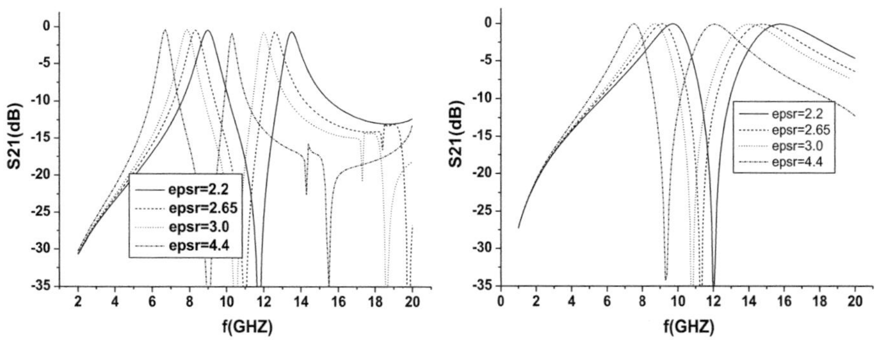

(a) CDSLS-FSS (b) DSLS-FSS

Fig. 7.25 Frequency responses of the CDSLS-FSS and DSLS-FSS with the dielectric constant

shifts to lower frequencies, whereas the bandwidth and transmission loss remain practically constant.

(5) Polarization and angle of incidence: Fig. 7.26 shows the frequency responses of the CDSLS-FSS and DSLS-FSS with different polarization modes and at different angles of incidence. The DSLS-FSS exhibits good performance for both *TE* polarization and *TM* polarization only at small angles of incidence. The performance of the CDSLS-FSS is better at any angle of incidence than that of the DSLS-FSS for *TM* polarization. However, although its *TE* polarization performance is not good, its performance is better when $\theta = 45°$.

3. **Experimental study of CDSLS-FSS**: To verify the accuracy of the simulation results, we also processed a 200 × 200 mm composite-type FSS, as shown in Fig. 7.29. The test was conducted using the same method used to test the Y-ring-type SIW FSS in the previous section. Figure 7.27 shows a comparison between the test and simulation results of the CDSLS-FSS at a normal angle of incidence. Owing to the limitations of the experimental conditions, only the

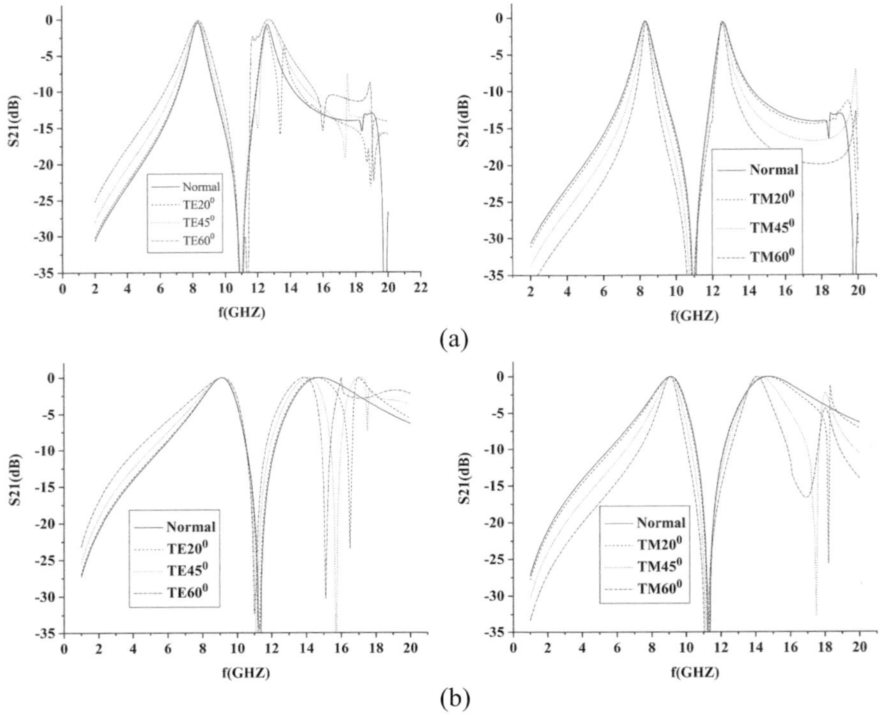

Fig. 7.26 Frequency responses of the CDSLS-FSS and DSLS-FSS for different polarization modes and different incident angles

transmission coefficients in the 7–14 GHZ band were tested. It is evident that the measured frequencies are offset from the simulation results, although the trends are generally the same.

Figure 7.28a and b show the test results at different angles for *TE* and *TM* polarizations, respectively. The trends are similar, although the measured frequency responses are offset.

Owing to the small area of the object, the limited experimental conditions, and small errors in the measurement process (especially when measuring different angles), the deflection angle could not be accurately fixed, so the measured results were not completely consistent with the simulation results. However, the measured and simulation results trends were similar, so the simulation results can be assumed to be relatively reliable (Fig. 7.29).

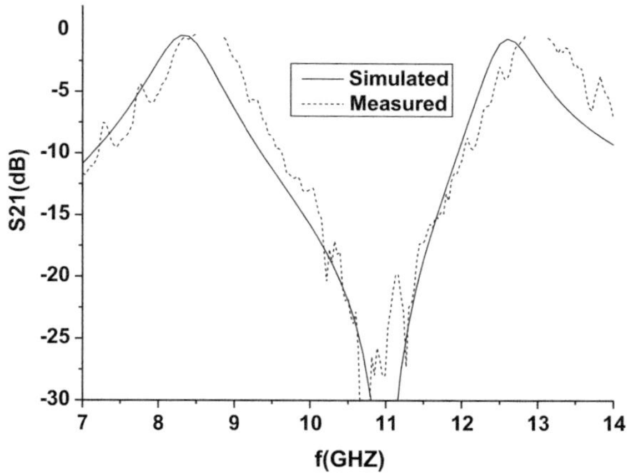

Fig. 7.27 Comparison of the test and simulation results of the CDSLS-FSS at normal incidence

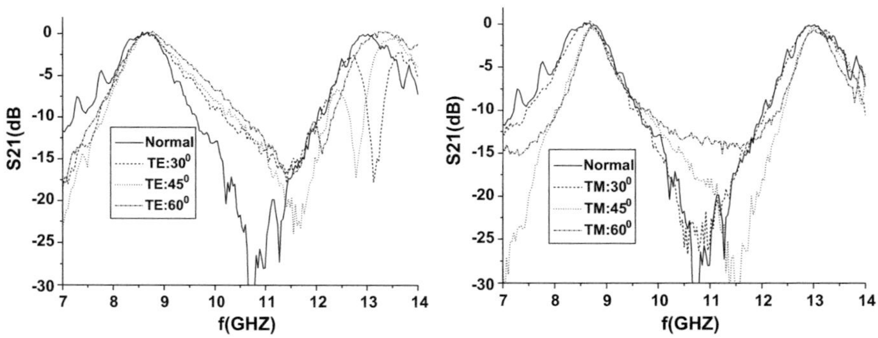

(a) TE (b) TM

Fig. 7.28 Test results of the transmission coefficient under different polarization conditions at different angles

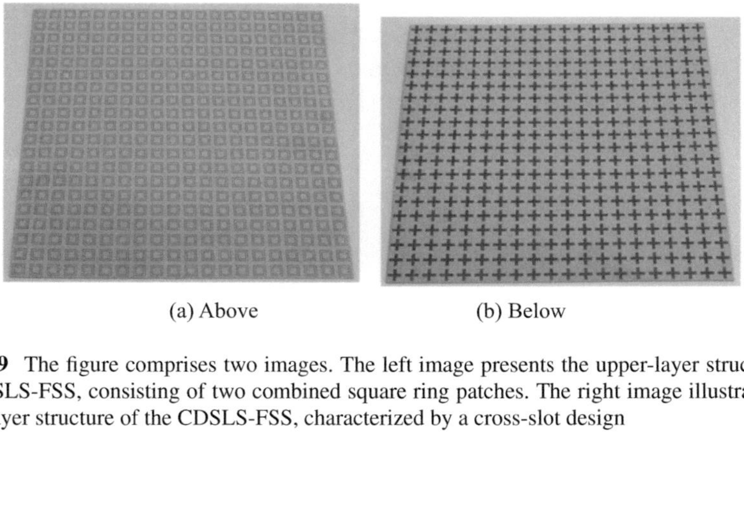

(a) Above (b) Below

Fig. 7.29 The figure comprises two images. The left image presents the upper-layer structure of the CDSLS-FSS, consisting of two combined square ring patches. The right image illustrates the lower-layer structure of the CDSLS-FSS, characterized by a cross-slot design

7.2 Research on Band-Stop FSS

To exert the FSS frequency selection feature on a limited size—that is, to arrange sufficient units within a limited space—it is necessary to reduce the unit area of the FSS. The ring FSS has the advantages of a simple structure and good oblique incident stability. Its resonant wavelength is approximately equal to the circumference of the ring. Consequently, an effective method of reducing the size of a ring FSS is to compress its volume while keeping the circumference constant. Because a fractal structure can effectively increase the perimeter of a structure within a fixed area, it can be used to design the ring FSS. Moreover, a fractal structure is a repetitive structure of a curve or graph with self-similarity, such as the Koch, Hilbert, and Peano curves. Zero-order, first-order, and second-order hierarchical structures are shown in Fig. 7.30.

An FSS with an isosceles array arrangement has the advantage of good angular stability. Consequently, the FSS structure involved a hexagonal honeycomb arrangement. Because the resonant wavelength of the ring FSS is roughly equivalent to the girth of the ring, increasing the perimeter of the metal belt within the existing area reduces its resonant frequency. At the same 3.2 GHz resonance, the existing area is effectively used owing to the second-order fractal FSS. Thus, this structure is smaller than other ring FSSs, which means more periodic units can be arranged in the same area to achieve a better resistance effect.

Size comparison of FSSs with the same resonant frequency:

- The area of a ring unit is 439 mm^2
- The area of the zero-order fractal FSS unit is 347 mm^2
- The area of the first-order fractal FSS unit is 134 mm^2
- The area of the second-order quasi-fractal FSS structure is 109 mm^2.

Thus, the area of the second-order quasi-fractal FSS structure is just 31.4% of that of the zero-order fractal FSS unit. An FSS unit with a hexagonal second-order fractal structure is shown in Fig. 7.31. The material's dielectric constant was 2.65, the thickness was 1 mm, and the incident wave was polarized in the x-direction.

The S_{21} obtained by means of HFSS software simulations is shown in Fig. 7.32. The results show that the FSS exhibits good resistance characteristics. The center frequency is 3.2 GHz, and the antenna bandwidth is 1.15 GHz (2.65–3.8 GHz) when S_{21} is less than -10 dB.

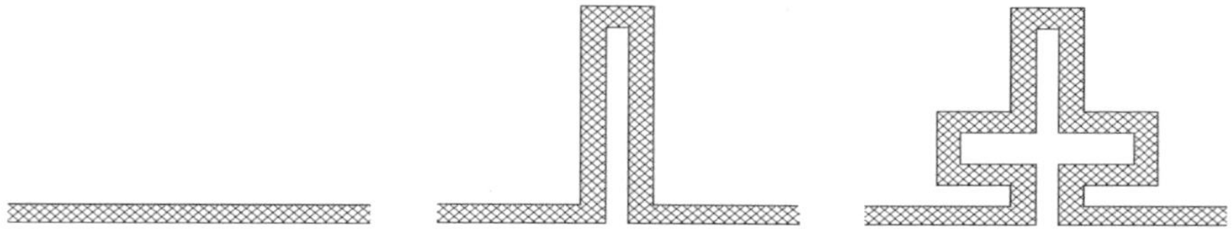

Fig. 7.30 Zero-order, first-order, and second-order hierarchical structures, from left to right

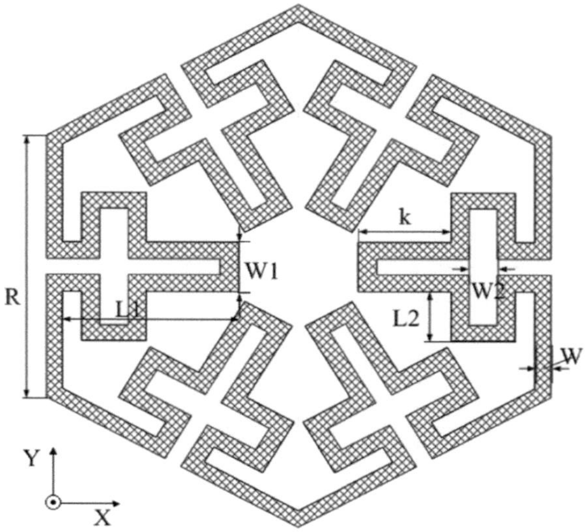

Fig. 7.31 Quasi-fractal structure FSS unit

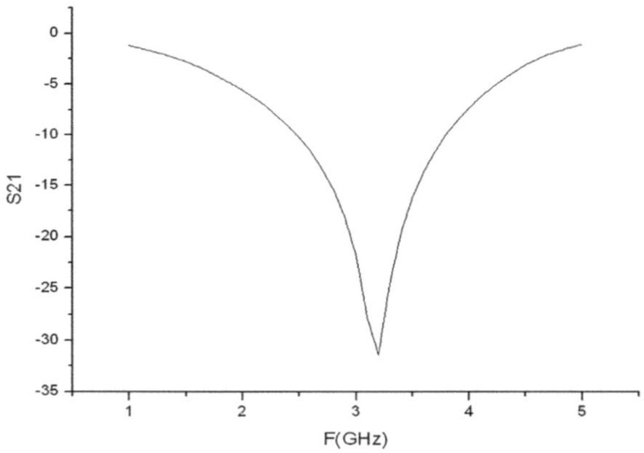

Fig. 7.32 S_{21} values of the quasi-fractal structure FSS

The S_{21} parameters at different angles are shown in Fig. 7.33. When the incident angle varies, the resonant frequency changes little, but the minimum S_{21} value changes. The results show that the multi-axisymmetric structure of the isosceles triangular element exhibits better angular stability. The designed FSS was then processed and tested, the test results of which are shown in Figs. 7.34 and 7.35. The results show that the unit exhibits the expected pass-through characteristic.

The first two sections of this chapter examined the bandpass FSS and band-resistance FSS. Based on the integrated waveguide, fractal structure, composite structure and other technical designs, various FSS were designed, and experimental tests were conducted. The results provide technical support for the design of the reflector and radome for array antennas.

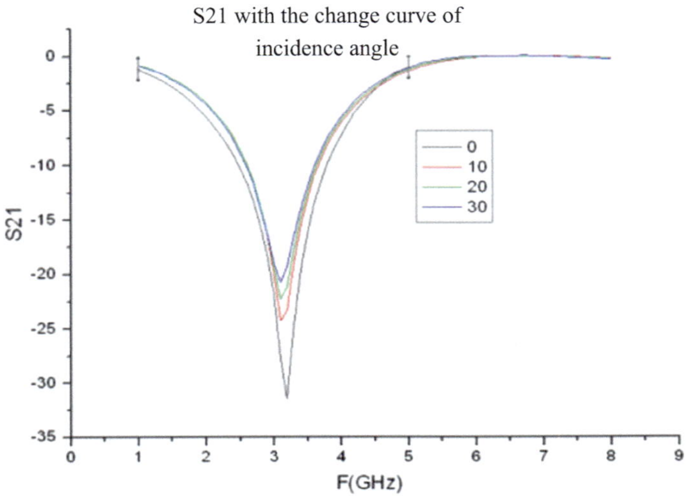

Fig. 7.33 S_{21} values of the quasi-fractal structure FSS with varying incidence angles

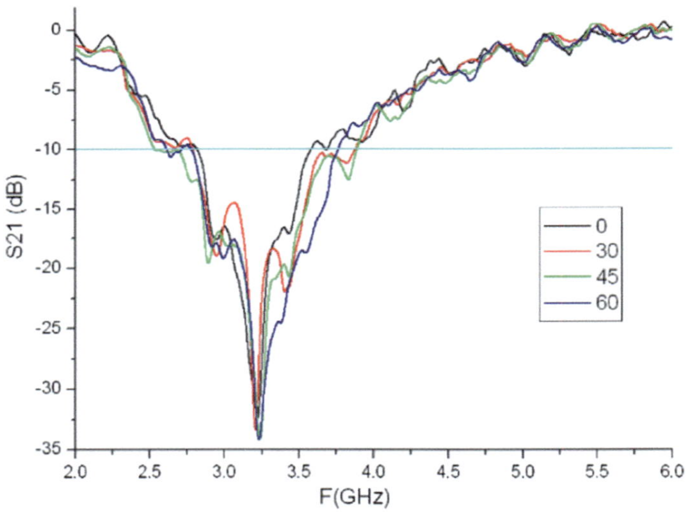

Fig. 7.34 S_{21} values of the quasi-fractal structure FSS with varying incident angles for vertical polarization

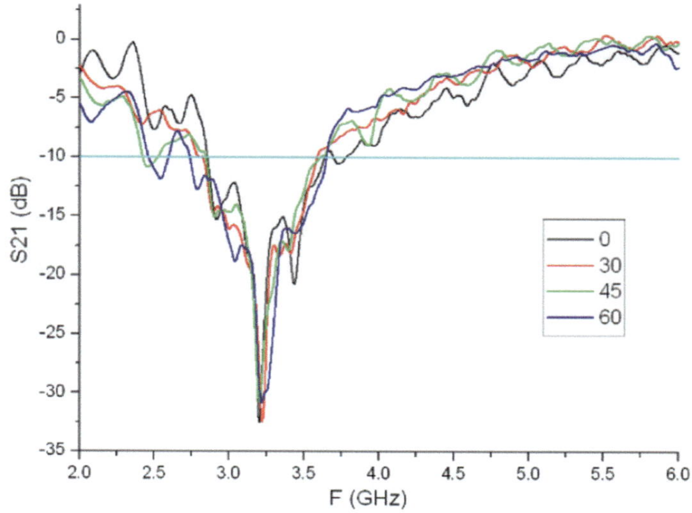

Fig. 7.35 S_{21} values of the quasi-fractal structure FSS with varying incident angles for horizontal polarization

7.3 Out-of-Band RCS Reduction of a Microstrip Antenna Based on a Band-Stop FSS

A typical microstrip printing unit is shown in Fig. 7.36. The working center frequency of the unit was 3 GHz. The thickness of the media substrate was 1 mm, and the relative dielectric constant was 2.65. The form of a coupled feed was adopted, and the feeder was connected to a 50-Ω SMA connector. Figure 7.37 shows the voltage standing wave ratios (VSWRs) of the antenna obtained by simulations. The bandwidth of the antenna was approximately 1.08 GHz when the standing wave was less than 1.5, and the gain of the antenna was 5.26 dB.

To increase the bandwidth and improve the antenna performance, a window-type PBG structure was added to the surface of the matrix, as shown in Fig. 7.38. The window-type structure was used to form a periodic resonance structure, thus improving the bandwidth characteristics of the antenna. The design dimensions of the window structure are shown in Fig. 7.39, the antenna using four window structures to form a periodic structure.

Based on the measured results shown in Fig. 7.40, the bandwidth of antenna can be considerably expanded by using the window-type periodic structure. The bandwidth of the typical printing array antennas is 1.08 GHz (2.52–3.6 GHz), whereas the bandwidth of the antennas with the PBG is 1.7 GHz (2.45–4.15 GHz). The reason for the broader bandwidth is that the periodic structure of the PBG suppresses the resonance peak at the 3.75 GHz frequency point, thereby broadening the antenna bandwidth.

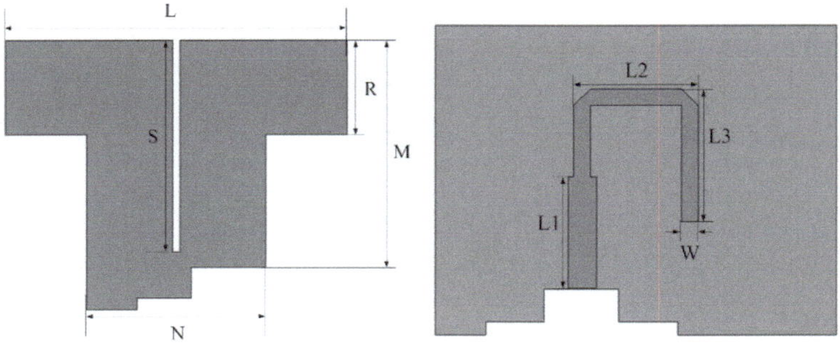

(a) Top view (b) Bottom view

Fig. 7.36 Geometry of a printed matrix

Fig. 7.37 Voltage standing wave ratio of the antenna

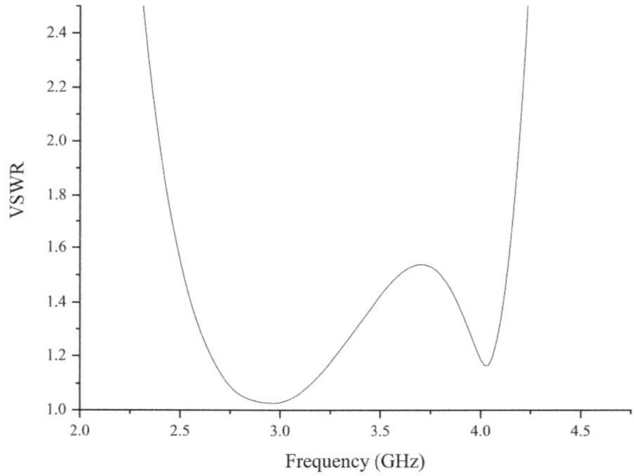

(a) Top view (b) Bottom view

Fig. 7.38 Improved printed matrix images

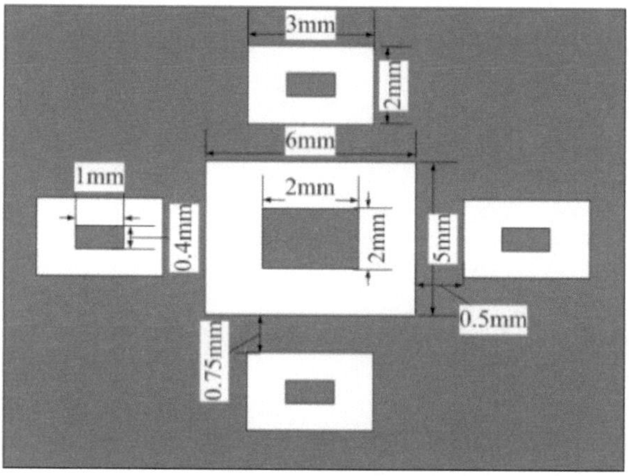

Fig. 7.39 Window-type structure unit

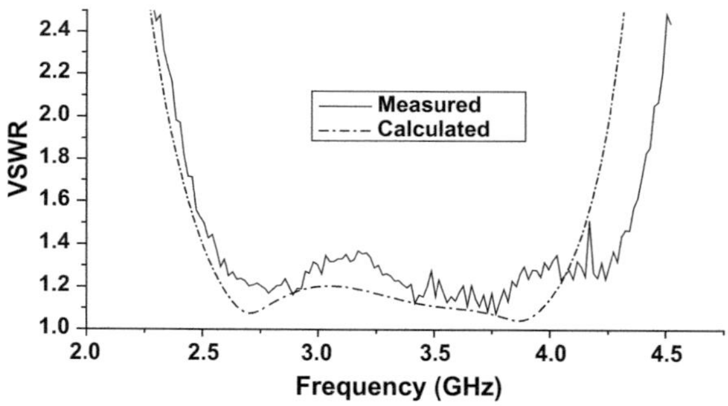

Fig. 7.40 Calculated and measured standing wave ratios of the improved printed microstrip

Figure 7.41 shows the antenna radiation patterns of the *E*-plane and *H*-plane of the modified microstrip printing system at frequencies of 2.5, 3.3, and 4.2 GHz. The maximum radiation direction of the antenna is defined to be in the direction of $\theta = 0°$. The results show that the main polarization direction diagram is consistent, and the cross-polarization is less than −30 dB.

Based on this analysis, the reflection board has a large influence on the RCS of the printed microstrip. In this work, the resistance FSS was introduced into the design of the microstrip printing system, thus controlling the RCS of the unit. The resistance band of the FSS was designed for the working band of the printing. Consequently, the FSS can be used to realize the effect of the reflector in the working band, which does not affect the radiation characteristics of the unit. Additionally, the FSS can project the incident plane wave out of the working band, effectively controlling the external RCS.

In the case of a dipole antenna, an FSS with a second-order structure can be used as the reflection of the floor. The matrix model with a baseplate is shown in Fig. 7.42, and the radiation impedance bandwidth of the antenna with the FSS reflector is shown in Fig. 7.43.

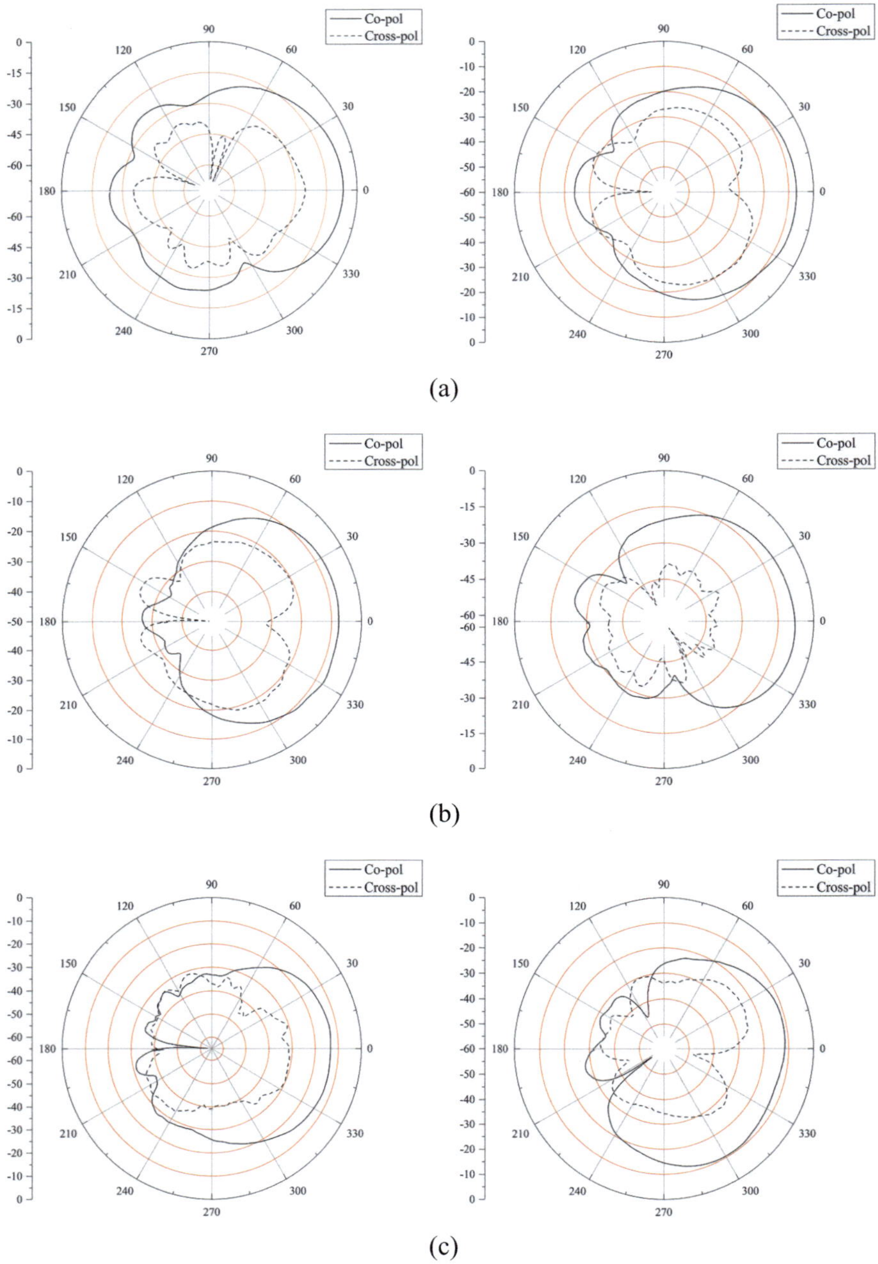

Fig. 7.41 Measured *E*-plane (left) and *H*-plane (right) radiation patterns

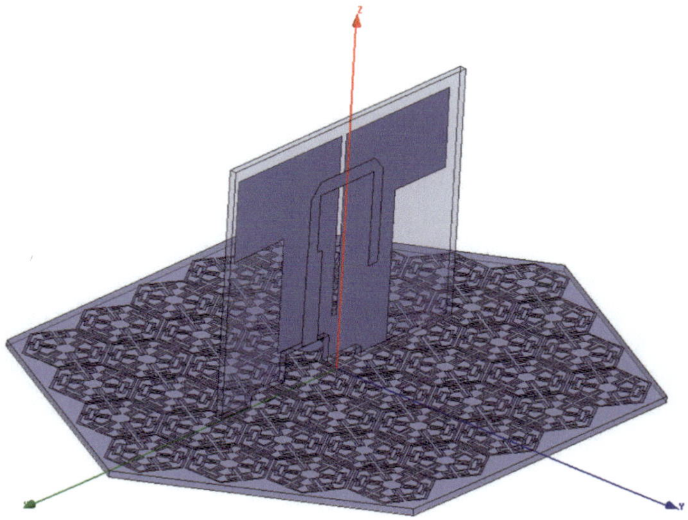

Fig. 7.42 Symmetrical array antenna and FSS model

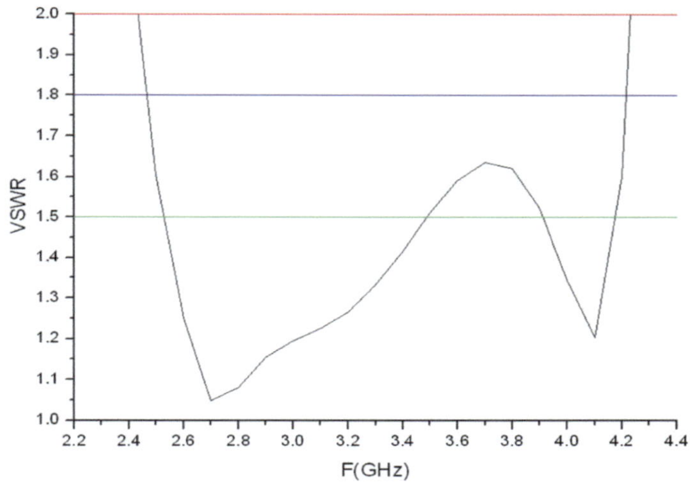

Fig. 7.43 VSWR of the antenna with the FSS reflector

The bandwidth of the antenna resonant VSWR < 1.8 is 2.45–4.2 GHz, which covers the resonance frequency of the second-order fractal structure FSS. A simulation was conducted for the unprinted array antenna with a PEC reflector as a reference formation. The results showed that the main lobe gain of 3.2 GHz increased from 7.1 to 7.3 dB, which was not significant. The unit antenna shown in Fig. 7.42 was processed and tested. The unprinted array antenna with the FSS as a reflected floor is shown in Figs. 7.44 and 7.45 shows the measured radiation patterns of the antenna.

As a comparison, the radiation patterns of the same oscillator using a good conductor as the reflector are presented and are in good agreement. The results show that the FSS exhibits good working characteristics in the pass-band.

Figure 7.46 shows the RCS measurement results of the antenna shown in Fig. 7.44 and the same oscillator antenna with a good conductor as the reflector. Figure 7.47 shows the simulation results of other angles. The measured and simulated results show that the second-order fractal FSS structure described above provides a good RCSR effect within its operating bandwidth, demonstrating the validity of the design concept. The most evident reduction of the monostatic RCS occurs at $\theta = 0°$, $\varphi = 0°$. Because the direction is the main scanning direction of the phase array antenna, the RCS of the antenna in this direction has the greatest threat. The RCSR in this direction is of great significance.

A 2 × 4 array antenna with a PBG structure without a printed array antenna is shown in Fig. 7.48.

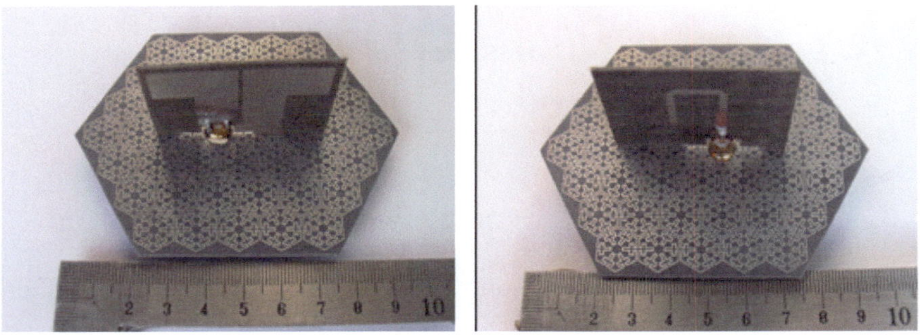

(a) Top view (b) Bottom view

Fig. 7.44 Microstrip with an FSS reflecting floor with a symmetrical array antenna

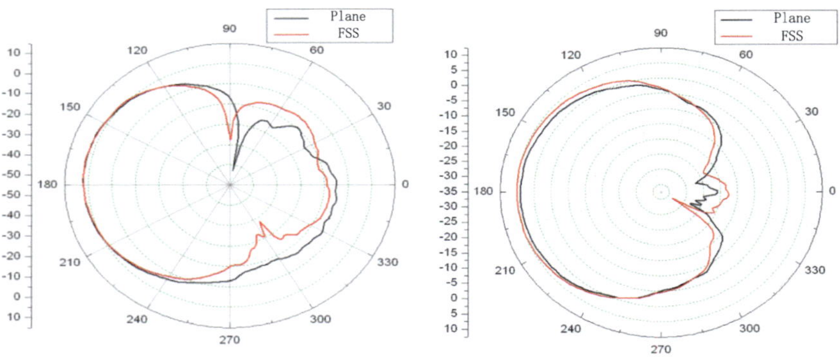

(a) *E*-plane (b) *H*-plane

Fig. 7.45 *E*- and *H*-plane radiation patterns of two different floors

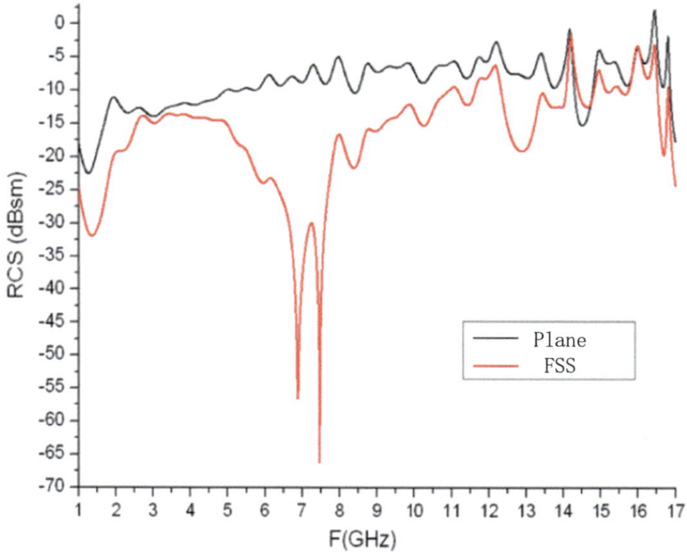

Fig. 7.46 Measured monostatic RCS of the antennas with all metal floor and FSS floor at $\theta = 0°, \varphi = 0°$

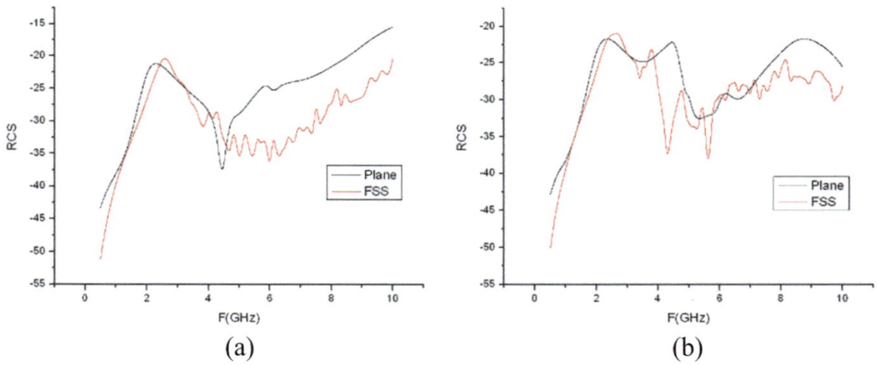

Fig. 7.47 Simulated monostatic RCS results of the antennas with full floor and FSS floor

The HFSS simulation software was used to simulate the antenna model. Figure 7.49 shows the simulation results of the antenna host. The results show that the input impedance bandwidth of the antenna is 2.2–4.14 GHz (VSWR < 1.8). The center frequency is 3.2 GHz (Fig. 7.50).

The antenna was fabricated and tested. Photos of the 1 × 4 array are shown in Fig. 7.51 (the total floor size being 280 × 175 mm). The spacing of the antenna unit was 60 mm, and the dielectric constant of the 1-mm-thick surface was 2.065, which was fed by two 1.4-mm microstrip power dividers. The energy was lost to the oscillator via the barren structure. A schematic diagram of a 2 × 4 symmetric matrix

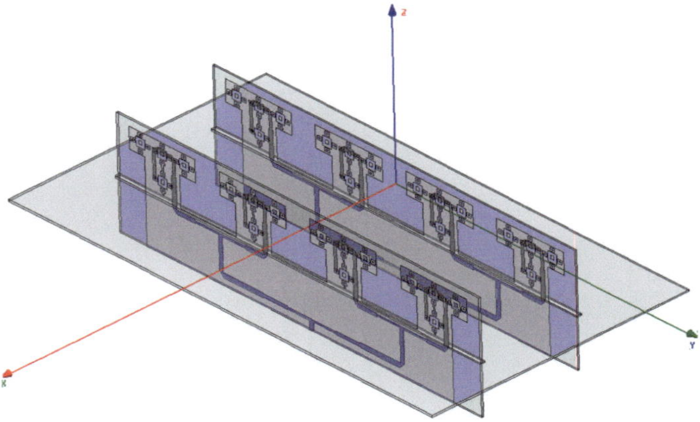

Fig. 7.48 Model of the dipole array

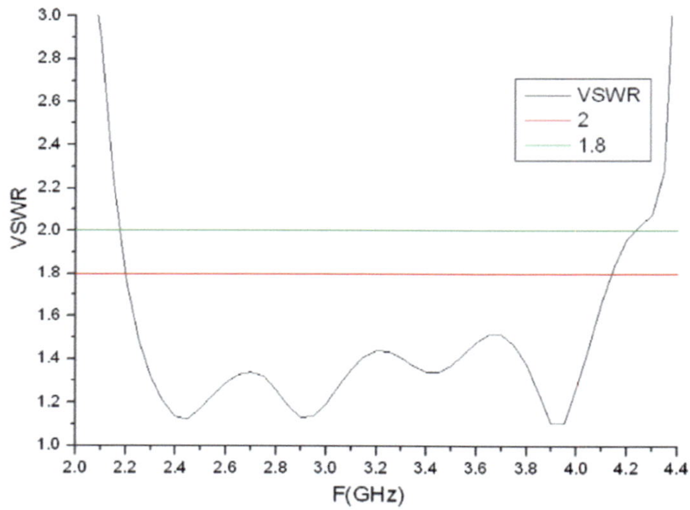

Fig. 7.49 VSWR curve of the dipole array (FSS floor)

constructed with an FSS surface as a reflector is shown in Figs. 7.52, and 7.53 shows a darkroom survey photograph.

As a comparison, the radiation pattern and the single station RCS curve of the 2×4 array antenna with the FSS and a good conductor as a reflector were tested. Figure 7.54 shows the measured E-plane and H-plane of the 4×2 array of the two floors. The radiation characteristics of the FSS floor are generally the same as those of the metal floor.

Figure 7.55 shows the monostatic RCS curves for a 2×4 array antenna using the two floors. The RCS of the antenna in the $\theta = 0°$, $\varphi = 0°$ direction is considerably

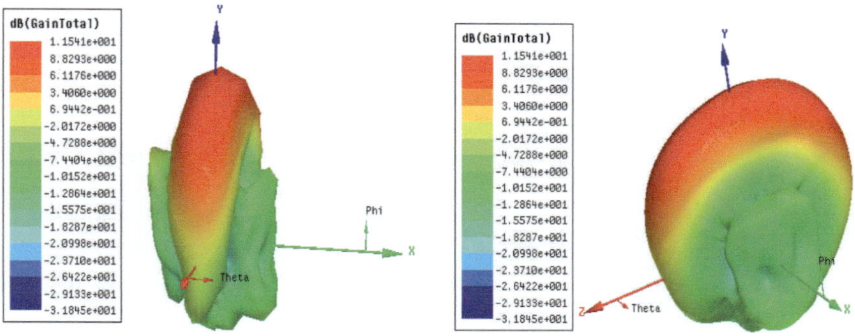

Fig. 7.50 The image depicts the three-dimensional gain radiation pattern of a symmetric dipole array antenna with an added FSS ground plane

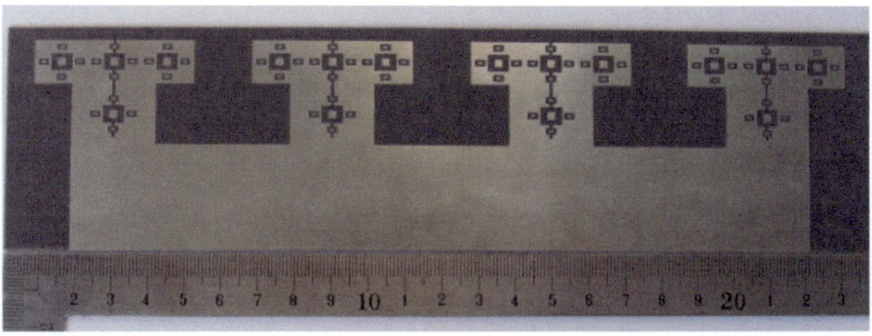

(a) Top view

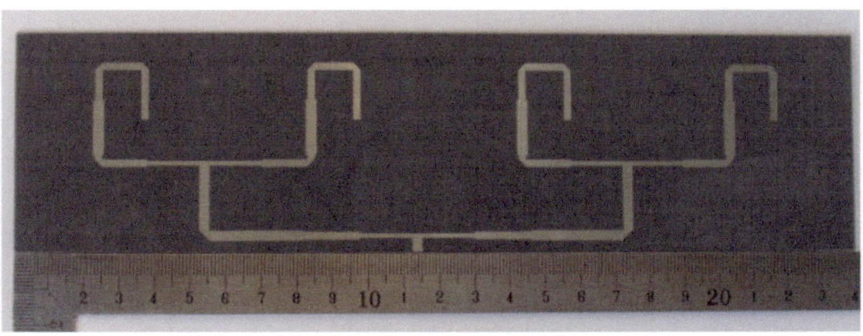

(a) Bottom view

Fig. 7.51 The 1 × 4 antenna array

reduced in the pass-band with the FSS floor. However, owing to the resistance characteristics of the reflector, there is no reduction in the pass-band near 3.3 GHz. In the $\theta = 60°$, $\varphi = 45°$ direction at 3.9, 7.4, and 14 GHz, the peaks are considerably reduced. In the other directions, the RCS of the antenna increases at some frequencies owing

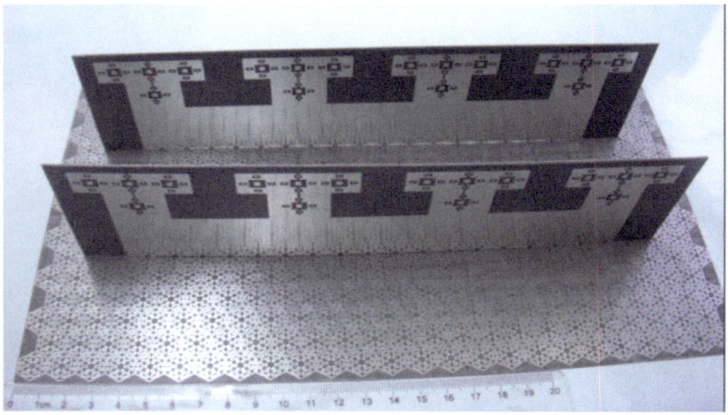

Fig. 7.52 The 2 × 4 dipole array with a reflecting floor FSS

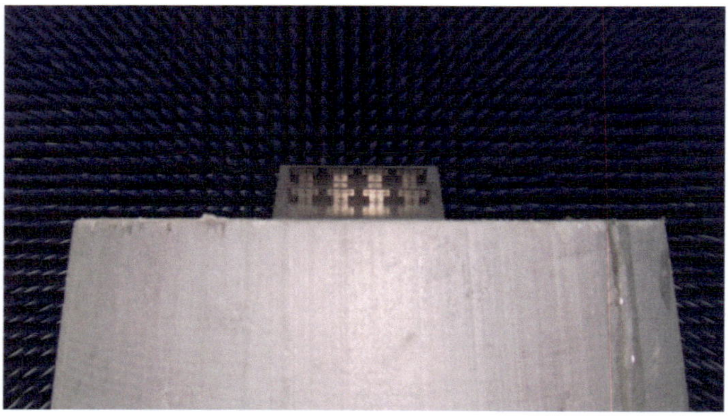

Fig. 7.53 Photograph of the measurement of the 2 × 4 array with a reflecting floor FSS

to the influence of the angle. However, because of the decrease in the peak antenna RCS (approximately 5 dB), the FSS floor exhibits some advantages in reducing the antenna RCS.

7.4 In-Band RCS Reduction of a Microstrip Antenna Based on a Polarization-Dependent FSS

The previous sections discussed methods for antenna RCSR using an FSS. Owing to the restrictions of the reduction mechanism, a traditional FSS can only be used to reduce the out-of-band RCS of the antenna. However, the in-band RCS is more important than the out-of-band RCS as the out-of-band RCS can also be reduced by many other methods, such as using artificial magnetic conductors (AMCs) and

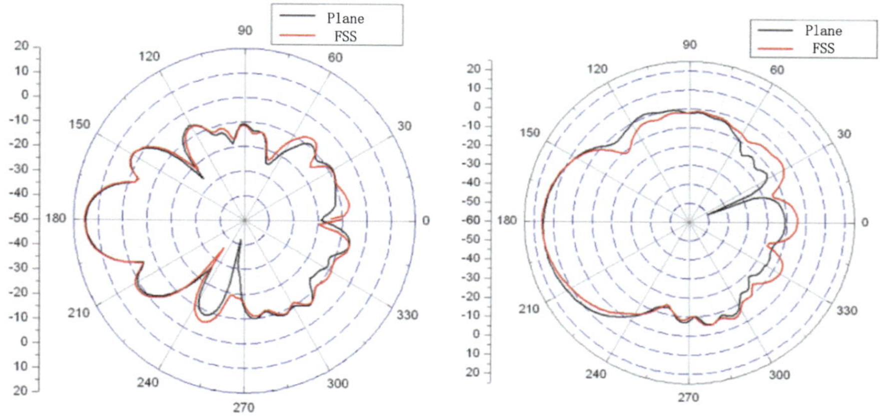

(a) E-plane (b) H-Plane

Fig. 7.54 Measured radiation patterns of the 2 × 4 antenna array

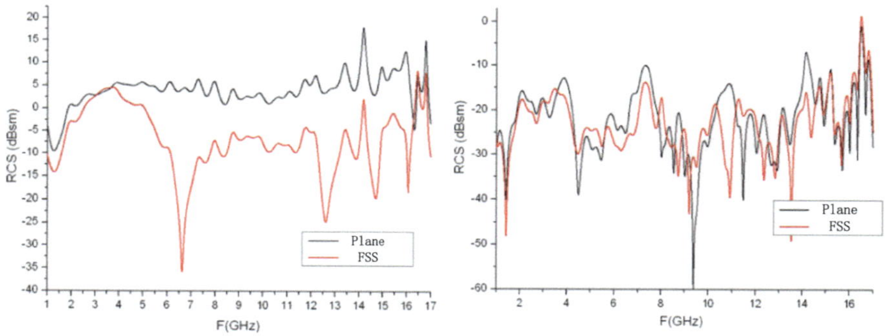

(a) $\theta = 0°, \varphi = 0°$ (b) $\theta = 60°, \varphi = 45°$

Fig. 7.55 Measured results of the 2 × 4 antenna array for two kinds of floors

RAM structures [23, 24]. Consequently, in this section, a polarization-dependent FSS (PDFSS) is proposed and applied to the in-band RCSR of a microstrip antenna.

When the electric field is normal to the split of a CSRR, the CSRR is excited [25] and exhibits transmission properties. By contrast, when the electric field is parallel to the split of the CSRR, the CSRR is not excited [25], and the entire incident wave is reflected. Taking advantage of the polarization-dependent properties, a polarization-dependent FSS was designed to transmit/reflect EM fields at the same frequency. The structure of the PDFSS unit cell is shown in Fig. 7.56a, the unit cell being a complementary split resonant ring (CSRR). The three pairs of slots in the CSRR can be used to make the structure of the unit cell more compact. The parameters of the unit cell were as follows: $L = 8$ mm, $L_1 = 1.25$ mm, $L_2 = 1$ mm, $L_3 = 0.5$ mm, $L_4 = 0.2$ mm, $L_O = 1$ mm, and $L_S = 0.5$ mm.

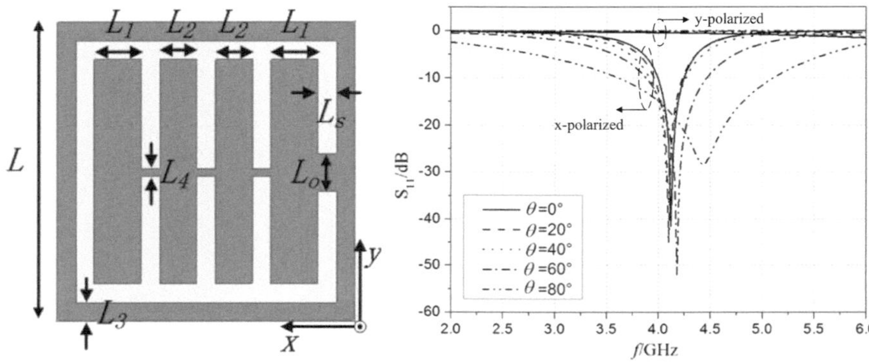

(a) Unit cell of the PDFSS (b) Reflection coefficients

Fig. 7.56 Geometry of the unit cell of the PDFSS and its reflection coefficients

Figure 7.56b shows the reflection coefficients of the PDFSS under oblique incident waves. The FSS exhibits bandpass and band-stop characteristics for *x*- and *y*-polarized incident waves, respectively. When the incidence angle increases from 0° to 80°, the resonant frequency of the PDFSS shifts to a higher frequency band. Because the bandwidth also increases with increasing incidence angle, the reflection coefficient is still below −17 dB at 4.1 GHz. Consequently, the proposed PDFSS exhibits stable bandpass/band-stop properties for oblique incident waves at 4.1 GHz, which can be used to realize in-band RCSR of the microstrip antenna over a wide range of incidence angles

The structures of the reference and proposed microstrip antennas are shown in Figs. 7.57a and b, respectively. The reference antenna operates at 4.1 GHz, the length of the antenna ground L_g is 29 mm, and the length of the patch L_p is 11.8 mm. The proposed low-RCS antenna has the same parameters except for the PDFSS structures and Γ-shaped slots on the ground plane.

Because the CSRR is a resonant structure and the transmission characteristics of the proposed PDFSS vary with the orientation of the unit cell, the arrangement of the unit cells on the ground plane of the microstrip antenna should be optimized to simultaneously achieve low RCS and good radiation performance. For the radiation case, the PDFSS should exhibit band-stop properties for the radiation fields. Figure 7.58 shows the current distribution on the metallic parts of the reference antenna at 4.1 GHz. The surface currents on the ground plane move in the *x*-direction, and the current intensity beside the radiation edges are stronger than in other positions. Consequently, the split of the CSRR beside the radiation edges should be parallel to the *x*-direction.

The electric field distribution on the metallic parts of the reference antenna with different orientations of the unit cell are shown in Fig. 7.59. Figure 7.59a and b show the electric field distribution when two CSRRs are arranged beside the radiation edge of the microstrip antenna with the split normal and parallel to the *x*-direction, respectively. The results show that the CSRRs with the split normal to the *x*-direction

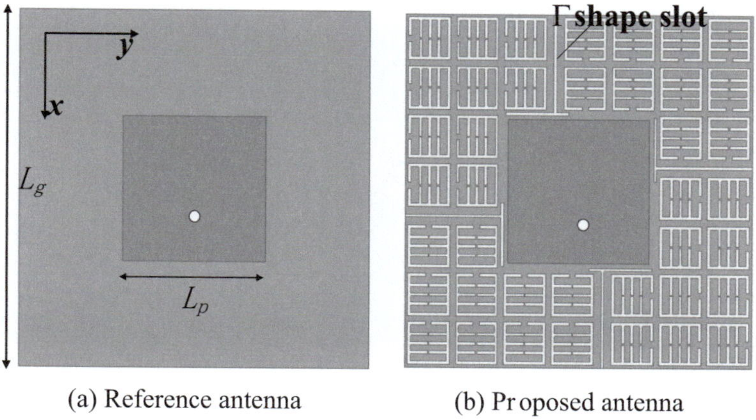

Fig. 7.57 Structures of the reference microstrip antenna and the proposed low-RCS antenna

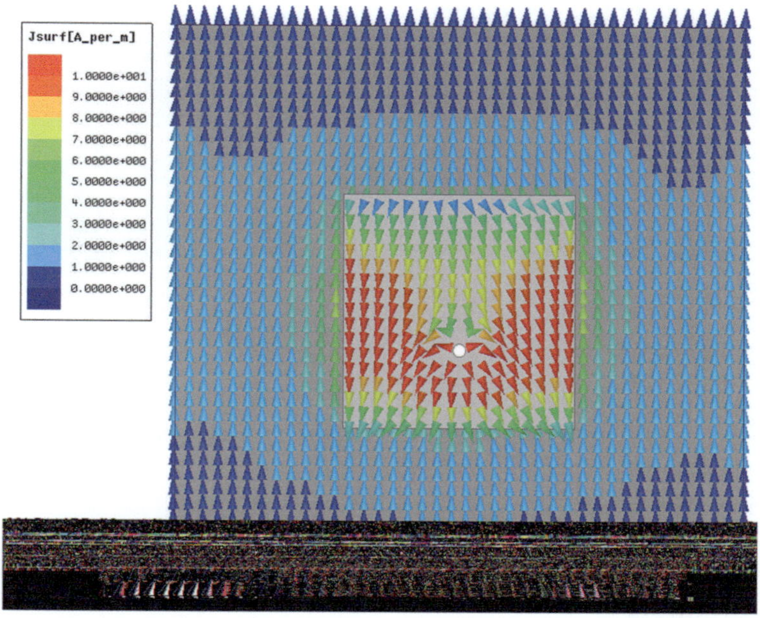

Fig. 7.58 Current distribution at 4.1 GHz

are excited by the field radiated from the antenna, while the CSRRs with the split parallel to the *x*-direction are not excited.

This occurs because the orientations of the unit cells of the PDFSS can influence the radiation characteristics of the microstrip antenna. To reduce the RCS in both the *xoz*-plane and *yoz*-plane and maintain the radiation characteristics of the proposed antenna, the unit cells etched on the ground must be in different orientations. The

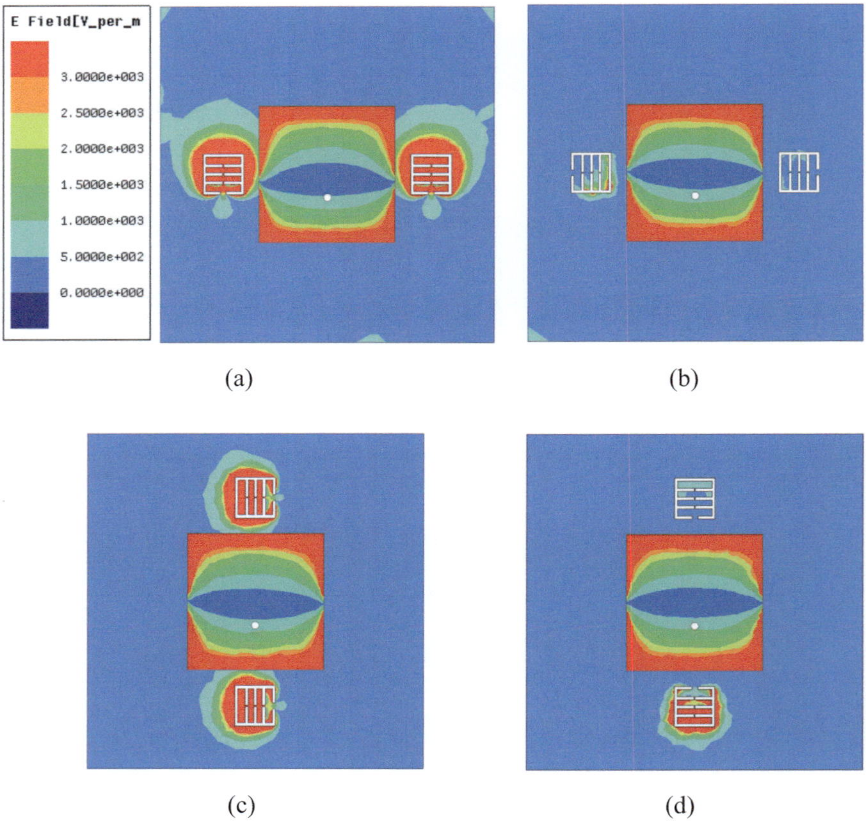

Fig. 7.59 Electric fields distribution at 4.1 GHz with different orientations of the CSRR

'Γ'-shaped slots between the unit cells with different orientations can be used to reduce their mutual coupling so the PDFSSs with unit cells of different orientations can operate well.

7.5 Summary

The characteristics of FSS structures were studied in this chapter, and applied to the design of low-RCS antennas. By applying the designed FSS structure to the antenna backplane and radiating element, low-RCS characteristics of the antenna were achieved. Prototypes of an antenna unit and antenna array were fabricated and tested, and the feasibility of this scheme was demonstrated by the simulation and test results.

References

1. C.H. Chan, R. Mittra, On the Analysis of Frequency Selective Surface Using Subdomain Basis Functions, IEEE Trans. Antennas Propagation, Vol. 38, No. 1, pp. 40–50, 1994.
2. W.L. Ko, R. Mittra, Scattering by a Truncated Periodic Array, IEEE Trans. Antennas Propagation, Vol. 36, No. 4, pp. 496–503, 1988.
3. T.K. Wu, Four-band Selective Surface with Double-Square-Loop Patch elements, IEEE Trans. Antennas Propagation, Vol. 42, No. 12, pp. 1659–1663, 1994.
4. T.K. Wu, S.W. Lee, Multiband Frequency Selective Surface with Multiring Patch Elements, IEEE Trans. Antennas Propagation, Vol. 42, No. 2, pp. 166–175, 1994.
5. Ben A. Munk, Frequency Selective Surface: Theory and Design, New York, Wiley, 2000.
6. Qiang Gao, Dunbao Yan, Naichang Yuan, Yunqi Fu, Research on the Characteristics of Loaded FSS, Chinese Journal of Radio Science, Vol. 21, No. 3, pp. 441–444, 2006.
7. Huanqing Wang, Mingyun Lv, Zhe Wu, Experimental Study of the Influence of Dielectric Loading on the Transmission Characteristics of FSS, Journal of Infrared and Millimeter Waves, Vol. 24, No. 1, pp. 27–30, 2005.
8. Hongyan Jia, Jinsong Gao, Yanhong Wang, Method of Stabilizing Center Frequency for FSS under wide angle incidence, Journal of Microwaves, Vol. 23, No. 6, pp. 7–9, 2007.
9. Hongyan Jia, XIaoguo Feng, Jinsong Gao, Effect of Changing Y-shaped Cell Arrangement on FSS Polarization Stability, Optics and Precision Engineering, Vol. 15, No. 6, pp. 978–982, 2007.
10. Xinyu Hou, Characteristic Analysis of Loaded FSS in Complex Media and Its Applications in Radome, Northwestern Polytechnical University, 1997.
11. Yu Lin Zhang, Wei Hong, Ke Wu, Ji Xin Chen, Hong Jun Tang, Novel Substrate Integrated Waveguide Cavity Filter with Defected Ground Structure, IEEE Transactions on Microwave Theory and Techniques, Vol. 53, No. 4, pp. 1280–1287, 2005.
12. Y. Cassivi, L. Perregrini, P. Arcioni, M. Bressan, K. Wu, G. Conciauro, Dispersion Characteristics of Substrate Integrated Rectangular Waveguide, IEEE Microwave Wireless Component Letters, Vol. 12, No. 9, pp. 333–335, 2002.
13. Xuping Li, Hao Liu, Xiaowei Shi, Multi-band FSS Design of Y Fractal Unit, Journal of Microwaves, Vol. 22, pp. 100–103, 2006.
14. Xinyu Hou, Peng Zhang, Jun Lu, Fragment Design of Frequency Selective Surface for a Double Curvature Radome, Journal of Projective Rockets, Missiles and Guidance, Vol. 26, No. 1, pp. 123–125, 2006.
15. Xiaoqiu Li, Jinsong Gao, Xiaoguo Feng, A Novel Composite Unit Frequency Selective Surface, Journal of Microwaves, Vol. 24, No. 1, pp. 7–14, 2008.
16. Xiaoqiu Li, Jun Lu, Hongyan Jia, Cross Shaped Composite Frequency Selective Surface with Frequency Band, Journal of Infrared and Millimeter Wave, Vol. 26, No. 2, pp. 146–148, 2007.
17. X.F. Luo, P.T. Teo, A. Qing, C.K. Lee, Design of Double-square-loop Frequency Selective Surfaces Using Differential Evolution Strategy Coupled with Equivalent-circuit Model, Microwave and Optical Technology Letters, Vol. 44, No. 2, pp. 159–162, 2005.
18. T.K. Wu, Double-square-loop FSS for Multiplexing Four(S/X/KU/KA) Bands, IEEE International Symposium Antennas and Propagation Society, Vol. 3, pp. 1885–1888, 1991.
19. Juan Yang, Zhongxiang Shen, A Thin and Broadband Absorber Using Double-square Loops, IEEE Antennas and Wireless Propagation Letters, Vol. 6, pp. 388–391, 2007.
20. Nana Qi, Shuxi Gong, Pengfei Zhang, Jianfu Liu, A Novel Y-loop Aperture Frequency Selective Surface Using Substrate-integrated Waveguide Technology, Microwave and Optical Technology Letters, Vol. 50, No. 12, pp. 3023–3027, 2008.
21. Na na Qi, Shuxi Gong, Yujie Zhang, Jianfu Liu, Reducing Bandwidth of FSS Using Substrate-integrated Waveguide Technology, Journal of Electromagnetic Waves and Application, Vol. 22, pp. 2087–2096, 2008.
22. Nana Qi, Research on Novel Frequency Selective Surface, Xidian University, Master Thesis, 2009.

23. Jiejun Zhang, Junhong Wang, Meie Chen, Zhan Zhang, RCS Reduction of Patch Array Antenna by Electromagnetic Band-Gap Structure, IEEE Antennas and Wireless Propagation Letters, Vol. 11, pp. 1048–1051, 2012.
24. Wenbo Pan, Cheng Huang, Po Chen, Xiaoliang Ma, Chenggang Hu, Xianggang Luo, A Low-RCS and High-Gain Partially Reflecting Surface Antenna, IEEE Trans. Antennas Propagation, Vol. 62, No. 2, pp. 945–949, 2014.
25. David R. Smith, Jonah Gollub, Jack J. Mock, Willie J. Padilla, David Schurig, Calculation and measurement of bianisotropy in a split ring resonator metamaterial, Journal of Applied Physics, Vol. 100, No. 2, pp. 024507–024507-9, 2006.

Chapter 8
Polarization Rotation Reflective Surfaces and Their Application for Antenna RCS Reduction

Metasurfaces are two-dimensional equivalents of metamaterials that take up less physical space than full three-dimensional metamaterial structures [1]. Metasurfaces can manipulate both the magnitude and phase of EM waves freely by optimizing the structure of the unit cells. Taking advantages of these properties, metasurfaces have been widely used for designing antennas, reflectarrays, lenses, radar absorbers, and RCSR materials [2–10]. Here, we only focus on metasurfaces that can used for RCSR.

There are two types of RCSR mechanisms—that is, absorbing incident waves, and reflecting incident waves in nonthreatening directions. The first mechanism is usually realized using radar-absorbing metamaterials (RAMs), which can simultaneously reduce both the monostatic and bistatic RCS as most of the incident waves are absorbed. However, lump resistors or resistive sheets have to be used when designing RAMs with wideband absorbing characteristics, which increases the complexity of the process. The other mechanism is commonly used for monostatic RCSR through passive cancelation. By arranging the unit cells of metasurfaces—such as AMCs, PGMs, and polarization rotation reflective surfaces (PRRSs)—in a chessboard-like configuration, the waves reflected by different parts are out of phase over a broad band in the backward direction. Consequently, the monostatic RCS can be reduced over a wide band. In this chapter, we introduce a design method for a wideband and high-efficiency PRRS and its mechanism for RCSR. Several low-RCS antennas using a PRRS are then presented.

Y. Liu et al., *Antenna Radar Cross Section: Theory and Design*,
https://doi.org/10.1007/978-981-96-1226-0_8

8.1 Design of a Wideband and High-Efficiency Polarization Rotation Reflective Surface

Owing to its broad application in RCSR and circularly polarized antennas, there has been increasing research interest in polarization rotators [11–15]. Generally, polarization rotators can be categorized into transmission- and reflection-type rotators. Transmission-type polarization rotators can exploit either the birefringence effect of anisotropic metamaterials [16–18] or the optical activity of chiral metamaterials [19–21]. Most of these structures operate in a narrow band. For reflection-type polarization rotators, plasmon resonances and asymmetric high-impedance surfaces can be used to realize the polarization rotation [22–29]. However, the PRRSs proposed in previous studies either work in a few separate narrow polarization rotation frequency bands or do not have sufficiently high polarization conversion ratios (PCRs) over the polarization rotation band. Consequently, in this section, we investigate several design methods for wideband and high-efficiency PRRSs.

8.1.1 Broadband PRRS with Asymmetric Vias [30]

In Yang et al. [28], a PRRS with two grounded vias loaded in the diagonal direction of an AMC patch was presented. By properly adjusting the distance between the two vias, which were symmetric with the center of the patch, a 29.1% polarization rotation band with high efficiency could be obtained. Based on this structure, a high-efficiency PRRS with a wider band could be achieved by using asymmetric vias. The unit cell of the proposed PRRS is shown in Fig. 8.1. It comprises a square metallic patch printed on a substrate backed by a metallic ground plane and two grounded vias loaded in a diagonal direction. It is worth noting that the two vias are not symmetric with the center of the PRRS unit cell and the square patch, which is critical for obtaining a high PCR across the entire operating frequency range. The period (p) of the PRRS unit cell is 6.0 mm, the width (w) of the patch is 4.0 mm, the diameters (D) of the two vias are 0.6 mm, and the offsets of the two vias from the center of the unit cell dt_1 and dt_2 are 2.5 and 1.2 mm, respectively. The patch was printed on a dielectric substrate with a permittivity of 2.65 and a thickness (t) of 2.0 mm ($\tan\delta = 0.001$).

Here, we can define $r_{TE/TM} = E_{TM}^{Ref}/E_{TE}^{Inc}$ and $r_{TE/TE} = E_{TE}^{Ref}/E_{TE}^{Inc}$ to denote the *TE*-to-*TM* reflection ratio and the *TE*-to-*TE* polarization conversions, respectively, where E_{TE}^{Inc} denotes the electric field of the *TE*-polarized incident EM wave, and E_{TE}^{Ref} and E_{TM}^{Ref} denote the electric fields of the *TE*- and *TM*-polarized reflected EM waves, respectively.

As reported in previous studies, an asymmetric high-impedance surface can be used to form the PRRS. An AMC structure with two grounded vias loaded along the diagonal direction was reported in [28], the two vias being symmetric with the center of the unit cell. Two narrow resonant frequency bands were generated, and a

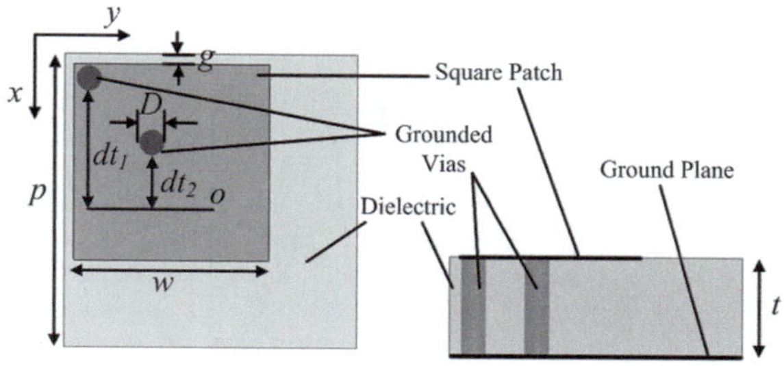

Fig. 8.1 Topological structure of the unit cell of the broadband PRRS with a high PCR

wide polarization rotation frequency band could be formed by adjusting the distance between the two vias. To understand the effect of the positions of the vias, we first make the two vias and the patch symmetric with the center of the unit cell, with the patch width (w) being 5.6 mm, while keeping the other parameters the same as in Fig. 8.1. Figure 8.2 shows that the two resonant frequency points move closer to each other as d decreases, which is consistent with the analysis in Menzel et al. [18].

We then kept the grounded via at the top left unchanged at 2.5 mm and moved the other in a diagonal direction. As shown in Fig. 8.3a, a third resonant point appears when the grounded via moves toward the center of the unit cell.

Figure 8.3b shows that when the two grounded vias are both in the upper left corner of the unit cell, the third resonant point appears. However, when the grounded via approaches the center of the unit cell, the two resonant frequency points at lower frequencies join together, which is similar to that of the symmetric case shown in Fig. 8.2. We can infer from Figs. 8.2 and 8.3 that the nonsymmetric distribution of the grounded vias generates one more polarization rotation frequency point than the symmetric structure

To produce a wider polarization rotation band with a high PCR, the width of the square patch was reduced, and the final design was optimized as shown in Fig. 8.1.

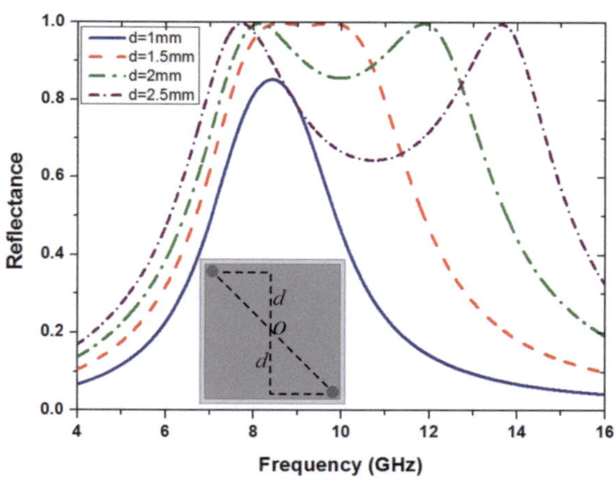

Fig. 8.2 Simulated magnitudes of the reflection coefficient $r_{TE/TM}$ of the PRRS with two symmetric vias

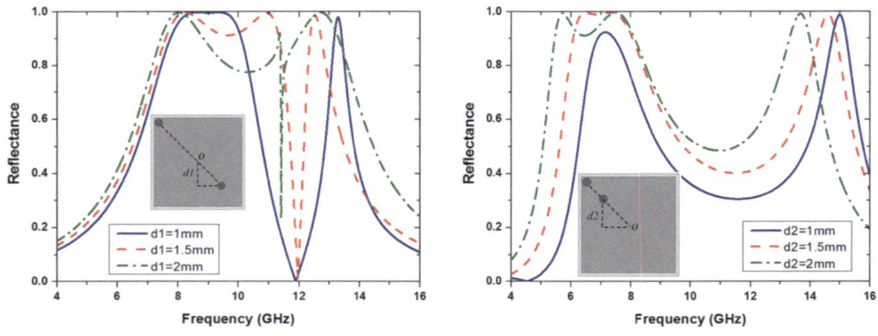

(a) (b)

Fig. 8.3 Simulated magnitudes of the reflection coefficient $r_{TE/TM}$ of the PRRS with two nonsymmetric vias **a** on different sides and **b** on the same side of the center of the PRRS unit cell in the diagonal direction

Figure 8.4 shows the magnitudes of the reflection coefficients of the proposed PRRS. There are three peak values that are approximately 1; this means that at the three frequency points—that is, 11.094, 14.098, and 16.798 GHz—almost all of the *TE*-polarized incident waves are converted to *TM*-polarized waves. Moreover, the minimum value of $r_{TE/TM}$ among the three peak values is 0.979, demonstrating high polarization conversion metasurface (PCM) stability at the operating frequency of the proposed PRRS.

The PCR can be defined as $\text{PCR} = |r_{TE/TM}|^2 / \left(|r_{TE/TM}|^2 + |r_{TE/TE}|^2\right)$. Figure 8.5 shows the simulated PCR of the proposed PRRS. The PCR is higher than 96.4% between the three resonant frequency points. The fractional bandwidth of the PCR higher than 96.4% is 49% (from 10.74–17.72 GHz).

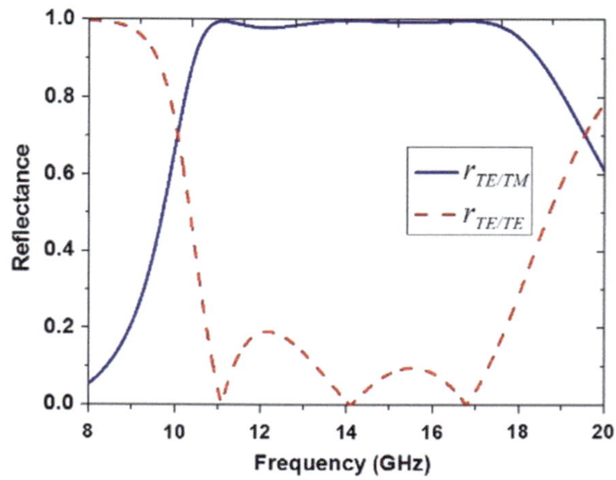

Fig. 8.4 Simulated reflection coefficients $r_{TE/TM}$ and $r_{TE/TE}$ of the proposed PRRS

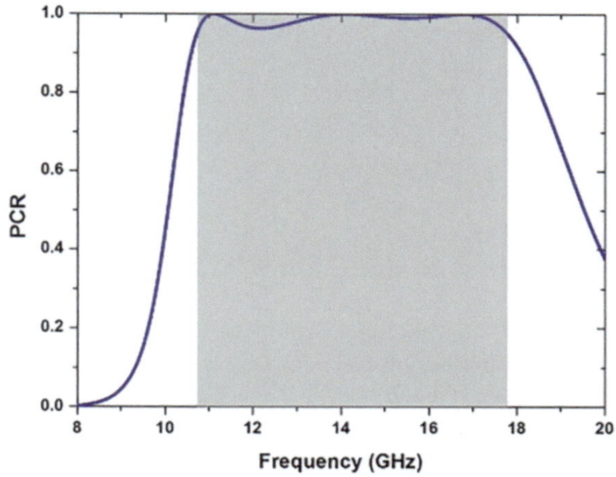

Fig. 8.5 Simulated PCR of the proposed PRRS

Figure 8.6 shows the reflection phase of the proposed PRRS under a *TE*-polarized incident wave. The reflection phases of $r_{TE/TE}$ are zero at the three polarization rotation frequency points—that is, 11.094, 14.098, and 16.798 GHz—owing to the proposed PRRS exhibiting high-impedance characteristics at these three points. Because the incident wave is *TE*-polarized and normal to the PRRS, its electric field is parallel to the *y*-axis. Moreover, the AMC can reflect the normally incident wave in phase. The induced current in the *y*-direction can be generated by the incident electric field (E_y). Owing to the asymmetric structure of the PRRS, a potential difference in the *x*-direction can be introduced by the nonuniform distribution of the surface impedance along the *x*-axis. Consequently, some current moves in the *x*-direction, which generates an electric field (E_x) in the *x*-direction.

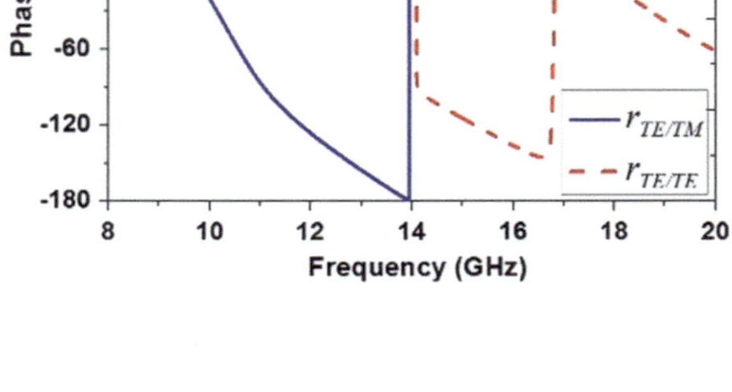

Fig. 8.6 Simulated reflection phases of $r_{TE/TM}$ and $r_{TE/TE}$ of the proposed PRRS

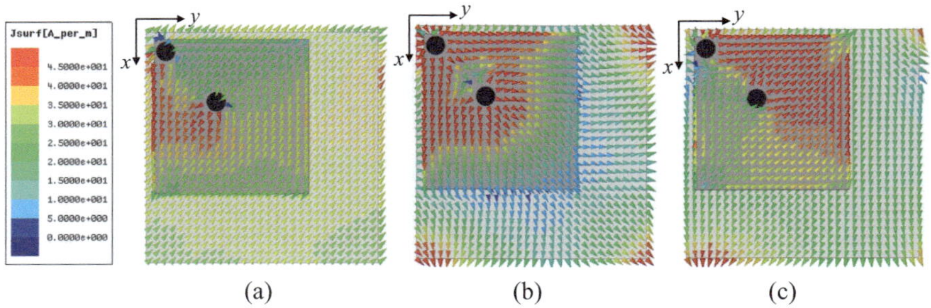

Fig. 8.7 Simulated surface current distributions on the metallic parts of the unit cell at three resonant points: **a** 11.094 GHz, **b** 14.098 GHz, and **c** 16.798 GHz

When the current moving in the *x*-direction is equal to that moving in the *y*-direction, the reflected field is orthogonal to the incident field with the same magnitude. The simulated surface current distributions on the metallic parts of the unit cell at the three polarization points are shown in Fig. 8.7. In Fig. 8.7a, the total surface current on the ground is stronger than that on the square patch. By contrast, as shown in Fig. 8.7b and c, the total surface current on the square patch is stronger. In all three cases, the intersection angle between the direction of the total surface current induced on the patch and the ground is a 45° angle from the *y*-axis at all three polarization rotation points, which suggests a 90° polarization rotation.

To investigate how the parameters influence the reflection rotation characteristics of the proposed PRRS, a series of parameter analyses was performed. First, dt_1 was changed while keeping the other parameters fixed. Figure 8.8 shows that with a reduction in dt_1, the bandwidth of the PRRS narrows. The third resonant frequency point moves to a lower frequency, and the second point shifts slightly to a higher frequency. When dt_1 is 2.1 mm, the second and third resonant frequency points combine, and only two resonant frequency points remain.

The parameter dt_2 was examined next. Figure 8.9 shows that the first resonant frequency point shifts to a higher frequency, and the other two remain unchanged or move only slightly with decreasing dt_2.

As shown in Fig. 8.10, when *w* decreases, the first two resonant frequency points both move to higher frequencies, while the third remains unchanged. Finally, to obtain a relatively high and stable PCR over the operating frequency band, the values of the parameters can be determined as presented above.

The stability of the PRRS under oblique incident waves has not been investigated in most previous studies. As shown in Fig. 8.11, when the incident angle (θ) varies from 0° to 30°, the frequency response is reasonably stable. However, as the incident angle continues to increase, the magnitudes of the reflection coefficients of the proposed PRRS are no longer stable. However, the reflection coefficient is still greater than 0.8 when the incident angle increases to 60°.

To validate the simulated results, a prototype with 80 × 50 cells (480 × 300 mm) was fabricated and measured, as shown in Fig. 8.12. Owing to restrictions associated

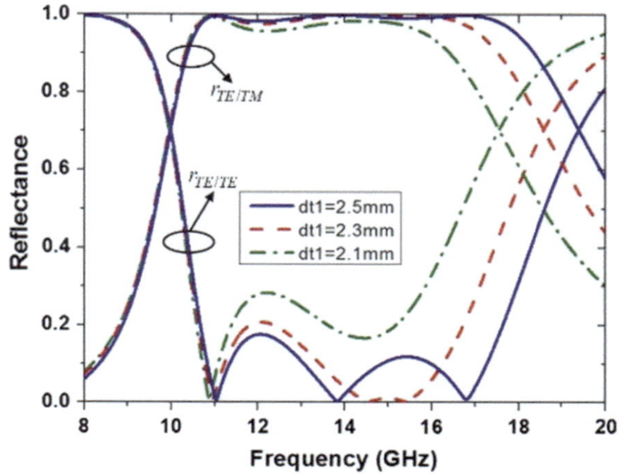

Fig. 8.8 Simulated magnitudes of the reflection coefficients $r_{TE/TM}$ and $r_{TE/TE}$ of the proposed PRRS as dt_1 decreases from 2.5 to 2.1 mm

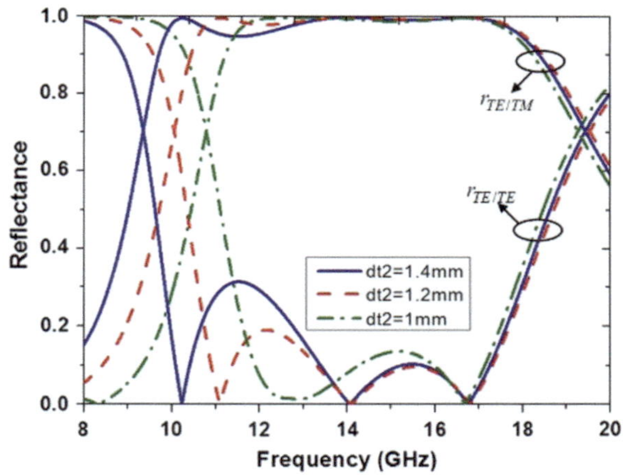

Fig. 8.9 Simulated magnitudes of the reflection coefficients $r_{TE/TM}$ and $r_{TE/TE}$ of the proposed PRRS as dt_2 decreases from 1.4 to 1 mm

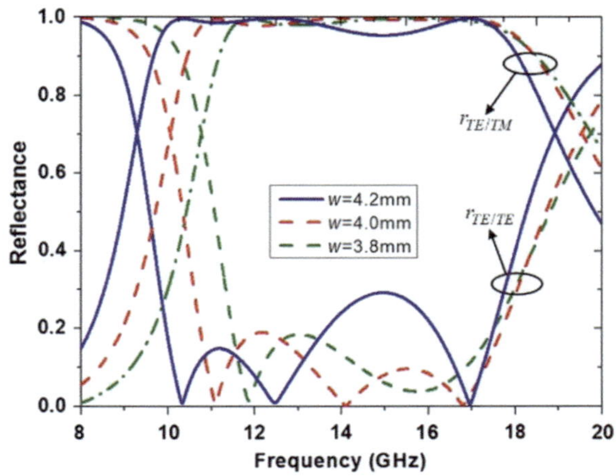

Fig. 8.10 Simulated magnitudes of the reflection coefficients $r_{TE/TM}$ and $r_{TE/TE}$ of the proposed PRRS as w decreases from 4.2 to 3.8 mm

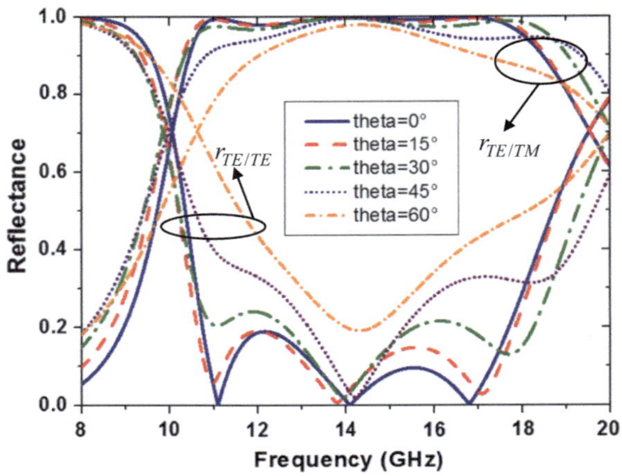

Fig. 8.11 Simulated magnitudes of the reflection coefficients $r_{TE/TM}$ and $r_{TE/TE}$ of the proposed PRRS under oblique incident waves

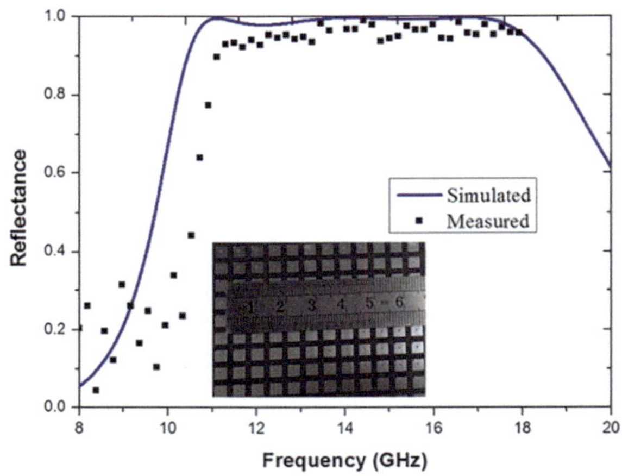

Fig. 8.12 Simulated and measured magnitudes of the reflection coefficients $r_{TE/TM}$ of the proposed PRRS

with the test conditions, the reflectance of the proposed PRRS was not measured at frequencies higher than 18 GHz. Nonetheless, it is evident that the measured reflection coefficients match the simulated coefficients well.

8.1.2 Dual-Patch Wideband PRRS with a Dielectric Superstrate [31]

In the previous section, 49% of the bandwidth of the PRRS was obtained with a high PCR of 96%. However, the polarization rotation bandwidth was still too narrow for most wideband applications. In this section, a novel ultrawideband and high-efficiency PRRS is designed by employing single-layer dual-patch unit cells. A unit

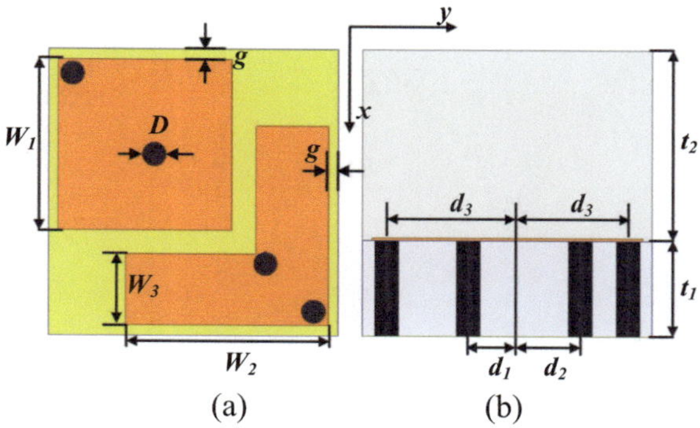

Fig. 8.13 Topological structure of the unit cell of the ultrawideband PRRS with a high PCR: **a** top view and **b** side view

cell comprises a square patch and an L-shaped patch printed on a substrate, which is covered by a superstrate and backed by a metallic ground. When the square and L-shaped patches are combined to form a dual-patch unit and covered by a superstrate, the electrical size of the unit cell is reduced, while the operating bandwidth is broadened to 97% with a high PCR of 96%.

The unit cell of the proposed PRRS is shown in Fig. 8.13. It comprises a square patch and an L-shaped metallic patch printed on the substrate backed by a metallic ground plane, both patches being connected to the metallic ground by two metallic vias loaded in a diagonal direction. The dual-patch unit cell is covered by a superstrate, the period of the unit cell being 6.0 mm, and the relative permittivity of the substrate and the superstrate being 2.65 ($\tan\delta = 0.001$). The other parameters are listed in Table 8.1.

Here, we can define $r_{xy} = E_x^{Ref}/E_y^{Inc}$ and $r_{yy} = E_y^{\mathrm{Ref}}/E_y^{Inc}$ to denote the reflection ratios of y-to-x and y-to-y polarization conversions, respectively, where E_y^{Inc} denotes the electric field of the y-polarized incident EM wave, and E_y^{Ref} and E_x^{Ref} denote the electric fields of the y- and x-polarized reflected EM waves, respectively.

To understand the response of the proposed PRRS to the incident EM wave with the y-polarized electric field, we can decompose the electric field of the incident EM wave ($\vec{E_i}$) into two perpendicular components, $\vec{E}_{iu}$ and $\vec{E}_{iv}$, as shown in Fig. 8.14a. For u-polarized and v-polarized incident EM waves, the magnitudes and phases of the reflected electric fields are shown in Fig. 8.14b. The two curves are similar, most of the energy being reflected with the same polarization as the incident EM wave. Moreover, the relative phase retardation is approximately 180° from 6.2–17.8 GHz. Using the frequency point where the reflected phase φ_u is 0° as an example, the electric vectors of the incident and reflected waves are shown in Fig. 8.14a. When φ_u

Table. 8.1 Unit cell dimensions (mm)

W_1	W_2	W_3	D	g	d_1	d_2	d_3	t_1	t_2
3.6	4.2	1.5	0.5	0.2	0.8	1.5	2.5	2	4

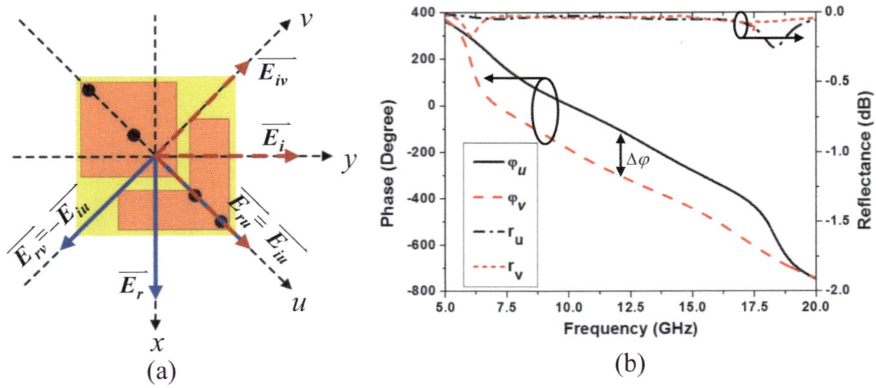

Fig. 8.14 **a** Schematic of y-to-x polarization conversion. **b** Amplitudes and phases of the reflected electric fields with the electric field of incident EM waves along the u-axis and v-axis

is 0°, φ_v is $-180°$ because the phase difference between φ_u and φ_v is approximately 180°. Thus, the reflected electric field ($\vec{E}_{ru}$) is equal to $\vec{E}_{iu}$, and $\vec{E}_{rv}$ is equal to $-\vec{E}_{iv}$.. Additionally, the magnitudes of the reflected electric fields are equal to those of the incident electric fields. Thus, the total reflected electric field of the reflected EM wave ($\vec{E}_r$) can be obtained. It has the same magnitude as $\vec{E}_i$ but is parallel to the x-axis. The same applies to the frequency range in which the phase difference is approximately 180°.

The simulated reflection coefficients (r_{xy} and r_{yy}) of the ultrawideband PRRS for oblique incidence of the y-polarized EM waves are shown in Fig. 8.15. Figure 8.15a shows that as the incident angle (θ) increases from 0° to 60°, the -3 dB bandwidth of the reflection coefficient (r_{xy}) is reduced by approximately 2 GHz. However, the frequency response of the proposed PRRS in a lower frequency range is reasonably consistent. The reflection coefficient (r_{xy}) is still greater than -2 dB from 6–16 GHz when the incident angle increases to 60°. Figure 8.15b shows that when $\theta = 0°$, there are five resonant frequencies, which decrease to four with oblique incident EM waves.

Next, we can use the surface currents of the PRRS unit cell to explain the polarization rotation mechanism. When the incident EM wave is y-polarized, a current can be induced in the y-direction. Owing to the asymmetric PRRS structure, the surface impedance in the x-direction is nonuniform. Consequently, some of the current moves in the x-direction, generating an electric field ($\vec{E}_x$) in the x-direction. When the current moving in the x-direction is equal to that moving in the y-direction, the reflected field becomes orthogonal to the incident field with the same magnitude. The simulated surface current distributions on the two patches of the unit cell at the five resonant frequencies (as shown in Fig. 8.15b) are shown in Fig. 8.16. The intersection angle between the direction of the total surface current induced on the patches and the y-axis is 45°, which indicates a 90° polarization rotation.

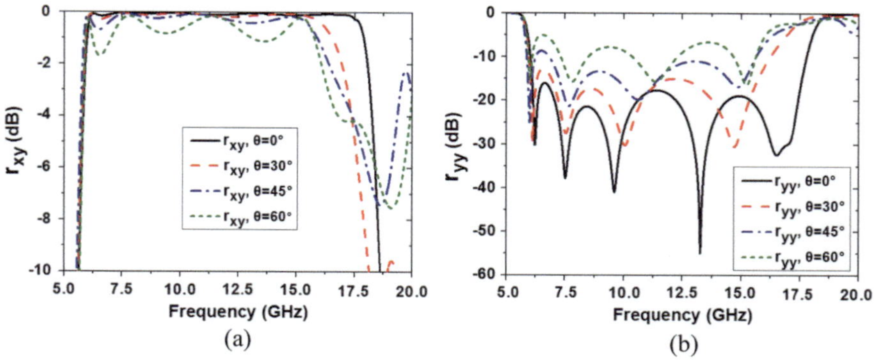

Fig. 8.15 Simulated reflection coefficients of the proposed PRRS for EM waves with oblique incidence: **a** r_{xy} and **b** r_{yy}

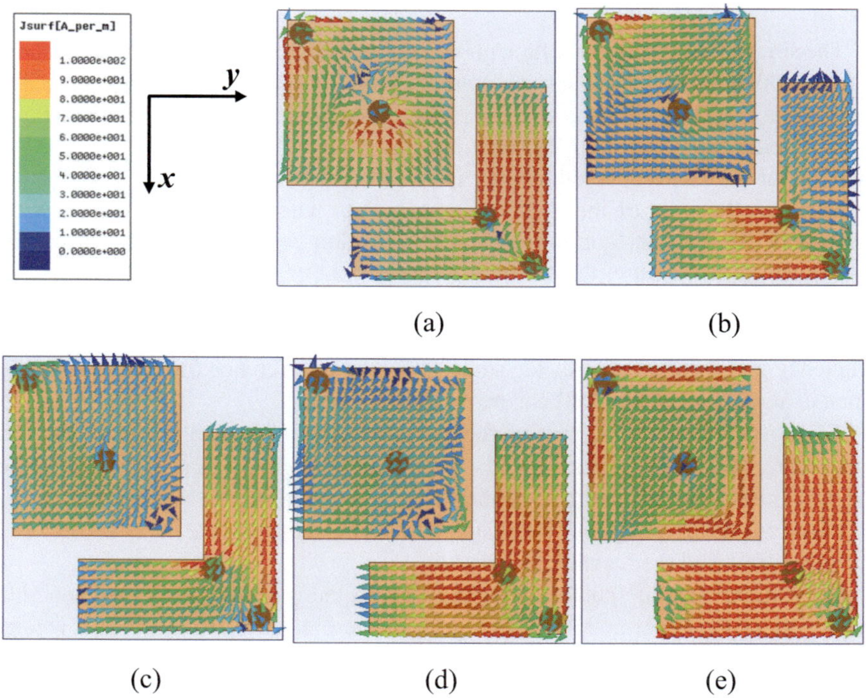

Fig. 8.16 Simulated surface current distributions on metallic patches of the unit cell at five resonant frequencies: **a** 6.42 GHz, **b** 7.68 GHz, **c** 9.38 GHz, **d** 13.3 GHz, and **d** 17.1 GHz

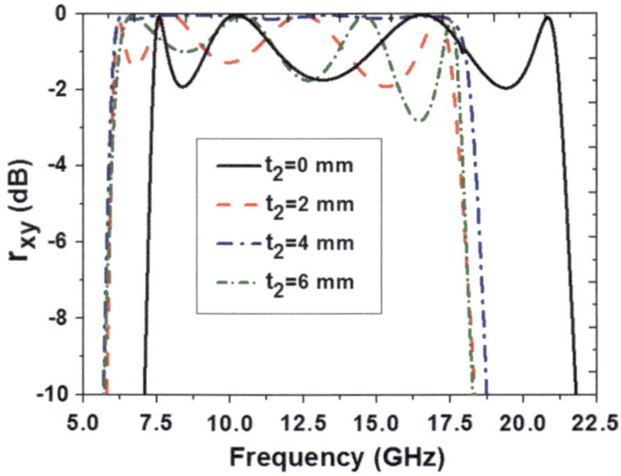

Fig. 8.17 Simulated reflection coefficients r_{xy} of the proposed PRRS for different thicknesses (t_2) of the superstrate

The simulated reflection coefficients (r_{xy}) of the proposed PRRS with and without different thicknesses (t_2) of the superstrate are shown in Fig. 8.17. The thickness of the superstrate has a large influence on the reflection coefficient (r_{xy}). When the PRRS is not covered by the superstrate, the operating band moves to a higher frequency range than that of the PRRS covered by the superstrate. When t_2 is 4 mm, the polarization conversion efficiency of the PRRS is greater than when t_2 is 2 and 6 mm. With the superstrate thickness optimized, the polarization conversion efficiency can be maximized, which ensures that the multiple reflections on the air-superstrate interface and superstrate-metallic patch interface enhance the PRRS polarization conversion efficiency. To further examine the relationship between the polarization conversion efficiency and the thickness of the superstrate, the interference theory presented in Chen et al. [32] and Chen [33] can be used.

As shown in Fig. 8.18, part of the y-polarized incident EM wave on the air-superstrate interface is reflected back into the air, and part of it is transmitted into the superstrate. The transmitted y-polarized EM wave continues to propagate in the propagation phase $\beta = \sqrt{\varepsilon} k_0 d$ in the superstrate until it reaches the interface between the superstrate and the metallic patches.

Because the metallic patches are connected to the ground sheet by the metallic vias, we can treat the metallic structures and the dielectric substrate as a whole, regardless of the multiple reflection process among them. Thus, the transmitted EM wave is reflected by the superstrate-metallic patch interface back into the superstrate. Because of the polarization conversion characteristics of the PRRS, the reflected waves comprise both y-polarized and x-polarized components. Consequently, there are two reflection coefficients—that is, $r_{23xy}\exp(i\emptyset_{23xy})$ and $r_{23yy}\exp(i\emptyset_{23yy})$. Partial reflection and transmission occur when the reflected EM waves reach the superstrate-air interface. When the EM waves reflected from the superstrate-air interface reach the superstrate-metallic patch interface again, there are two additional reflection coefficients—that is, $r_{23xx}\exp(i\emptyset_{23xy})$ and $r_{23yx}\exp(i\emptyset_{23yx})$. Because the proposed

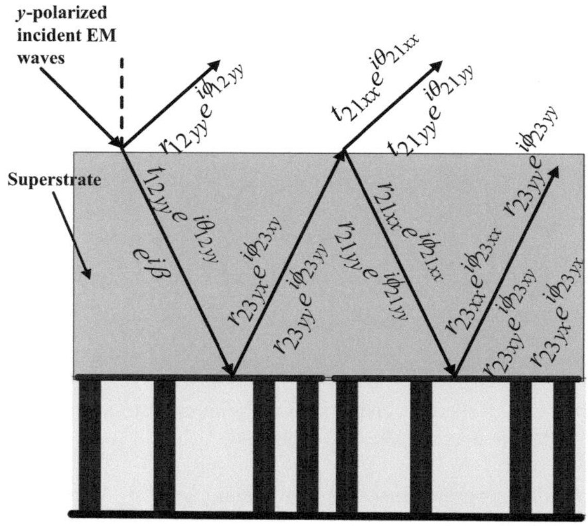

Fig. 8.18 Schematic of the multiple reflection-transmission process described by the reflection and transmission coefficients

structure is symmetric with respect to the diagonal, $r_{23xy} = r_{23yx}$, $r_{23xx} = r_{23yy}$, $\emptyset_{23xy} = \emptyset_{23yx}$, and $\emptyset_{23xx} = \emptyset_{23yy}$.

The magnitude and phase of the reflection coefficient at the superstrate-metallic patch interface are shown in Fig. 8.19a and b, which were obtained using full-wave EM simulation software. The reflection and transmission coefficients at the air-superstrate interface are constants because both the superstrate and air are homogeneous and isotropic. Thus, the total reflection of the x-polarized EM wave can be calculated by superposition of all multiple reflections. The PCR of the PRRS can be expressed as $\mathrm{PCR} = |r_{xy}|^2 / \left(|r_{xy}|^2 + |r_{yy}|^2\right)$. It is evident that the calculated reflection coefficients and PCR are in good agreement with those from the simulation, as shown in Fig. 8.20a and b, respectively. The PCR is higher than 96% in 97% of the bandwidth from 6.1–17.7 GHz.

Theory of PRRSs for RCSR: This section investigates the theory of the PRRS for RCSR. The dual-patch PRRS proposed above can be used to form a checkerboard-like surface to reduce the RCS of a metallic sheet. Moreover, the relationship between the RCSR of this surface and the PCR of the PRRS is analyzed. Figure 8.21 shows the proposed low-RCS checkerboard-like surface, which can be obtained by arranging the PRRS unit cells in four orthogonal directions. Each of the four areas is formed by 16 × 16 unit cells with a total area of 96 × 96 mm^2, the total area of the checkerboard-like surface being 192 × 192 mm^2. The RCS of a metallic ground-backed substrate with the same permittivity and size as the proposed checkerboard-like surface was selected as the reference.

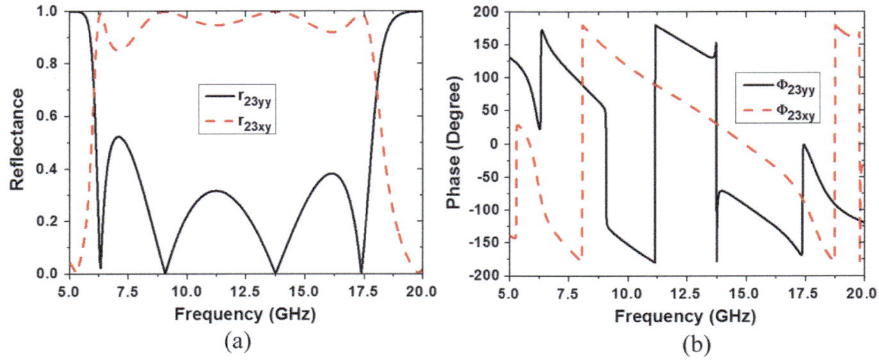

Fig. 8.19 Reflection coefficients at the superstrate-metallic patch interface obtained by numerical simulation: **a** amplitudes and **b** phases

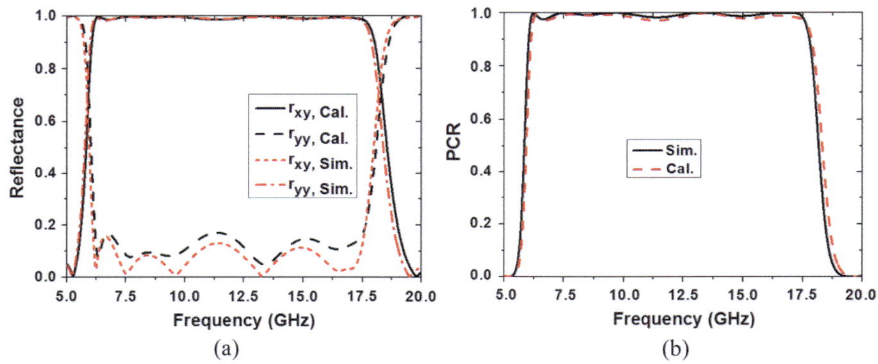

Fig. 8.20 Calculated and simulated **a** reflection coefficients r_{xy} and r_{yy}, and **b** PCR of the proposed PRRS

Fig. 8.21 Structure of the checkerboard-like surface formed by the proposed ultrawideband PRRS

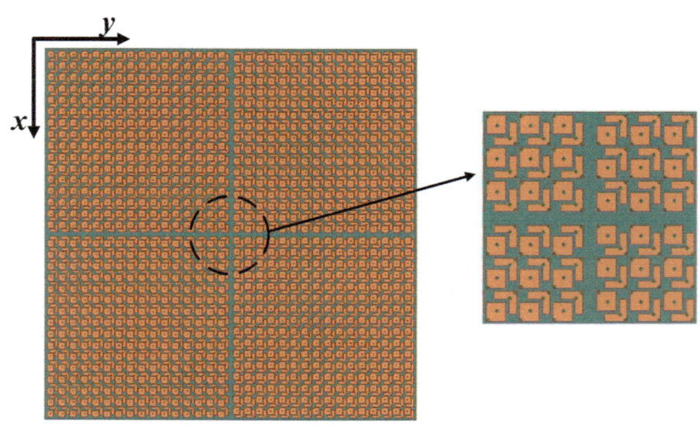

The RCS of a target σ can be expressed as follows [34]:

$$\sigma = 4\pi \lim_{R\to\infty} R^2 \frac{\left|\vec{E}_s\right|^2}{\left|\vec{E}_i\right|^2} \tag{8.1}$$

where $\vec{E}_i$ and $\vec{E}_s$ denote the electric field intensities of the incident and scattered EM waves, respectively, and R denotes the distance between the target and detection radar.

Using Eq. (8.1), we can analyze the RCS of the PRRS using the electric fields of the incident and reflected EM waves. Without loss of generality, assume that the structure is impinged by a normally incident EM wave with a y-polarized electric field ($\vec{E}_{iy}$), and its RCS can be analyzed using the reflected electric field as shown in Fig. 8.22. Because the proposed PRRS has the characteristics of polarization rotation, the electric field of the reflected EM wave has two components—that is, the x-polarized electric field ($\vec{E}_{rx}$) and the y-polarized electric field ($\vec{E}_{ry}$). The magnitudes of $\vec{E}_{rx}$ and $\vec{E}_{ry}$ can be expressed as follows:

$$\left|\vec{E}_{rx}\right| = \left|r_{xy}\right| \cdot \left|\vec{E}_{iy}\right| \tag{8.2}$$

$$\left|\vec{E}_{ry}\right| = \left|r_{yy}\right| \cdot \left|\vec{E}_{iy}\right| \tag{8.3}$$

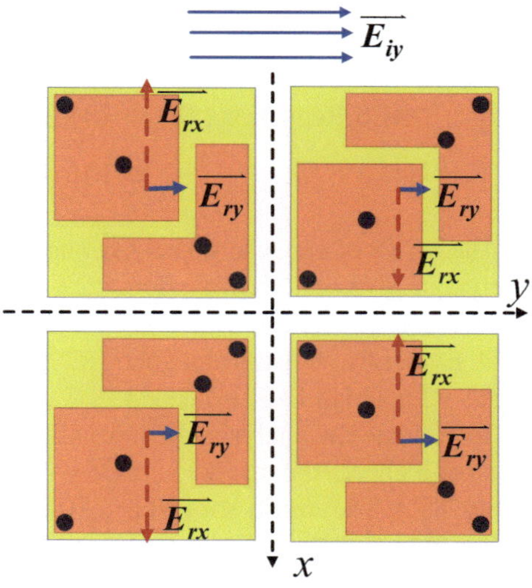

Fig. 8.22 Schematic of the electric field of the reflected wave with a y-polarized incident EM wave

Figure 8.22 shows that the x-polarized electric fields of the EM waves reflected from different parts of the checkerboard-like surface have a 180° phase difference. Moreover, the four orthogonally arranged PRRSs cover exactly the same area. Thus, the reflected EM waves with the x-polarized electric field cancel each other out. Consequently, the total reflected EM wave has only a y-polarized electric field. The RCS of the proposed checkerboard-like surface σ_c can be expressed as follows:

$$\sigma_c = 4\pi \lim_{R\to\infty} R^2 \frac{\left|\vec{E}_{ry}\right|^2}{\left|\vec{E}_{iy}\right|^2} \tag{8.4}$$

The RCSR of the checkerboard-like surface, compared to that of a PEC with the same dimensions, can be approximated by:

$$\begin{aligned}\text{RCS Reduction (dB)} &= 10\log_{10}\left[\frac{\lim\limits_{R\to\infty} 4\pi R^2 \frac{\left|\vec{E}_{ry}\right|^2}{\left|\vec{E}_{iy}\right|^2}}{\lim\limits_{R\to\infty} 4\pi R^2 |1|^2}\right] = 10\log_{10}\left[\frac{\left|\vec{E}_{ry}\right|^2}{\left|\vec{E}_{iy}\right|^2}\right] \\ &= 10\log_{10}\left[\frac{\left(\left|r_{yy}\right|\left|\vec{E}_{iy}\right|\right)^2}{\left|\vec{E}_{iy}\right|^2}\right] = r_{yy}(\text{dB})\end{aligned} \tag{8.5}$$

Regardless of the dielectric loss, the relationship between the RCSR of the checkerboard-like surface and the PCR of the PRRS can be represented by:

$$\begin{aligned}\text{RCS Reduction (dB)} &= 10\log_{10}\left[\left|r_{yy}\right|^2\right] = 10\log_{10}\left[1 - \left|r_{xy}\right|^2\right] \\ &= 10\log_{10}[1 - \text{PCR}]\end{aligned} \tag{8.6}$$

The RCSR of a checkerboard-like surface versus the PCR of the PRRS obtained by applying Eq. (8.6) is shown in Fig. 8.23. In the ideal case of an infinite checkerboard-like surface, a 10-dB RCSR can be obtained when the PCR of the PRRS is greater than 90%. Evidently, to achieve a good RCSR over an ultrawide band, an ultrawideband PRRS with a high PCR is needed.

To validate the analysis, the normalized monostatic RCS for the proposed ultra-wideband checkerboard-like surface with respect to the reference surface is shown in Fig. 8.24. The results show that a 10-dB RCSR is achieved over more than 103% of the frequency bandwidth (from 5.6–17.7 GHz). Moreover, the RCSR curve is consistent with that of r_{yy} of the proposed PRRS, which is consistent with the results

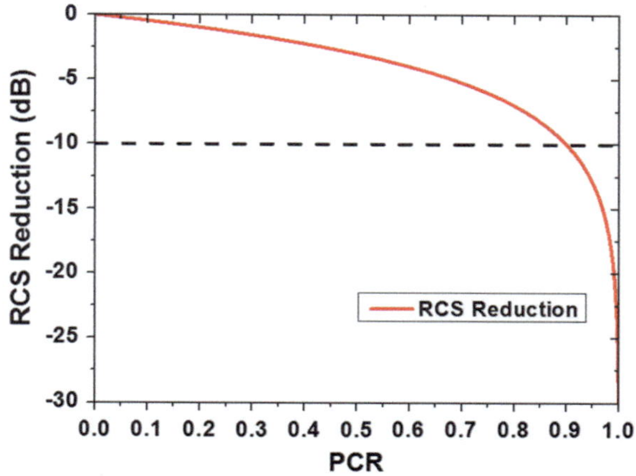

Fig. 8.23 RCSR of the proposed checkerboard-like surface versus the PCR of the PRRS

of Eq. (8.5). The difference between the two curves can be attributed to diffraction from the edges.

A comparison of the monostatic RCS between the checkerboard-like surface and the reference with oblique incident EM waves at four different frequencies is given in Fig. 8.25. The monostatic RCS of the checkerboard-like structure is smaller than that of the reference at most angles of the incident EM waves.

In the case of RCSR using nonabsorptive materials, the total scattered energy is not reduced owing to the principle of energy conservation. Consequently, the reduction in RCS at certain angles inevitably leads to an increase in the RCS at other angles. To examine the distribution of the scattered energy, we compared the bistatic RCS of the checkerboard-like surface with that of the reference at 12 GHz for a

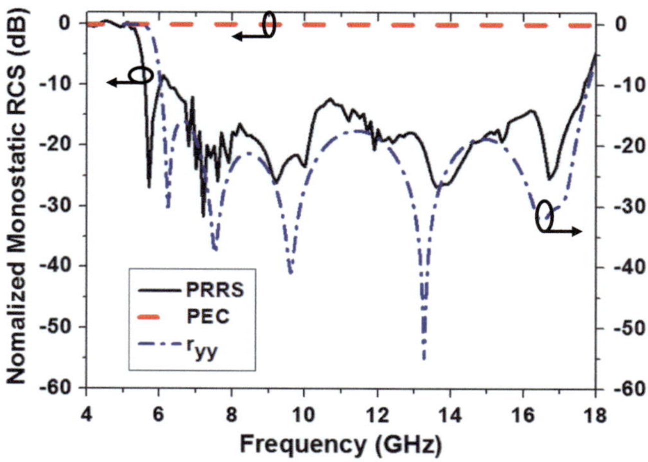

Fig. 8.24 Normalized monostatic RCS of the proposed ultrawideband checkerboard-like surface with respect to the reference and r_{yy} of the PRRS

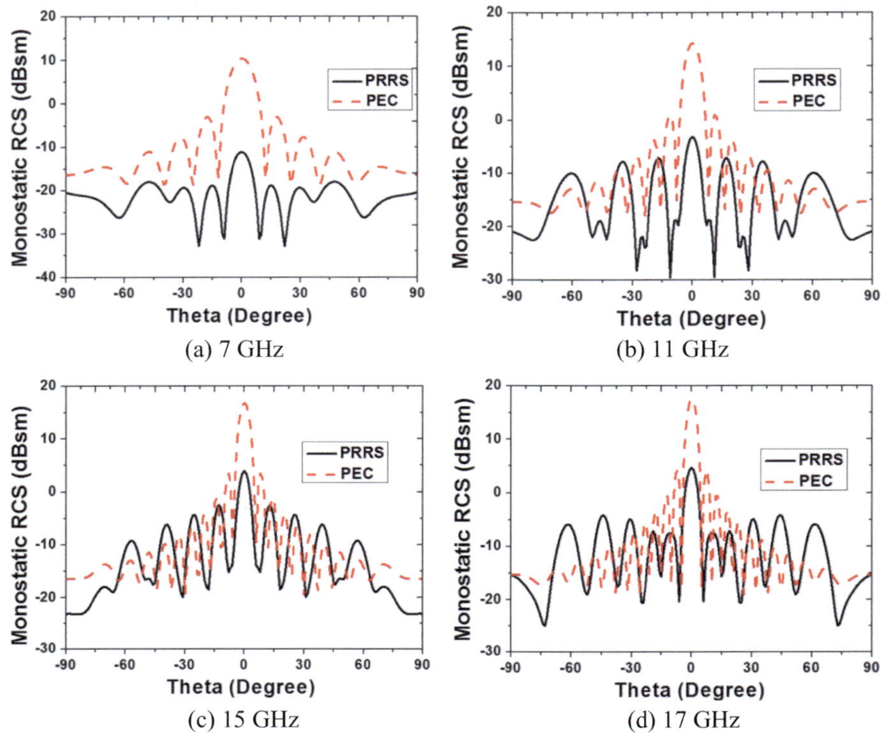

Fig. 8.25 Monostatic RCSs of the checkerboard-like surface and the reference for oblique incident EM waves at **a** 7 GHz, **b** 11 GHz, **c** 15 GHz, and **d** 17 GHz

normally incident EM wave, as shown in Fig. 8.26. Compared with the bistatic RCS of the reference, the main lobe of the bistatic RCS of the checkerboard-like surface is considerably reduced, and four side lobes are generated. This comparison clearly indicates where the energy reduced in the backward direction is redirected. To further examine the bistatic RCSR of the proposed checkerboard-like surface, the figure of merit for bistatic reduction (FMB) is shown in Fig. 8.27, which was first defined in Liu et al. [35], as follows:

$$\text{FMB}\,(f) = \frac{\text{Max}\,\{\text{Bistatic RCS of the proposed structure}\}}{\text{Max}\,\{\text{Bistatic RCS of the reference}\}} \tag{8.7}$$

where the Max{·} function domain covers all bistatic scattering angles (θ_s, $\emptyset_s$) ($0 \leq \theta_s \leq 90°$ and $0 \leq \emptyset_s \leq 360°$) for a given frequency (f) and incident angle (θ_i, $\emptyset_i$).

The FMBs for both *TE*- and *TM*-polarized incident waves were calculated over a range of incident angles from normal to 40° off-normal incidence. As shown in Fig. 8.27a, a 6-dB RCSR bandwidth of 95% is evident for normal incidence. For oblique incidence, the RCSR value and bandwidth decrease. A 5.5-dB RCSR or

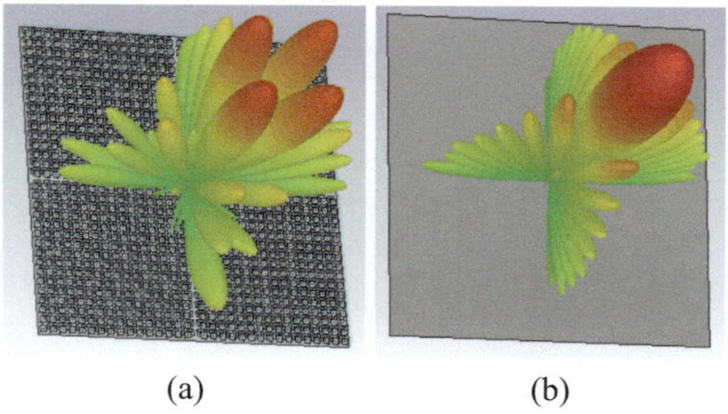

Fig. 8.26 Bistatic RCS at 12 GHz for a normally incident EM wave: a checkerboard-like surface and b reference

higher is obtained with a bandwidth of 88% and up to a 40° angle of incidence, as shown in Fig. 8.27b and c.

The checkerboard-like surface was fabricated as shown in Fig. 8.28, and its scattering properties were measured in a microwave anechoic chamber. Figure 8.29 shows a comparison between the simulated and measured RCSs of the checkerboard-like surface. A reasonable agreement between the measured and simulated results is evident in the designed frequency band.

8.2 Application of PRRS to Antenna RCS Reduction

The PRRS can be used to achieve RCSR and can also be applied to antenna design. In this section, design methods for both linearly-polarized and circularly-polarized low-RCS antennas using PRRSs are introduced. Additionally, the effect of the PRRS on the radiation performance of these antennas is discussed.

8.2.1 Out-of-Band RCS Reduction of a Slot Array Antenna Using a PRRS [36]

A 2 × 2 slot array antenna with an operating frequency band from 4.47–4.83 GHz was selected as the reference antenna to validate the RCSR effect of using a PRRS, as shown in Fig. 8.30a. As shown, a ground with four slots is located on the top surface of a 1-mm-thick FR4 dielectric substrate ($\varepsilon_r = 4.4$, $\tan\delta = 0.02$). The feeding network is printed on the bottom surface of the FR4 substrate.

A metal-backed substrate slab is added to the bottom of the feeding network, which helps to suppress the undesired back lobe. The FR4 slab is 3 mm thick. To reduce the RCS of the slot array antenna, the ground of the antenna is covered with a fishbone-shaped PRRS, which is shown in Fig. 8.30b. To alleviate the interference caused by the PRRS, the copper ground of the PCM is eliminated, the antenna and

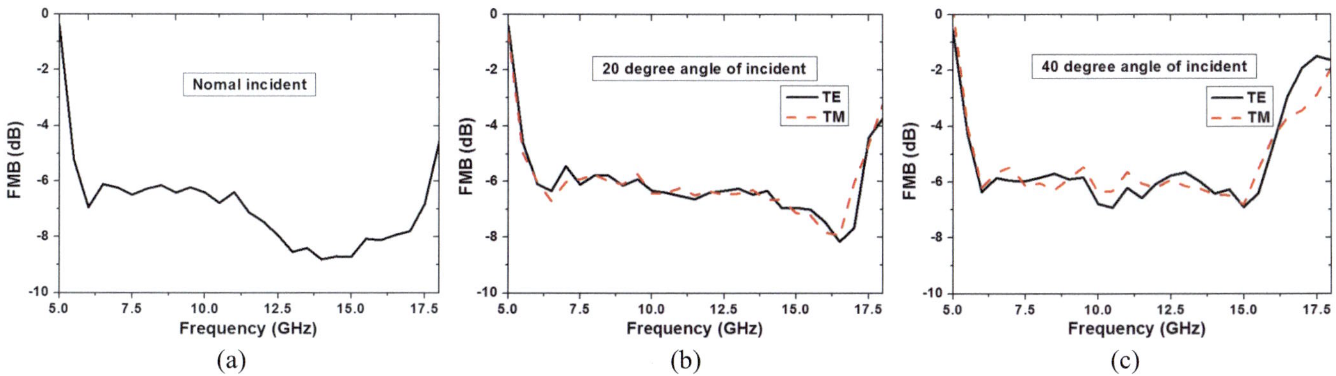

Fig. 8.27 Figures of merit bistatic RCSR (FMB) of the checkerboard-like surface versus frequency at different angles of incidence: **a** $\theta = 0^{\circ}$, **b** $\theta = 20^{\circ}$, and **c** $\theta = 40^{\circ}$

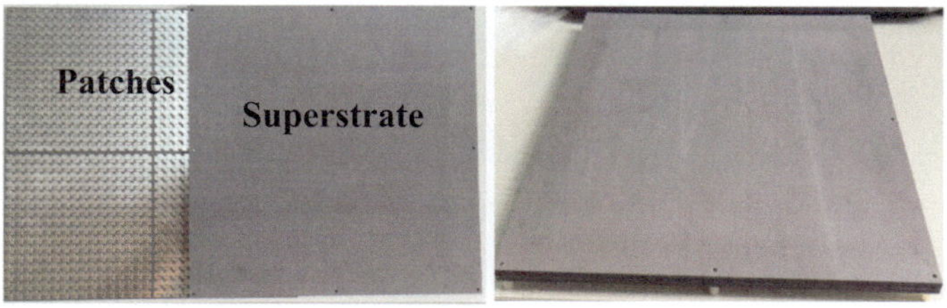

Fig. 8.28 Photographs of the proposed checkerboard-like surface

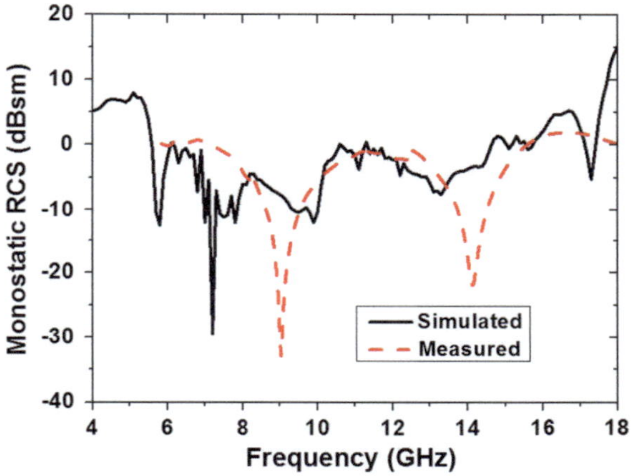

Fig. 8.29 Comparison between the simulated and measured monostatic RCSs of the proposed checkerboard-like surface

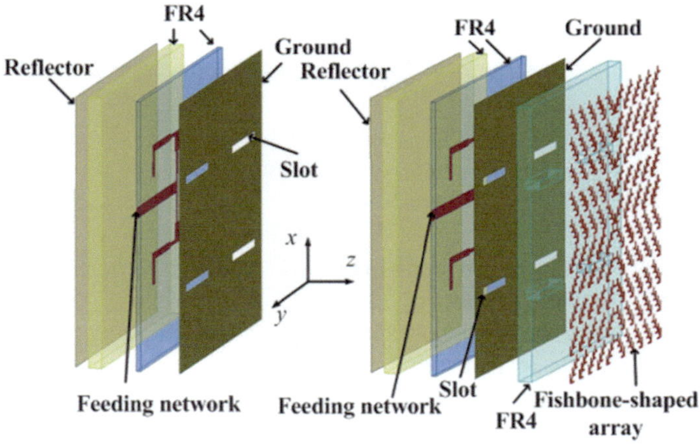

Fig. 8.30 3D sketches of **a** the reference slot array antenna and **b** the proposed low-RCS antenna with a PRRS

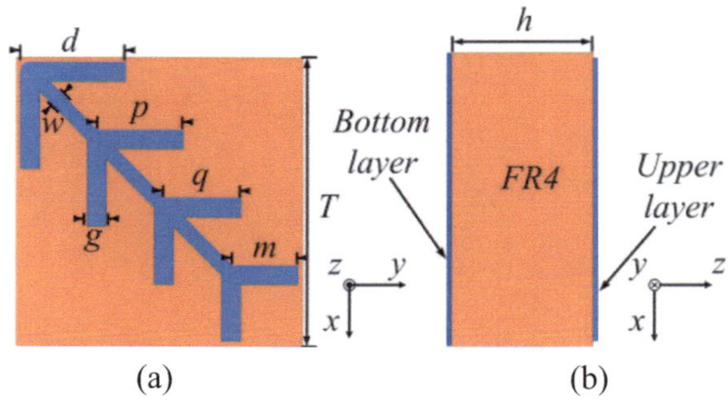

Fig. 8.31 Geometry of the fishbone-shaped unit cell: **a** top view, **b** side view

the PRRS sharing the same ground. Covering the antenna with the PRRS results in resonant frequency shifting and gain loss. To make the EM wave radiate into space as much as possible and reduce the coupling between the PRRS and the slots, the parts of the PRRS above the slots are cut, as shown in Fig. 8.30b.

Figure 8.31 shows the geometry of the fishbone-shaped unit cell, which comprises two metallic layers separated by a lossy substrate FR4 ($\varepsilon_r = 4.4$, $\tan\delta = 0.02$). The top layer comprises an etched fishbone-shaped pattern, the bottom layer being a copper sheet. To ensure the polarization insensitivity of the unit to the impinging wave, the geometry was designed to be symmetrical with respect to the + 45° diagonal. The dimensions of the unit cell were as follows: $T = 6$ mm, $d = 2$ mm, $p =$ 1.8 mm, $q = 1.6$ mm, $m = 1.4$ mm, $w = 0.3$ mm, $g = 0.4$ mm, and $h = 3$ mm.

Figure 8.32 presents the co- and cross-polarization reflections of the infinite periodic fishbone-shaped unit or the mirror unit under a normally impinging plane wave. A 5-dB bandwidth of the polarization conversion ranging from 5.5–17.5 GHz is evident. The curve of the cross-polarization shows three peaks at 6.0, 10.5, and 16.2 GHz, where the incident wave is effectively converted to a cross-polarized wave. Ideally, the polarization should be completely converted orthogonally after reflection.

Based on the discussion in Sect. 8.2, the RCSR can be obtained when the fishbone-shaped unit cells are arranged in a chessboard-like structure. To verify the proposed method, the antennas shown in Fig. 8.30 were fabricated and measured. Photographs of both antennas are shown in Fig. 8.33. The radiation patterns were measured in a Satimo SG 24-L chamber, and an Agilent E5071C vector network analyzer was used to measure the reflection coefficients.

Figure 8.34 shows the simulated and measured reflection coefficients of both antennas. The impedance bandwidth was 300 MHz and extended from 4.45–4.75 GHz. The simulated and measured gains of both antennas agree very well at 4.60 GHz (Fig. 8.35). As is evident, the measured gains of both antennas are 0.2 dB lower than the simulated gains for the *xoz*- and *yoz*-planes. The measured back lobe levels of both antennas are 7 dB lower than those of the simulations. On the *yoz*-plane, the measured sidelobe levels of both antennas at 120° and 240° are 7 dB higher than the simulated levels. The discrepancies between the measurements and simulations are caused by the following:

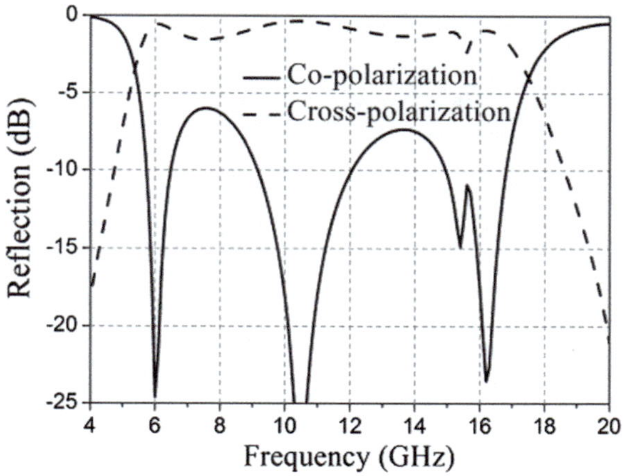

Fig. 8.32 Co- and cross-polarization reflections of the infinite periodic fishbone-shaped PRRS

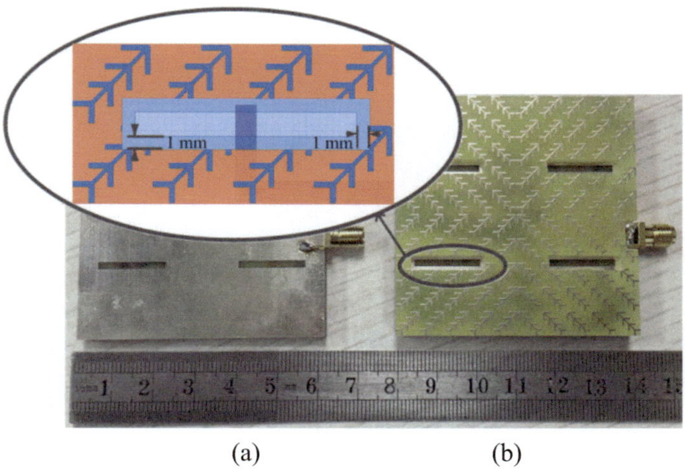

Fig. 8.33 Photographs of the fabricated **a** antenna without a PRRS and **b** antenna with a PRRS

(1) the background noise in the measurement setup;
(2) the SMA and cable loss that are not accounted for in the simulations;
(3) the misalignment between the PRRS and the antenna; and
(4) the fabrication tolerance.

For most applications, the gain loss achieved for the RCSR is acceptable.

Figure 8.36 shows the frequency responses of the simulated and measured monostatic RCSs for normally impinging plane waves with *x*- and *y*-polarizations. Good agreement between the simulated and measured results is evident. The experimental results indicate that a 5-dB RCSR for the antenna with the PRRS occurs from

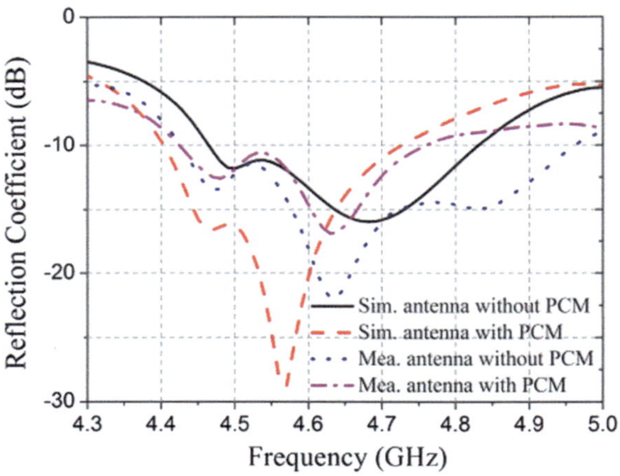

Fig. 8.34 Simulated and measured reflection coefficients of the antenna without the PCM and the antenna with the PCM

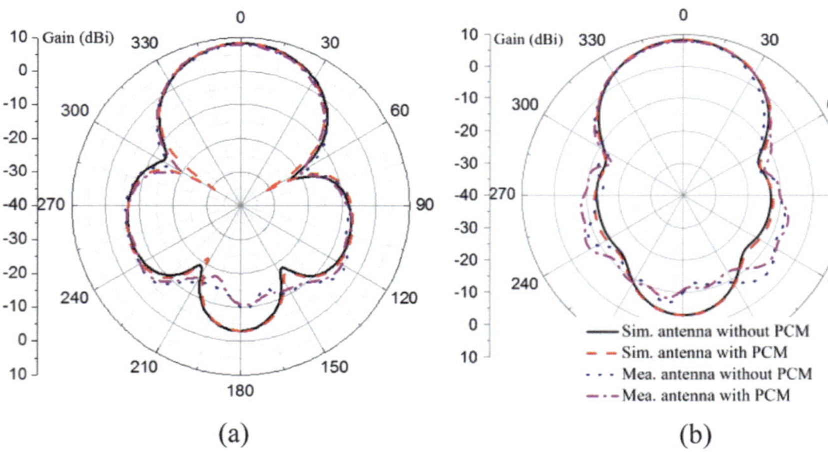

Fig. 8.35 Simulated and measured radiation patterns of both antennas at 4.60 GHz. **a** *xoz*-plane. **b** *yoz*-plane

6–18 GHz. However, discrepancies are also evident between the simulated and measured results. The reasons are because:

(1) the impinging wave in the simulation is a plane wave, whereas the far-field radiation from the horn aperture cannot be regarded as a standard plane wave;
(2) the imperfect nature of the absorbing material in the chamber; and
(3) the fabrication tolerance and misalignments in the measurement setup.

As shown in Fig. 8.37, the monostatic RCSs of both antennas with oblique impinging plane waves for *TE*- and *TM*-polarizations at 14.0 GHz are presented.

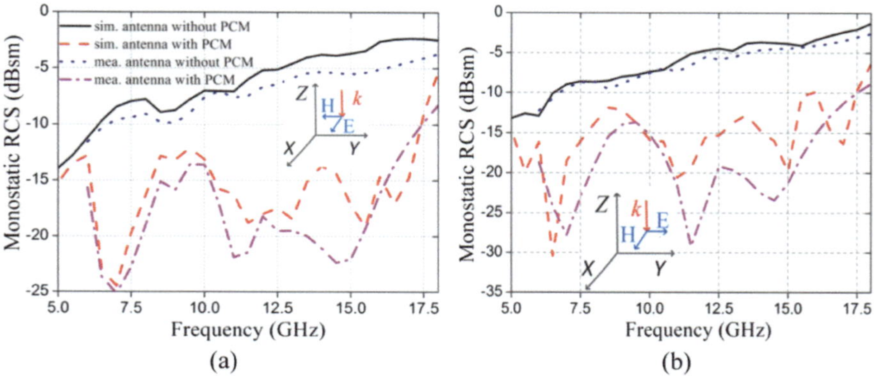

Fig. 8.36 Simulated (Sim.) and measured (Mea.) RCSs of both antennas versus frequency for normal incidence with **a** *x*-polarization (*E*-incident field along the *x*-axis) and **b** *y*-polarization (*E*-incident field along the *y*-axis)

A notable RCSR of the antenna with the PCM is achieved at angles from $-10°$ to $+10°$ for both polarizations. Moreover, after reaching a certain angle, the PRRS loses its ability to reduce the RCS.

The measured results show that the PRRS can effectively achieve a 5-dB RCSR over a wide band ranging from 6–18 GHz. Moreover, the RCSR method can be applied to other frequencies by changing the substrate properties or adjusting the dimensions of the fishbone-shaped unit cell. Consequently, the proposed design provides a new and general way to reduce the RCS of antennas across a wideband range.

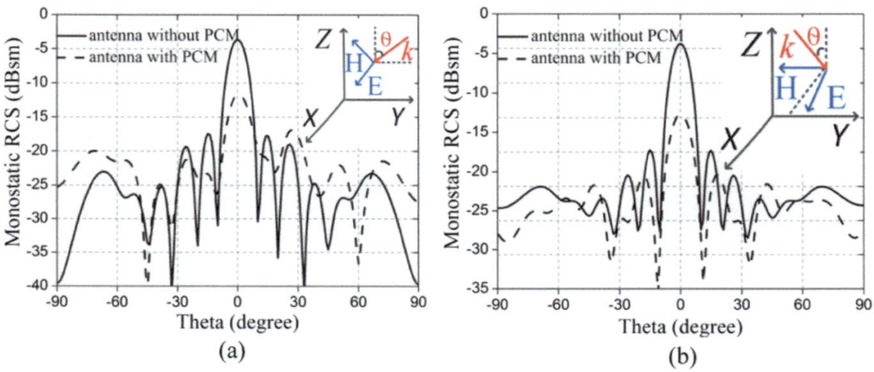

Fig. 8.37 Simulated monostatic RCSs of both antennas at 14.0 GHz with various angles of impinging waves with **a** TE polarization (*E*-incident field parallel to the *xoy*-plane, $\varphi = 90°$) and **b** *TM* polarization (*H*-incident field parallel to the *xoy*-plane, $\varphi = 0°$)

8.2.2 *In-Band RCS Reduction of a Slot Array Antenna Using a PRRS [37]*

The PRRS used for RCSR presented above operates out of the antenna's operating band, which results in an out-of-band RCSR. However, in-band RCSR is more desirable in practical applications. Consequently, the method of in-band RCSR using a PRRS is discussed in this section.

Here, we can use a slot array antenna resonating at 8.42 GHz as the reference, as shown in Fig. 8.38a. It comprised a 4 × 4 slot array, which is backed by a metallic ground and covered by a dielectric superstrate of thickness 2 mm. The length of the slot ls is 10 mm, and the distances between the adjacent slots in the *x*- and *y*-directions are both 12 mm. The relative permittivities of the substrate and superstrate are 2.65 and 2.2, respectively. The proposed low-RCS slot array antenna has the same parameters as those of the proposed antenna except for the PRRS structure, as shown in Fig. 8.38b.

The unit cell of the PRRS used in the proposed antenna is shown in Fig. 8.39. It comprised two pairs of L-shaped metallic patches printed on a dielectric substrate, which is backed by a metallic ground and covered by a superstrate. The superstrate comprises a dielectric layer and a pair of L-shaped metallic patches printed on the dielectric layer. The parameters of the unit cell of the proposed PRRS are listed in Table 8.2.

As shown in Fig. 8.40a, the working bandwidth of this PRRS is 7.8–34.7 GHz, in which the cross-polarization reflection coefficient is higher than 0.96. The calculated PCR versus frequency is shown in Fig. 8.40b. The calculated PCR is larger than 0.9 in the frequency range from 7.8–34.7 GHz.

The proposed configuration of the PRRS array used as the superstrate is shown in Fig. 8.41. The unit cells of the PRRS are arranged in four triangular areas with

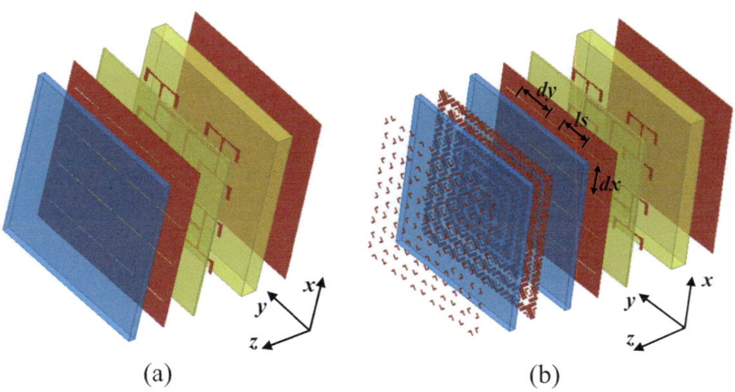

Fig. 8.38 Geometries of the **a** reference slot array antenna and **b** the proposed low-RCS slot array antenna

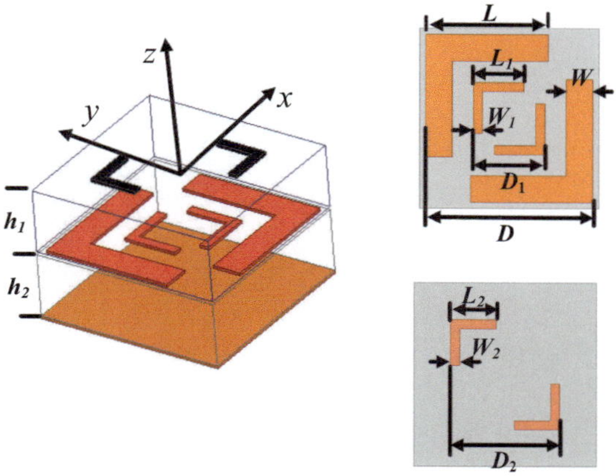

Fig. 8.39 3D schematic of a unit cell of the PRRS

Table. 8.2 Unit cell dimensions of the proposed PRRS (mm)

L	W	L_1	W_1	L_2	W_2	D	D_1	D_2	h_1	h_2
4.1	0.9	2.7	0.3	1.5	0.3	5.6	2.4	3.6	2	2

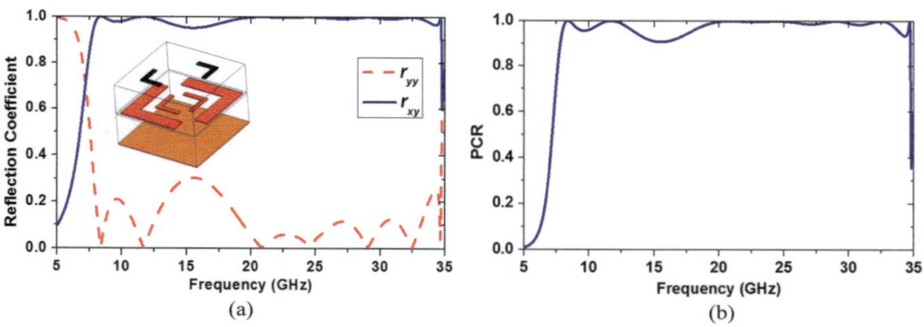

Fig. 8.40 **a** Simulated reflection coefficients r_{xy} (solid blue curve) and r_{yy} (dashed red curve) of the proposed PRRS at normal incidence of y-polarized EM waves. **b** Calculated PCR

orthogonal directions. When the PRRS is used as the superstrate, the metallic ground is removed, and the slot array antenna serves as the PRRS ground.

To investigate how the PRRS configurations influence the radiation and scattering performance of the slot array antenna, we designed four PRRS configurations, as shown in Fig. 8.42. The copolarization RCSs of the slot array antennas covered by the four different PRRS configurations are shown in Fig. 8.43. The result show that the smaller the period of the PRRS, the lower the RCSR obtained in the low-frequency range. When the period of the PRRS arranged in the same direction is two (Conf. 3), the RCSR is less than 2 dB at frequencies from 7–10 GHz. Consequently, to obtain good RCSR in the operating frequency band of the slot array antenna, the period of the PRRS should be at least four.

Fig. 8.41 Configuration of the proposed PRRS array

Fig. 8.42 Different superstrate configurations

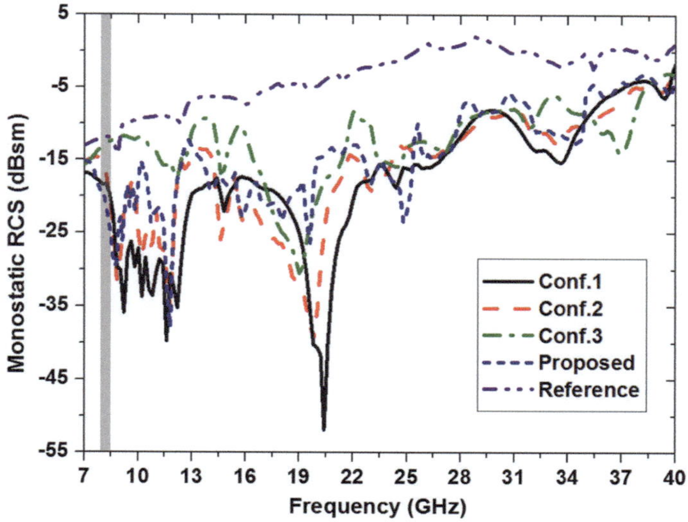

Fig. 8.43 Copolarization RCSs of the reference antenna and the slot array antennas covered by the four different configurations

The $|S_{11}|$ values and gains of the slot array antennas covered by the four PRRS superstrates are shown in Fig. 8.44. Figure 8.44a shows that all of the PRRS configurations increase the impedance bandwidth of the slot array antenna as the PRRS is generated by the slot, and an additional resonant frequency appears. However, the gain of the antenna is greatly affected by the superstrate, as shown in Fig. 8.44b. The gain of the antenna covered by Conf. 1 is higher than that of the reference antenna. It is worth noting that the antenna covered by Conf. 1 is circularly polarized (CP), and its axial ratio (AR) is less than 3 dB at 8.42 GHz.

The surface current distribution on the PRRS unit cell at 8.42 GHz when the PRRS is generated by the x-polarized incident wave is shown in Fig. 8.45. There are two orthogonal surface currents with equal magnitudes when the PRRS is generated by

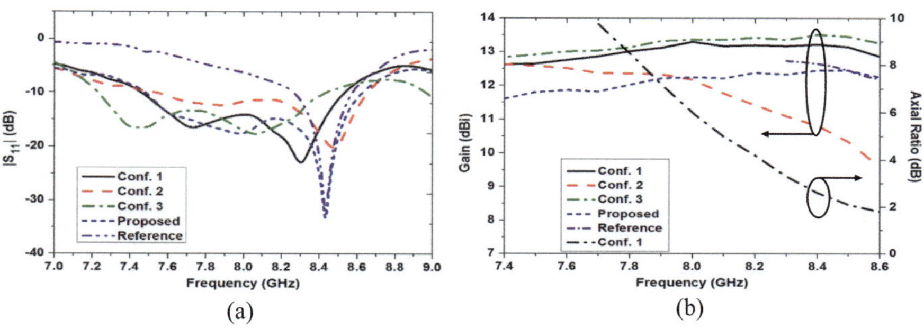

Fig. 8.44 Radiation performance of the reference antenna and the slot array antenna covered by different PRRS superstrates: **a** $|S_{11}|$, **b** gain and AR

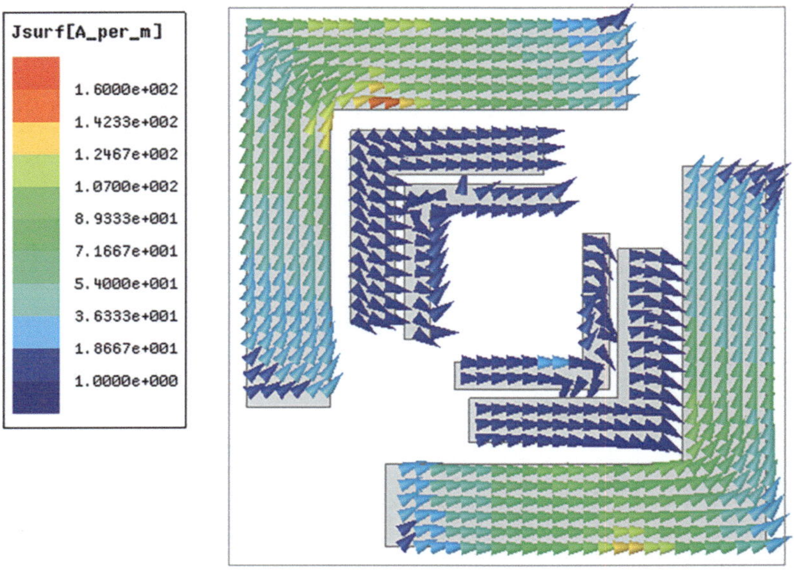

Fig. 8.45 Surface current distribution on metallic parts of the PRRS unit cell at 8.42 GHz

the x-polarized EM field, which generate two orthogonal electronic fields of equal magnitude. When the phase difference between the two orthogonal electronic fields is 90°, a CP EM wave is generated.

When the PRRS is arranged similar to Conf. 2, the left-handed and right-handed CP waves that radiate from the four different parts cancel each other out. As a result, the gain of the antenna covered by Conf. 2 is reduced considerably (by 2.41 dB) compared with that of the antenna covered by Conf. 1. Figure 8.42 shows that the PRRS cannot work at 8.42 GHz when the period is two as the RCS of the antenna covered by Conf. 3 is not reduced. Consequently, no CP waves are generated, and the gain does not decrease. For the proposed PRRS configuration, the x-polarized EM field radiated from the slot is in the diagonal direction of the PRRS unit cells. In this case, only an x-polarized electronic field can be generated. Thus, the gain of the antenna is not reduced much.

A comparison of the radiation and scattering performance of the reference antenna and the slot array antenna covered by different PRRS configurations at 8.42 GHz is listed in Table 8.3. Both the antenna covered by Conf. 1 and proposed antenna exhibit good radiation performance and in-band copolarization RCSR. However, the structure of Conf. 1 can only reduce the copolarization RCS under x-/y-polarized incident waves, while the proposed structure can reduce both the copolarization and total RCS of the antenna under arbitrarily polarized incident waves. Furthermore, the superstrate arranged as Conf. 1 converts the linearly polarized (LP) EM waves to CP EM waves, which is undesirable. The gain of the antenna covered by Conf. 2 is substantially reduced, although the in-band RCS of Conf. 3 is only reduced by

0.55 dB. Consequently, the proposed PRRS structure was selected as the final solution to balance the radiation and scattering performances of the slot array antenna.

The monostatic RCS under oblique incident waves and the bistatic RCS of the proposed antenna at 8.42 GHz were also investigated. A comparison of the monostatic RCS between the reference antenna and the proposed antenna is shown in Fig. 8.46. The monostatic RCS is reduced at angles of $\pm 15°$ for both x- and y-polarized incident waves. Regardless of the dielectric loss, the PRRS is lossless, and the scattered energy is not reduced. Consequently, to investigate the energy distribution of the scattered energy, a comparison of the bistatic RCSs of the reference antenna and the proposed antenna at 8.42 GHz was conducted, as shown in Fig. 8.47.

The comparison shows that the main lobe of the scattering pattern is considerably reduced, and two side lobes form in the x-direction, which indicates that the energy is scattered in nonthreatening directions.

The reference and proposed antennas were fabricated and measured to validate the simulated results. Photographs of the reference and proposed antennas are shown in Fig. 8.48a. The simulated and measured $|S_{11}|$ values are shown in Fig. 8.48b. The measured $|S_{11}|$ of the proposed antenna are at higher frequencies, which is caused by installation errors and the air gap between the superstrate and slot array antenna. The

Table. 8.3 Comparison of the radiation and scattering performance between the reference antenna and slot array antenna covered by PRRS

Configuration	Gain (dBi)	Relative bandwidth (%)	In-band copolarization RCS (dBsm)
Conf. 1	13.22	13.39	−19.67
Conf. 2	10.81	13.7	−20.30
Conf. 3	13.51	15.3	−12.50
Reference	12.65	3.5	−11.95
Proposed	12.45	14.20	−24.32

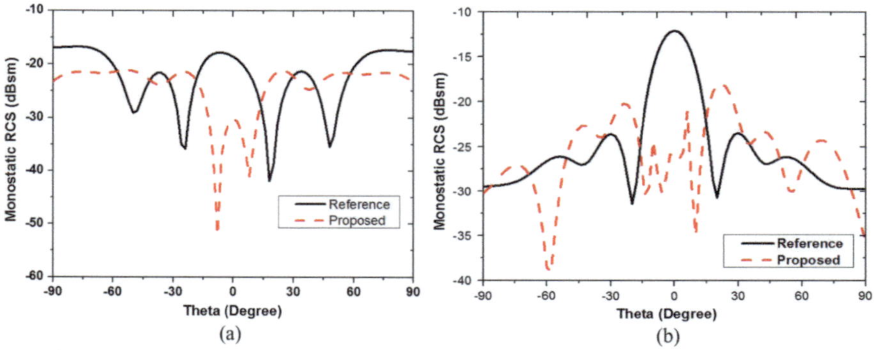

Fig. 8.46 Monostatic RCSs of the reference and proposed antennas for oblique impinging plane waves in the xoz-plane with **a** x-polarization and **b** y-polarization

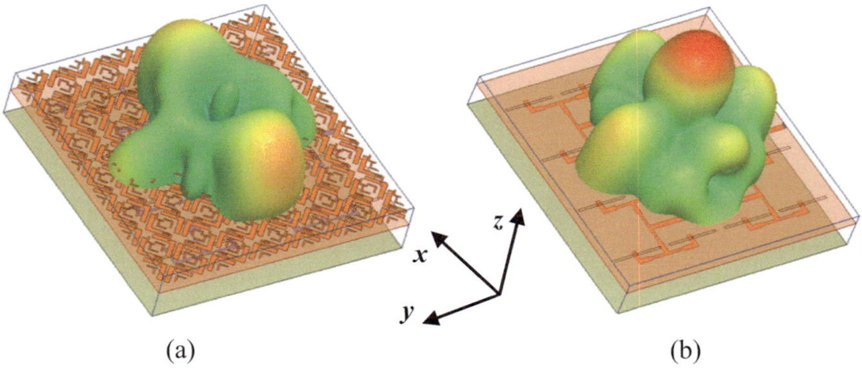

Fig. 8.47 Bistatic RCSs at 8.42 GHz: **a** proposed antenna and **b** reference antenna

radiation patterns in the *xoz*- and *yoz*-planes are shown in Fig. 8.49. The measured gain of the proposed antenna is 0.3 dB lower than the simulated results (caused by the SMA connector and dielectric loss).

The simulated and measured RCSs of the reference and proposed antennas for normally impinging plane waves are shown in Fig. 8.50. Owing to restrictions in the test conditions, the RCSs were not measured at frequencies higher than 18 GHz. The results show that the RCS is considerably reduced from 7–40 GHz and that a reasonable agreement between the simulated and measured results is obtained, which indicates good scattering performance of the proposed antenna.

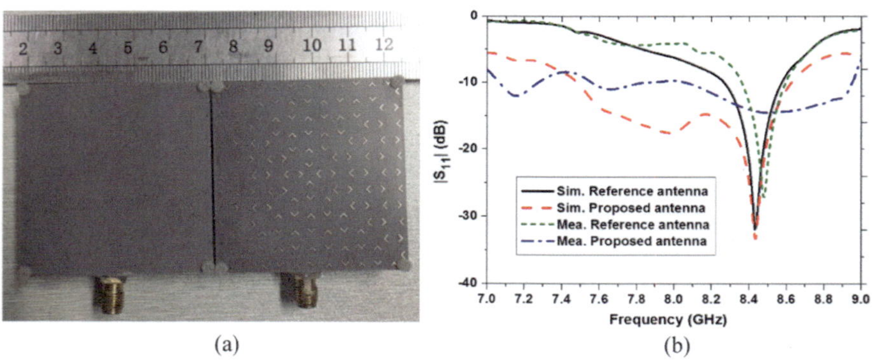

Fig. 8.48 **a** Photographs of the reference and proposed antennas, **b** simulated and measured $|S_{11}|$ values of the antennas

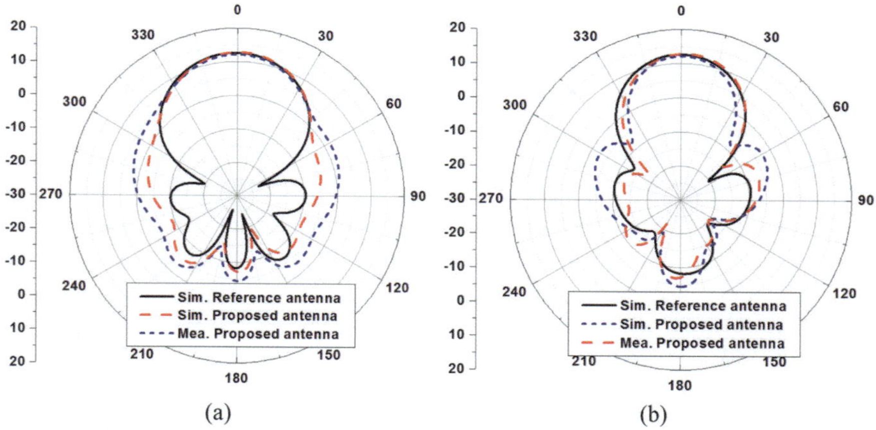

Fig. 8.49 Simulated (Sim.) and measured (Mea.) radiation patterns at 8.42 GHz in **a** the *xoz*-plane and **b** the *yoz*-plane

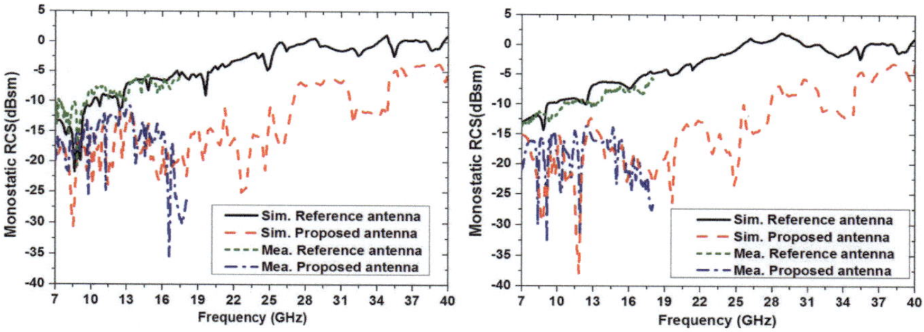

Fig. 8.50 Simulated (Sim.) and measured (Mea.) RCSs of the reference and proposed antennas for normally impinging plane waves with **a** *x*-polarization and **b** *y*-polarization

8.2.3 Low-RCS CP Antenna Using a PRRS [38]

By using the phase-modulating characteristics of a PRRS, the design of a CP antenna with low RCS is proposed in this section. The structure of the proposed CP antenna and the unit cell of the PCM are shown in Fig. 8.51. The parameters of the proposed antenna are listed in Table 8.4. The upper and lower dielectric substrates were composed of Arlon AD450 with a relative permittivity of 4.5 and a loss tangent of 0.0035. Three types of patterned artificial magnetic conductors were printed on the upper substrate. The unit cell comprised a square patch, a ground plane, and two asymmetric metallic vias loaded in the diagonal direction.

Here, we briefly describe the CP radiation mechanism of the proposed CP antenna again. The new unit cell of the PCM is enlarged and shown in Fig. 8.51b. The

Table. 8.4 Dimensions of the CP antenna and polarization conversion metasurface (unit: mm)

L	t_1	t_2	W_S	L_S	W_m	L_m	W_p	p	D	d_1	d_2	g
64	3	0.5	2	36	0.94	37	39	10	0.6	3	1.8	0.5

x-polarized electric field of the EM wave coupled from the feeding slot can be decomposed into two orthogonal components—that is, $\overrightarrow{E_1}$ and $\overrightarrow{E_2}$.

For the $\overrightarrow{E_1}$ component, the PCM can be regarded as an RLC circuit, as shown in Fig. 8.52a. However, for the $\overrightarrow{E_2}$ component, the PCM can be considered to be another RLC circuit because the two metallic vias produce a parallel inductance (L_g), as shown in Fig. 8.52b. The parallel inductance (L_g) results in the equivalent circuit shown in Fig. 8.52b being less inductive than that shown in Fig. 8.52a. Consequently, the phase of $\overrightarrow{E_1}$ leads that of $\overrightarrow{E_2}$ after passing through the PCM. Thus, when the magnitudes of the two induced orthogonal currents are equal and the phase difference between them is 90°, the proposed CP antenna radiates a right-handed circular polarization wave. By contrast, if the two metallic vias are loaded in the other diagonal direction, the proposed antenna radiates a left-handed circular polarization wave.

The simulated $|S_{11}|$ values and boresight gain of the proposed CP antenna are shown in Fig. 8.53a. The proposed CP antenna achieves a −10-dB impedance bandwidth of 21%, and the peak boresight gain is 9.3 dBi at 5.42 GHz. Figure 8.53b shows

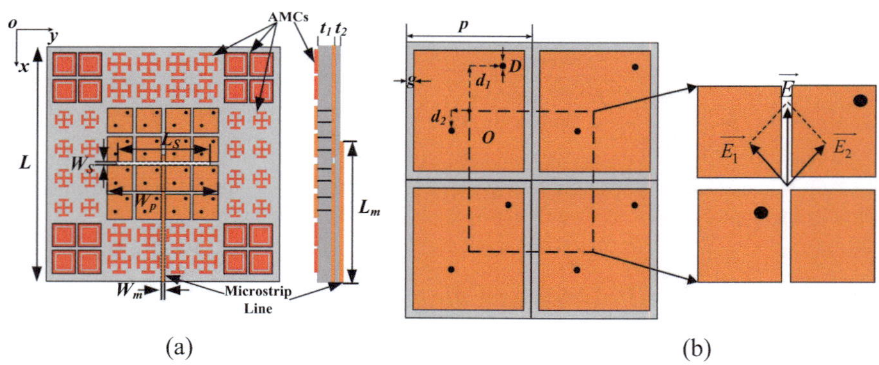

Fig. 8.51 Geometry of the proposed CP antenna and the polarization conversion metasurface

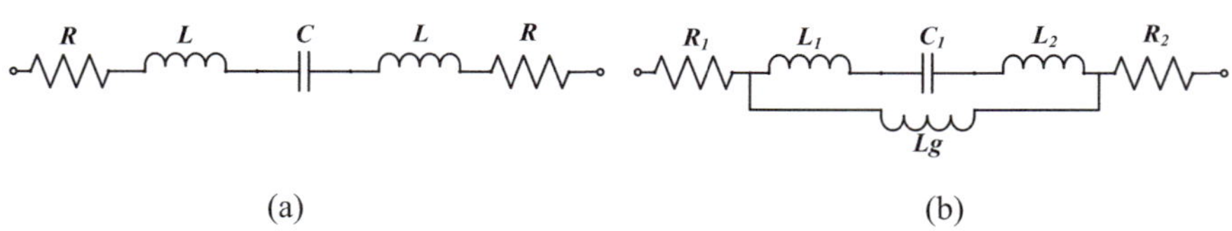

Fig. 8.52 Equivalent circuits of the new unit cell seen by **a** $\overrightarrow{E_1}$ and **b** $\overrightarrow{E_2}$

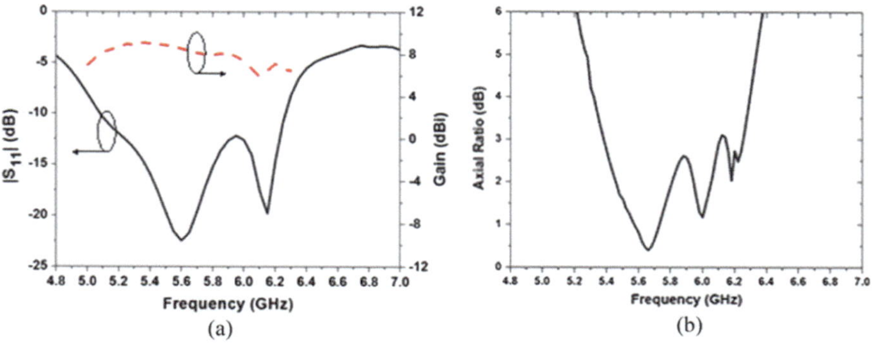

Fig. 8.53 Simulated radiation performance of the proposed antenna. **a** $|S_{11}|$ and boresight gain. **b** Axial ratio

that the simulated 3-dB AR bandwidth of the proposed antenna is 14.9%. The −10-dB impedance bandwidth and the 3-dB AR bandwidth overlap from 5.37–6.25 GHz.

By combining an artificial magnetic conductor and a PEC or two different types of artificial magnetic conductors with a 180° phase difference in a checkerboard-like configuration, the reflections from the two different surfaces cancel each other out under a normal incident plane wave, thus reducing the monostatic RCS in the normal direction. The RCSR for the combination of an artificial magnetic conductor and a PEC structure can be obtained only in a narrow frequency band. The reflection phase of the proposed PCM is similar to that of a PEC outside its operating band. Thus, if only one type of artificial magnetic conductor structure is used, the bandwidth of the RCSR will be narrow. Accordingly, to achieve a wideband RCSR, three different types of artificial magnetic conductors and the PCM are used.

The monostatic RCSR principle using four different structures is shown in Fig. 8.54. The directions and lengths of the arrows represent the reflection phases and magnitudes of the four different structures. Because the areas occupied by the four structures are the same, the magnitudes of the reflected waves are equal, and the four arrows have the same lengths. As shown in Fig. 8.54a, when the phase differences among the reflected waves are 90°, the reflected waves cancel each other out, and there is no reflection in the backward direction. However, it can be difficult to keep the phase difference at 90° over a wide frequency band. Consequently, we can assume that the reflection phases vary in the adjacent 90° ranges. Under these conditions, the total magnitude of the reflected waves reaches a maximum value when the reflection phases those as shown in Fig. 8.54b. Thus, the monostatic RCS can be reduced by at least 6 dB when the reflection phases of the artificial magnetic conductors and PCM vary in adjacent 90° ranges.

Based on the principle of monostatic RCSR, three different artificial magnetic conductors were designed, the reflection phases of the artificial magnetic conductors and PCM being as shown in Fig. 8.55. The reflection phase of the proposed PCM structure changes considerably at frequencies from 5–7 GHz because the proposed PCM structure resonates in this frequency range.

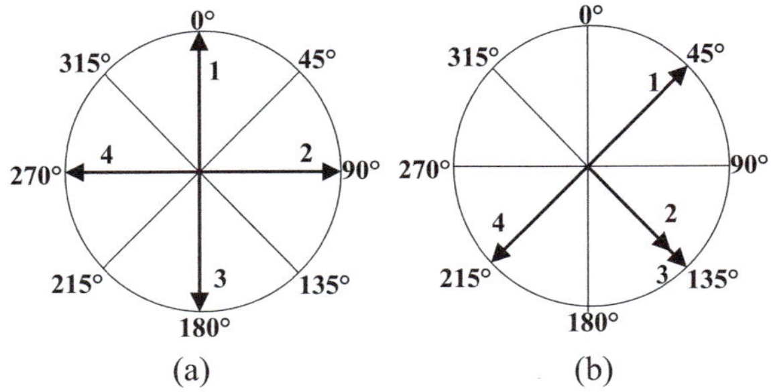

Fig. 8.54 Intuitive schematic of RCSR when the reflection phases of the artificial magnetic conductors and polarization conversion metasurface vary in adjacent 90° ranges: **a** the case in which the RCS reaches a minimum and **b** the case in which the RCS reaches a maximum

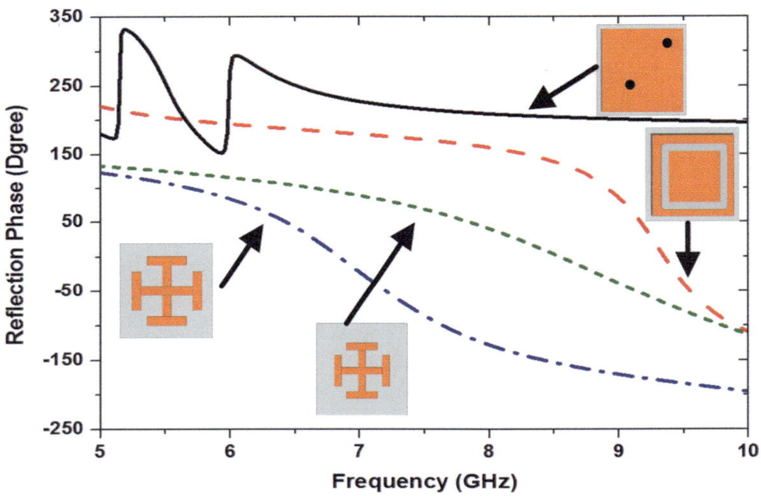

Fig. 8.55 Reflection phases of the artificial magnetic conductors and polarization conversion metasurface

RCS comparisons between the proposed CP antenna, the proposed CP antenna without AMCs, and a conventional CP patch antenna were conducted. Figure 8.56 shows the RCS comparison between the three antennas under normally-incident waves. The RCS of the CP antenna without AMCs is lower than that of the conventional CP antenna because there is a phase difference between the waves reflected from the PCM and the PEC of the antenna. However, the maximum reflected phase difference between them is less than 90°. Consequently, the RCSR is small. The RCS of the proposed CP antenna is considerably lower because the waves reflected from the four different areas cancel each other out.

To further investigate the principle of RCSR, a comparison of the bistatic RCSs of the three antennas is shown in Fig. 8.57. Compared with the reference antenna, the

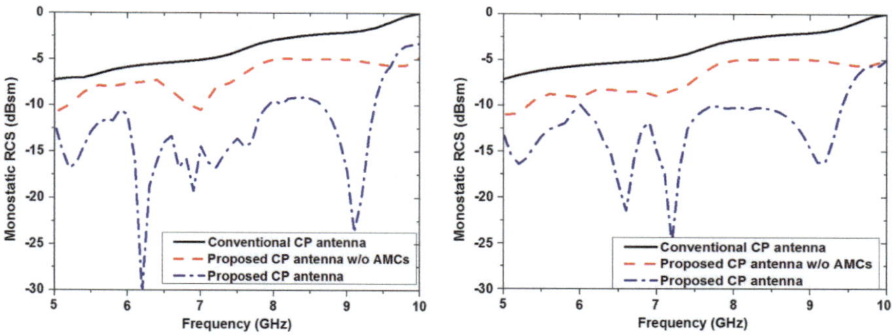

(a) (b)

Fig. 8.56 Monostatic RCS of the conventional CP patch antenna, the proposed CP antenna, and the proposed CP antenna without AMCs for normal impinging plane waves with **a** *x*-polarization and **b** *y*-polarization

main lobe of the proposed antenna's scattering pattern is considerably reduced, and the RCS in the other directions increases, which indicates that the energy is scattered in nonthreatening directions.

A prototype of the proposed CP antenna was fabricated and measured to validate the simulated results. Figure 8.58 shows a comparison between the measured and simulated radiation performance of the antenna. Reasonable agreement is evident between the measured and simulated results.

In summary, the proposed antenna could simultaneously achieve a high gain, low profile, and low RCS because its aperture efficiency was higher than the other antenna types. The 4×4 cell array of the PC cells increased the radiation area of the antenna. Additionally, the radiator of the antenna acted as part of the structure for RCSR rather than only as a radiator. Consequently, it could achieve a higher gain with a smaller aperture and a lower profile.

8.2.4 High-Gain and Low-RCS CP Antenna Based on a CPCM [39]

A higher gain is usually desired when designing antennas. This section introduces a design method for high-gain and low-RCS antennas using a Fabry–Perot (FP) cavity and chessboard polarization conversion metasurfaces (CPCMs).

Figure 8.59 shows a schematic view of the proposed antenna with a CPCM. A CP slot array antenna excited by a sequentially-rotated feeding network was chosen as the source antenna, and a CPCM was placed over it at a distance of H. The operating frequency of the antenna was chosen to be from 8.5–9.5 GHz, and the low-RCS band was from 6–14 GHz. As shown in Fig. 8.59, the CPCM comprises four parts—that is, Parts I and III are PCMs, and parts II and IV are their mirror images.

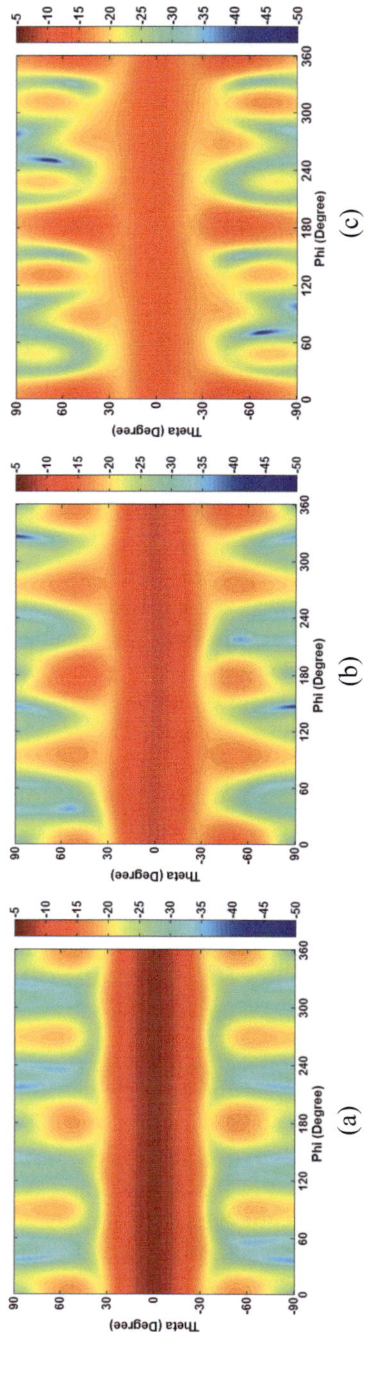

Fig. 8.57 Bistatic RCSs of **a** the conventional CP patch antenna, **b** the proposed CP antenna without AMCs, and **c** the proposed CP antenna at 5.2 GHz

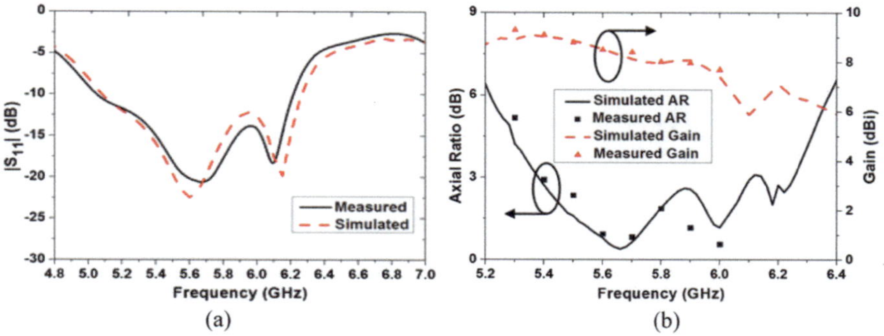

Fig. 8.58 **a** Measured and simulated $|S_{11}|$ values. **b** Boresight gain and AR

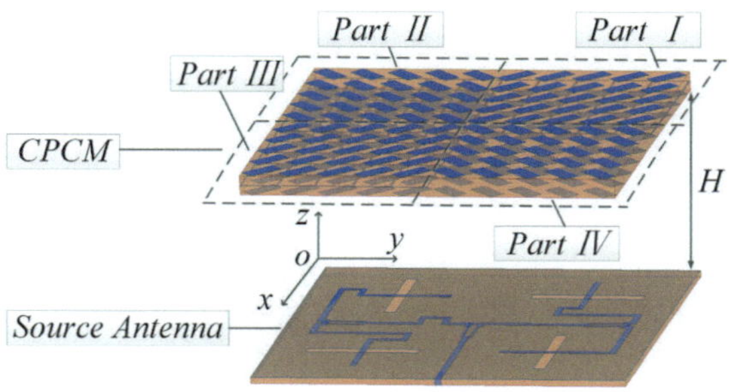

Fig. 8.59 Schematic view of the antenna with a CPCM

Figure 8.60a shows an enlarged view of the PCM with an LP source. The proposed PCM unit cell is a sandwich structure inspired by Muhammad et al. [40]; it has two adjoining substrates with three metallic patterns (Fig. 8.60b). As shown in Fig. 8.60c, the top and bottom metallic layers are both rectangular pieces of copper, while the middle metallic layer is a square ring. To ensure the same characteristics for both *x*- and *y*-polarized waves, the geometry of the unit cell was chosen to be symmetrical with respect to the +45° diagonal (Fig. 8.60c). An infinite periodic boundary condition was simulated by using a floquet port and master/slave boundaries. Taconic TLA-35 ($\varepsilon_r = 3.5$, $\tan\delta = 0.003$) was used as the substrate.

Each PCM comprises 5 × 5 unit cells that perform three functions. First, a CP wave is generated when an LP wave passes through the PCM. Second, gain enhancement is achieved by the FP resonant cavity formed by the PCM and the ground of the slot antenna. Third, the PCM backed by the ground of the slot antenna can achieve a reflective wideband polarization conversion.

However, it can be difficult to optimize the three functions simultaneously. Thus, the objective is to achieve a tradeoff between these features. Accordingly, the PCM can be designed as follows:

(1) Design the unit cell of the PCM (Fig. 8.60b) to cover the operating band of the antenna (8.5–9.5 GHz), including the initial parameters.

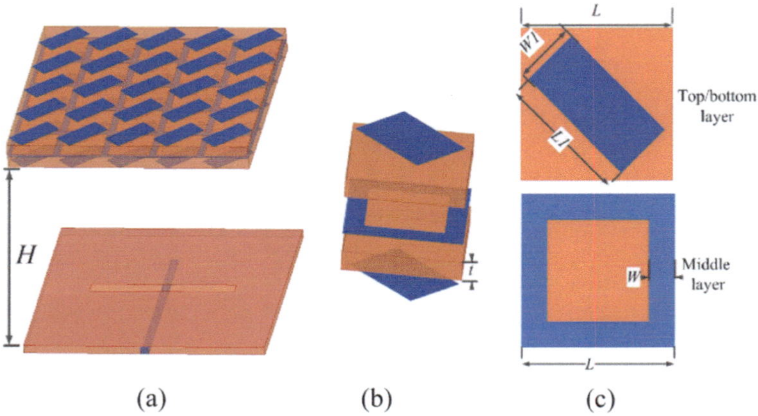

Fig. 8.60 **a** A slot antenna with a PCM; **b** 3D topology of the PCM unit cell; and **c** top view of the PCM unit cell

(2) Optimize the co- and cross-polarization transmission coefficients of the unit cell to convert an LP wave into a CP wave.
(3) Optimize the x- and y-components of the reflection coefficients to satisfy the FP resonant condition for circular polarization.
(4) Optimize the unit cell backed with the PEC to achieve broadband polarization conversion.
(5) Repeat steps 2, 3 and 4 until a desired tradeoff is achieved.

The optimization process shows that the parameters L_1, W and W_1 are most sensitive to the three features discussed above. Based on the design process, the final geometric parameters of the unit were obtained: $L = 6$ mm, $W_1 = 2.6$ mm, $L_1 = 5.8$ mm, $t = 1.524$ mm, and $W = 0.4$ mm.

Two prerequisites should be considered to obtain a CP wave after an LP wave passes through a PCM [41]. First, the phase difference between the copolarization transmission coefficient T_{xx} and the cross-polarization component T_{xy} should be approximately 90° from 8–10 GHz (Fig. 8.61a). Second, the difference between them should be less than 3 dB (Fig. 8.61b).

Figure 8.62 shows the simulation results of the reflection coefficients, the boresight left-hand circularly polarized (LHCP)/LP gain and AR versus frequency of the slot antennas with and without the PCM. In the impedance bandwidth from 8.75–9.35 GHz, the boresight LHCP gain of the slot antenna with the PCM is higher than the LP gain of the standalone slot antenna. Moreover, the AR of the slot antenna with the PCM is less than 3 dB from 8.75–9.25 GHz. The simulated results demonstrate that the PCM has the ability to convert an LP wave into an LHCP wave, which is based on the transmission characteristics of the PCMs in Orr et al. [42]. Moreover, the gain of the antenna also improves; the reasons for which are explained below.

Gain enhancement of the slot antenna with PCM can be realized by the FP resonance effect. Based on the design guidelines of the FP antenna [43], the circularly

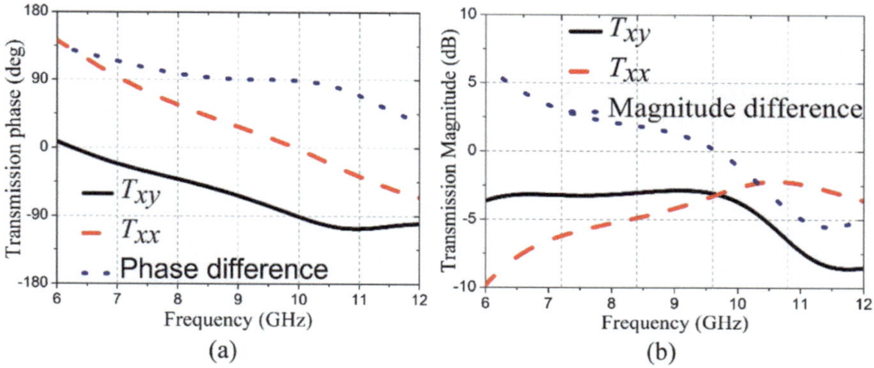

Fig. 8.61 Co- and crosspolar transmission coefficients T_{xx} and T_{xy}: **a** phase and phase difference, and **b** magnitude and difference in magnitude

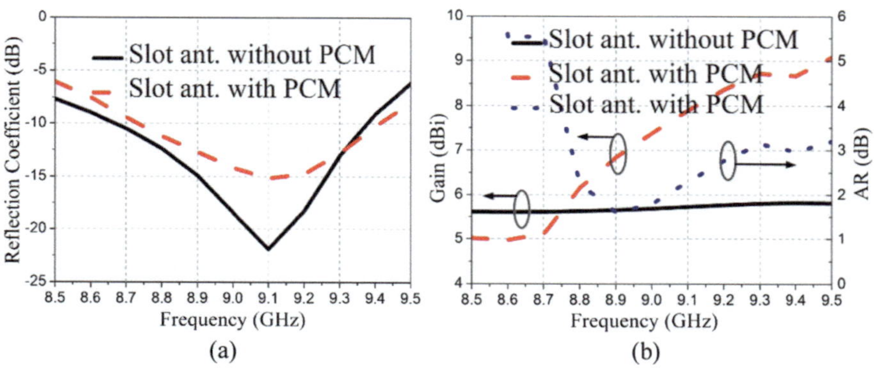

Fig. 8.62 Simulated **a** reflection coefficients and **b** boresight LHCP/LP gain and AR versus frequency of the slot antenna (ant.) with and without the PCM

polarized FP (CPFP) resonance condition should satisfy the resonance requirement for both the x- and y-polarization components. As described in Zhu et al. [41], the reflection coefficients of the PCM in the x- and y-directions should be similar to meet the same resonance conditions. Moreover, the resonance cavity height (H) can be expressed as follows:

$$H = \frac{c}{f_0}\left(\frac{\Phi_{1x} + \Phi_{2x} + 2n\pi}{4\pi}\right) = \frac{c}{f_0}\left(\frac{\Phi_{1y} + \Phi_{2y} + 2n\pi}{4\pi}\right) \tag{8.8}$$

where H denotes the height, c *denotes* the speed of light in a vacuum, n denotes an integer ($n = 0, 1, 2, \ldots$), Φ_{1x} and Φ_{1y} denote the reflection coefficient phases of the antenna ground in the x- and y-directions, respectively, and Φ_{2x} and Φ_{2y} denote the corresponding phases of the PCM.

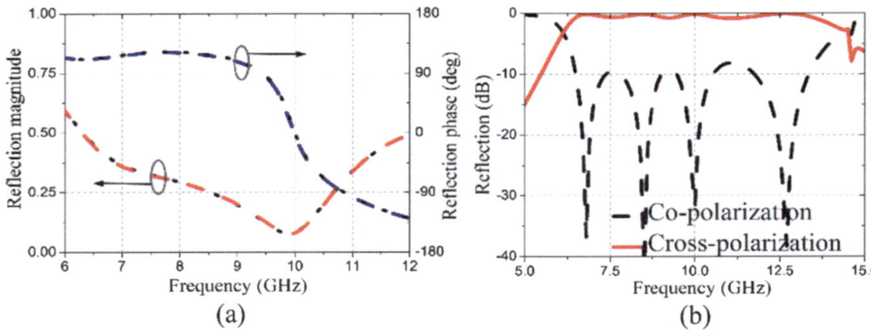

Fig. 8.63 **a** Simulated reflection magnitudes and phases of the *x*- and *y*-components and **b** simulated co- and cross-polarization reflections of the PCM

Assuming that n for both polarizations has the same value, then Eq. (8.8) implies that:

$$\Phi_{1x} + \Phi_{2x} = \Phi_{1y} + \Phi_{2y} \tag{8.9}$$

Equation (8.9) indicates that the sum of the reflection coefficient phases must be equal in the two directions to meet the same resonance conditions for both polarizations. Furthermore, the reflection coefficient phases of the antenna ground in the *x*- and *y*-directions are equal to π, which means $\Phi_{1x} = \Phi_{1y} = \pi$. Consequently, Φ_{2x} should be equal to Φ_{2y} to achieve a high-gain CP antenna using the FP resonance cavity principle.

Figure 8.63a shows the simulated reflection coefficient magnitudes and phases of the PCM. The reflection magnitudes in the *x*- and *y*-directions and the reflection phases share the same curve, which can be explained by the symmetrical structure with respect to the +45° diagonal. The same curves for both reflection phases and magnitudes demonstrate that the PCM has the ability to construct the CPFP resonant cavity, which is demonstrated by the radiation improvement in Fig. 8.62b.

The CPCM is the transmission structure, which enables a much easier way to analyze the antenna's backside scattering feature compared to other models. The PCM unit cell was simulated with the PEC ground. The distance between the unit cell and the PEC was optimized as 11 mm. Figure 8.63b shows the co- and cross-polarization reflections of the infinite PCM unit cell backed with the PEC ground. It is evident that the cross-polarization component of the reflection coefficients becomes the major component, while the copolarization component loses its dominance from 6–14 GHz. Consequently, the chessboard configuration of the PCM was chosen because of its wideband low-RCS feature.

As shown in Fig. 8.62b, LHCP radiation is generated when the PCM is placed over the slot antenna (Fig. 8.62a). Similarly, right-hand circular polarization (RHCP) radiation can be generated by replacing the PCM with a mirror PCM. Accordingly, considering the arrangement of the CPCM, each antenna element beneath its corresponding PCM/mirror PCM should not be arranged in the same orientation or fed by in-phase excitation. If two types of PCMs are excited by slots of the same orientation, the addition of the LHCP and RHCP radiation from the neighboring PCMs results in a decrease in gain in the boresight direction and creates four grating lobes, which can be explained using the array antenna theory [43].

To mitigate this type of interference, a CP slot array antenna based on a sequentially rotated feeding network was chosen as the source antenna. The sequentially rotated feeding network, which excited the four orthogonal slots, was printed on the bottom surface of the 1-mm-thick substrate. The neighboring slots were excited with the same magnitude and 90° phase difference. Only a LHCP wave is produced after each LP wave passes through the PCM or mirror PCM. Thus, there is no gain degradation in the boresight direction. Consequently, good scattering features can be achieved without sacrificing the radiation characteristics by using such a source.

Considering the CP radiation performance of the reference antenna, it appears to be unnecessary to use the CPCM to produce LHCP radiation. However, the antenna with the CPCM has a broader AR bandwidth because better polarization purity can be achieved by replacing the four LP elements with four CP elements. Considering the scattering features, the objective of achieving the wideband low RCS can be realized by using the CPCM. Because of the symmetrical structure of the CPCM, the RCSR is similar for x- and y-polarized incidence waves. Thus, the radiation performance and scattering feature tradeoff can be maintained by codesigning the radiation source and suitable arrangement of the CPCM.

To achieve the high-gain and wideband low-RCS CP antenna, the CPCM was placed above the source antenna, and four nylon pillars were used to support it. The CPCM and the source antenna shared the same (metal) dimensions of 60 $\times$ 60 mm. However, considering some space for the nylon pillars, the substrate dimensions were 66 $\times$ 66 mm, as shown in Fig. 8.64. The optimized distance (H) between the CPCM and the ground of the antenna was 11 mm.

Figure 8.65a shows that the simulated and measured reflection coefficients for both antennas are similar. The impedance bandwidth is 1.25 GHz (13.8%), and it extends from 8.25–9.5 GHz. Figure 8.65b shows the simulated and measured boresight gain versus frequency. The LHCP gain of the antenna with the CPCM is higher than that of

Fig. 8.64 Photograph of the fabricated antennas with and without the CPCM

the antenna without the CPCM within the operating band from 8.5–9.5 GHz, which covers the 3-dB gain decrease. The simulated LHCP gain of the antenna with the CPCM is approximately 4.8 dBi higher than that of the antenna without the CPCM at 9.2 GHz. The simulated normalized LHCP and RHCP radiation patterns for both antennas at 9.2 GHz are shown in Fig. 8.66, and the measured radiation patterns of both antennas are shown in Fig. 8.67.

The discrepancies between the simulations and measurements are due to the background noise in the measurement setup, fabrication tolerances, and assembly errors. The radiation characteristics shown in the figures are consistent with each other, demonstrating that the presence of the CPCM improves the radiation performance.

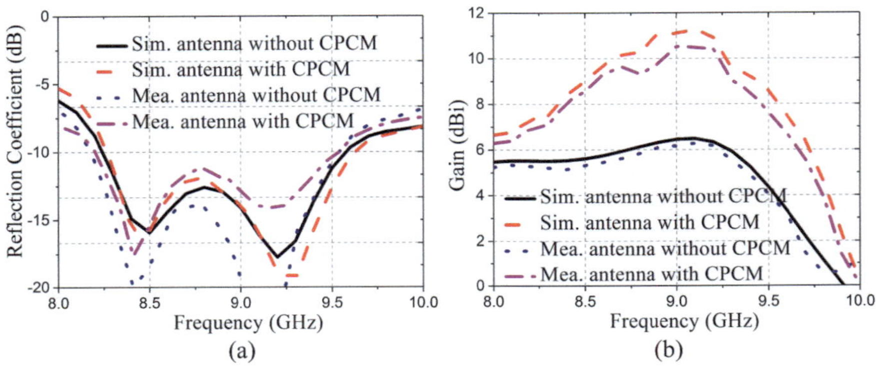

Fig. 8.65 Simulated (Sim.) and measured (Mea.) reflection coefficients and gains versus frequency of both antennas

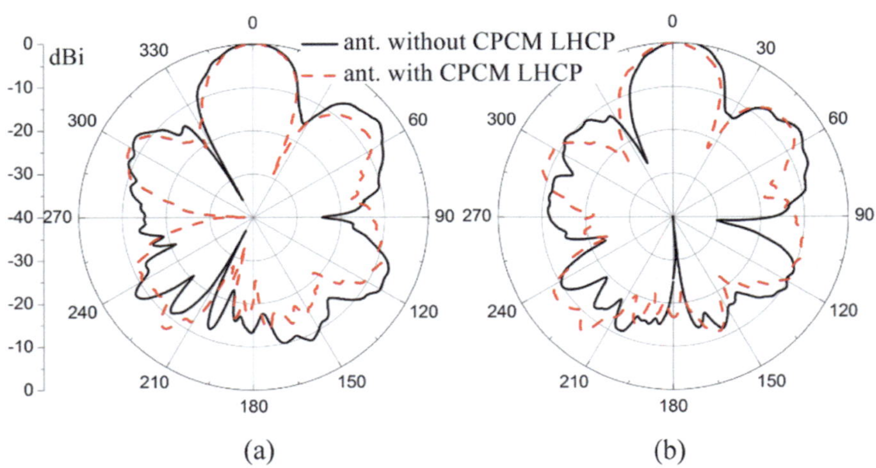

Fig. 8.66 Simulated normalized radiation patterns of both antennas (ant.) at 9.20 GHz: **a** *xoz*-plane and **b** *yoz*-plane

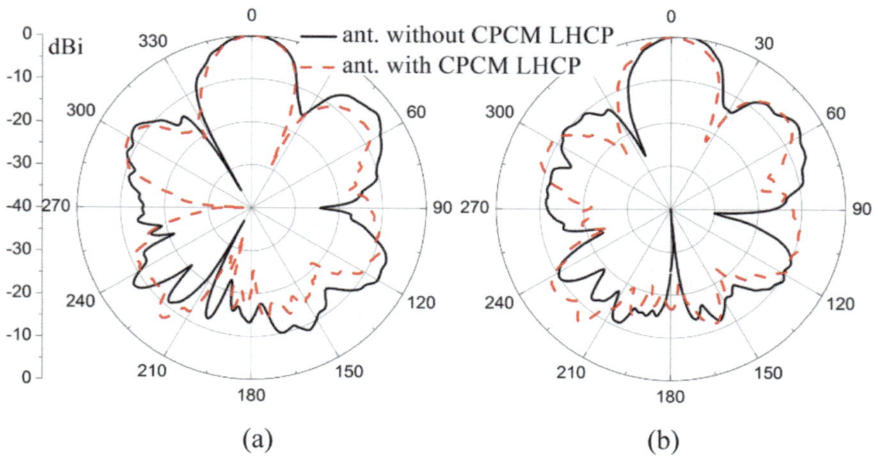

Fig. 8.67 Measured normalized radiation patterns of both antennas (ant.) at 9.20 GHz: **a** *xoz*-plane and **b** *yoz*-plane

Figure 8.68a shows the measured and simulated boresight AR curves of both antennas, which indicate that the 3-dB AR bandwidth can be broadened by using the CPCM. A combined operating bandwidth (3-dB AR, 3-dB gain drop, and −10-dB impedance matching) from 8.5–9.5 GHz with a maximum realized boresight LHCP gain of 11.2 dBi is achieved. The measured results are consistent with those of the simulations. Figure 8.68b and c show the simulated 3D bistatic scattered fields of both antennas at 13.0 GHz under normal incidence. The energy in the normal direction is redistributed in the four grating lobes with the directions $(\varphi, \theta) = (45°, 24°)$, $(135°, 24°)$, $(225°, 24°)$ and $(315°, 24°)$.

Figure 8.69 shows the frequency response of the simulated and measured monostatic RCSs for normally impinging plane waves with *x*- and *y*-polarizations. The

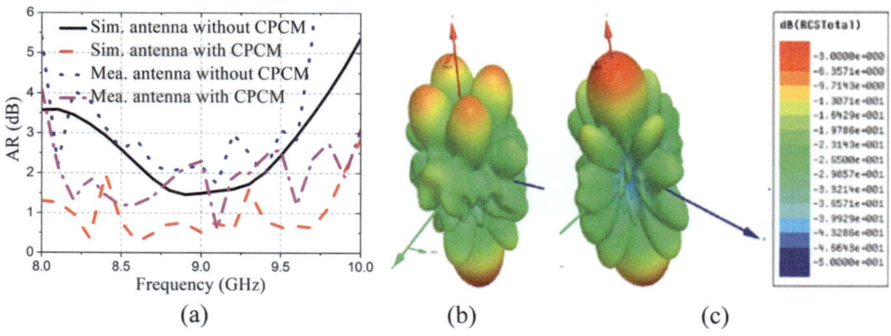

Fig. 8.68 **a** Simulated (Sim.) and measured (Mea.) AR versus frequency of both antennas; 3D bistatic scattered field at 13.0 GHz under normal incidence for **b** the antenna with the CPCM and **c** the antenna without the CPCM

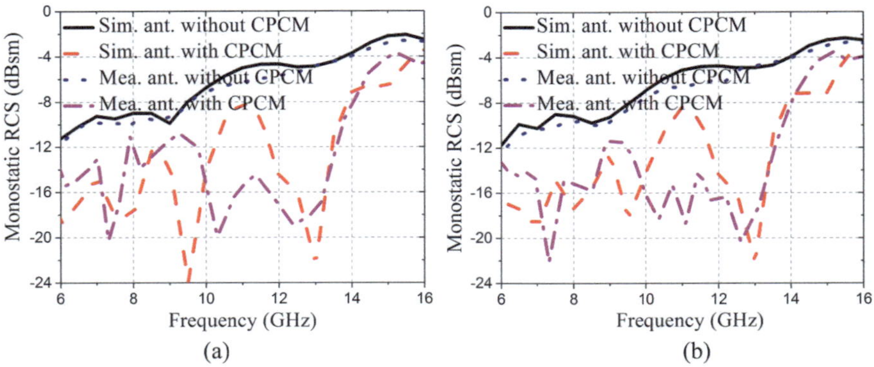

Fig. 8.69 Simulated (Sim.) and measured (Mea.) RCSs of both antennas (ant.) as a function of the frequency for normal incidence with **a** x-polarization (E-incident field along the x-axis) and **b** y-polarization (E-incident field along the y-axis)

realized RCSR bandwidth is from 6–14 GHz, while the simulated RCSR values peak at 7.4, 9.5, 10.2, and 13 GHz. The RCSR band and the maximum RCSR frequencies are consistent with the polarization conversion band and polarization resonant frequencies in Fig. 8.63b, respectively. Moreover, the maximum RCSR at 13 GHz is at least 14 dB for both polarizations.

8.3 Summary

This chapter introduced a method for designing wideband PRRSs. By loading a dielectric superstrate and optimizing its thickness, the bandwidth and polarization rotation efficiency could be considerably enhanced. The PRRS could be used for wideband RCSR when arranged in a chessboard-like configuration. In contrast to the AMC, only one kind of unit cell was needed for wideband RCSR, simplifying the design procedure. Additionally, the PRRS could be used for wideband RCSR. By using the phase-modulating characteristics of the PRRS, CP antennas with low RCS could also be achieved.

References

1. C L Holloway, E F Kuester, J A Gordon, *et al.* An Overview of the Theory and Applications of Metasurfaces: The Two-Dimensional Equivalents of Metamaterials. IEEE Antennas and Propagation Magazine, Vol. 54, No. 2, pp. 10–35, 2012.
2. W. Liu, Z. Chen, X. Qing, Metamaterial-Based Low-Profile Broadband Aperture-Coupled Grid-Slotted Patch Antenna. IEEE Transactions on Antennas and Propagation, Vol. 63, No. 7, pp. 3325–3329, 2015.

3. T H Hand, S A. Cummer, Reconfigurable Reflectarray Using Addressable Metamaterials. IEEE Antennas and Wireless Propagation Letters, Vol. 9, No. 1, pp. 70–74, 2015.
4. N. Kundtz, D. R. Smith, Extreme-angle broadband metamaterial lens. Nature Materials, Vlo. 9, No. 2, pp. 129–132, 2010.
5. M J Freire, R Marques, L. Jelinek, Experimental demonstration of a $\mu = -1$ metamaterial lens for magnetic resonance imaging. Applied Physics Letters, Vol. 93, No. 23, pp. 3966–640, 2008.
6. N. I. Landy, S. Sajuyigbe, J J Mock, *et al.* Perfect Metamaterial Absorber. Physical Review Letters, Vol. 100, No. 20, pp. 207402, 2008.
7. F. Ding, Y. Cui, *et al.* Ultra-broadband Microwave Metamaterial Absorber. Applied Physics Letters, Vol. 100, No.10, pp. 103506–103506–4, 2012.
8. T. Liu, X. Cao, *et al.*, RCS Reduction of Waveguide Slot Antenna with Metamaterial Absorber. IEEE Transactions on Antennas and Propagation, Vol. 61, No. 3, pp. 1479–1484, 2013.
9. S. Li, X. Cao, T. Liu, *et al.* Double-layer Perfect Metamaterial Absorber and Its Application for RCS Reduction of Antenna. Radioengineering, Vol. 23, No. 1, pp. 222–228, 2014.
10. Y. Zhao, C. Yu, J. Gao, *et al.* Broadband Metamaterial Surface for Antenna RCS Reduction and Gain Enhancement. IEEE Transactions on Antennas and Propagation, (Early Access), pp. 1–1, 2018.
11. L. Ren, Y. Jiao, F. Li, J. Zhao, and G. Zhao, A dual-layer T-shaped element for broadband circularly polarized reflectarray with linearly polarized feed, IEEE Antennas and Wireless Propagation Letters, Vol. 10, pp. 407–410, 2011.
12. M. Zhao, G. Zhang, X. Lei, J. Wu, and J. Shang, Design of new single layer multiple-resonance broadband circularly polarized reflectarrays, IEEE Antennas and Wireless Propagation Letters, Vol. 12, pp. 356–359, 2013.
13. S. A. Muhammad, R. Sauleau, G. Valerio, L. L. Coq, and H. Legay, Self-polarizing Fabry–Perot antennas based on polarization twisting element, IEEE Transactions on Antennas and Propagation, Vol. 61, No. 3, pp. 1032–1040, 2013.
14. W. Yang, K.-W. Tam, W.-W. Choi, W. Che, and H. Hui, Polarization rotation reflective surface based on artificial magnetic conductor and its application, Electronic Letters, Vol. 50, No. 21, pp. 1500–1502, 2015.
15. F. Costa, S. Genovesi, and A. Monorchio, Reading chipless RFID located on metallic platforms by using cross-polar scattering, General Assembly and Scientific Symposium, Beijing, China, 2014.
16. J. Y. Chin, M. Lu, and T. J. Cui. Metamaterial polarizers by electric-field-coupled resonators. Applied Physics Letters, Vol. 93, pp. 251903, 2008.
17. R. Singh *et al.*, Terahertz metamaterial with asymmetric transmission, Physical Review B, Vol. 80, pp. 153104, 2009.
18. C. Menzel *et al.*, Asymmetric Transmission of Linearly Polarized Light at Optical Metamaterials. Physical Review Letters, Vol. 104, pp. 253902, 2010.
19. Y. Ye and S. He, 90∘ polarization rotator using a bilayered chiral metamaterial with giant optical activity, Applied Physics Letters, Vol. 96, pp. 203501, 2010.
20. Z. Wei, Y. Cao, Y. Fan, X. Yu, and H. Li, Broadband polarization transformation via enhanced asymmetric transmission through arrays of twisted complementary split-ring resonators. Applied Physics Letters, Vol. 99, pp. 221907, 2011.
21. H. Shi, A. Zhang, S. Zheng, J. Li, and Y. Jiang, Dual-band polarization angle independent 90° polarization rotator using twisted electric-field-coupled resonators. Applied Physics Letters, Vol. 104, pp. 034102, 2014.
22. D. Yan, Q. Gao, C. Wang, C. Zhu, and N. Yuan, A novel polarization convert surface based on artificial magnetic conductor. Microwave Conference, Suzhou, China, pp. 1–2, 2005.
23. X.-C. Zhu et al, A Novel Reflective Surface With Polarization Rotation Characteristic. IEEE Antennas and Wireless Propagation Letters, Vol. 12, pp. 968–971, 2013.
24. M. Feng, J. Wang, H. Ma, W. Mo, H. Ye, and S. Qu, Broadband polarization rotator based on multi-order plasmon resonances and high impedance surfaces. Journal of Applied Physics, Vol. 114, pp. 074508, 2013.

25. Y. Z. Cheng *et al.*, Broadband terahertz reflective linear polarization convertor, 39th IEEE International Conference on Infrared, Millimeter and Terahertz Waves, pp. 1–2, 2014.
26. H. Chen, *et al.* Ultra-wideband polarization conversion metasurfaces based on multiple plasmon resonances. Journal of Applied Physics, Vol. 115, pp. 154504, 2014.
27. X. Huang, D. Yang, and H. Yang, Multiple-band reflective polarization converter using U-shaped metamaterial. Journal of Applied Physics, Vol. 115, pp. 103505, 2014.
28. YW. Yang, K.-W. Tam, W.-W. Choi, W. Che, and H. T. Hui, Novel Polarization Rotation Technique Based on an Artificial Magnetic Conductor and Its Application in a Low-Profile Circular Polarization Antenna. IEEE Transactions on Antennas and Propagation, Vol. 62, No. 12, pp. 6206–6216, 2014.
29. L. Zhang, P. Zhou, H. Lu, H. Chen, J. Xie, and L. Deng, Ultra-Thin Reflective Metamaterial Polarization Rotator Based on Multiple Plasmon Resonances. IEEE Antennas and Wireless Propagation Letters, Vol. 14, pp. 1157–1160, 2015.
30. Y. Jia, Y. Liu, Y. J. Guo, *et al.* Broadband Polarization Rotation Reflective Surfaces and Their Applications to RCS Reduction. IEEE Transactions on Antennas and Propagation, Vol. 64, No. 1, pp. 179–188, 2015.
31. Y. Jia, Y. Liu, Y. J. Guo, *et al.* A Dual-Patch Polarization Rotation Reflective Surface and Its Application to Ultra-Wideband RCS Reduction. IEEE Transactions on Antennas and Propagation, Vol. 65, No. 6, pp. 3291–3295, 2017.
32. H. T. Chen, J. Zhou, J. F. O'Hara, F. Chen, A. K. Azad, and A. J. Taylor, Antireflection coating using metamaterials and identification of its mechanism. Physical Review Letters, Vol. 105, pp. 073901-1-073901-4, 2010.
33. H. T. Chen, Interference theory of metamaterial perfect absorbers. Optics Express, Vol. 20, pp. 7165–7172, 2012.
34. E. F. Knott, J. F. Shaeffer, and M. T. Tuley, Radar Cross Section. Raleigh, NC, USA: SciTech Pub., 2004. a. Edalati, and K. Sarabandi, Wide Angle, Polarization Independent RCS Reduction Using Nonabsorptive Miniaturized-Element Frequency Selective Surfaces. IEEE Transactions on Antennas and Propagation, Vol. 62, No. 2, pp. 747–754, 2014.
35. Y. Liu, K. Li, Y. Jia, *et al.* Wideband RCS Reduction of a Slot Array Antenna Using Polarization Conversion Metasurfaces. IEEE Transactions on Antennas and Propagation, Vol. 64, No. 1, pp. 326–331, 2016.
36. Y. Jia, Y. Liu, W. Zhang, *et al.* In-Band Radar Cross Section Reduction of Slot Array Antenna. IEEE Access, Vol. 99, pp. 1–1, 2018.
37. Y. Jia, Y. Liu, *et al.* A low-RCS and high-gain circularly polarized antenna with a low profile. IEEE Antennas and Wireless Propagation Letters, Vol. 16, pp. 2477–2480, 2017.
38. K. Li, Y. Liu, Y. Jia and Y. J. Guo, A Circularly Polarized High-Gain Antenna with Low RCS over a Wideband Using Chessboard Polarization Conversion Metasurfaces. IEEE Transactions on Antennas and Propagation, Vol. 65, No. 8, pp. 4288–4292, 2017.
39. L. Li, Y. Li, *et al.* Novel Polarization-Reconfigurable Converter Based on Multilayer Frequency-Selective Surfaces. Proceedings of the IEEE, Vol. 103, No. 7, pp. 1057–1070, 2015.
40. S. A. Muhammad, R. Sauleau, and H. Legay, Self-generation of circular polarization using compact Fabry–Perot cavity antennas. IEEE Antennas and Wireless Propagation Letters, Vol. 10, pp. 907–910, 2011.
41. H. L. Zhu, S. W. Cheung, K. L. Chung, and T. I. Yuk, Linear-to-Circular Polarization Conversion Using Metasurface. IEEE Transactions on Antennas and Propagation, Vol. 61, No. 9, pp. 4615–4623, 2013.
42. R. Orr, G. Goussetis, and V. Fusco, Design Method for Circularly Polarized Fabry–Perot Cavity Antennas. IEEE Transactions on Antennas and Propagation, Vol. 62, No. 1, pp. 19–26, 2014.
43. C. A. Balanis, Antenna Theory: Analysis and Design, 3rd ed. Hoboken, NJ, USA: Wiley, 2005.

Chapter 9
Low-RCS Antennas Based on Metasurfaces

A phase gradient metasurface (PGM) is a type of two-dimensional structure with subwavelength dimensions. PGMs were originally used to produce spiral waves to demonstrate their ability to control light [1]. In Sun et al. [2], H-shaped unit cells were used to form a PGM to determine the relationship between propagating waves and surface waves. Owing to the phase change on the surface of PGMs, the classic Snell's Law—in which the angle of incidence is equal to the angle of reflection for a plane wave incident on an isotropic surface—does not apply. The incident wave is either coupled to the surface wave or anomalously reflected. This characteristic makes PGMs good candidates for RCSR.

Many structures have been proposed to couple the incident wave to a surface wave or to reflect it [3–5]. However, the operating bandwidths of these structures can be quite narrow. Reflectarray elements have been used to form the gradient surface to achieve a 10-dB RCSR from 8.9–12 GHz in Kim et al. [6]. A two-dimensional PGM formed by combined split-ring resonators was designed in Li et al. [7], the RCSR bandwidth being 7.8–17 GHz. However, few antenna RCSR studies based on PGMs have been published.

9.1 Phase Gradient Metasurfaces and Their Application to Slot Array Antenna RCS Reduction [8]

9.1.1 Design of the PGM

The blue plane in Fig. 9.1 represents the surface of a PGM. The parameters $d\Phi/dx$ and $d\Phi/dy$ denote the phase gradients on the x-axis and y-axis, respectively, and their synthetic vector is $d\Phi/dr$. When the propagating wave illuminates the PGM, two

Y. Liu et al., *Antenna Radar Cross Section: Theory and Design*,
https://doi.org/10.1007/978-981-96-1226-0_9

Fig. 9.1 Schematic diagram of the PGM

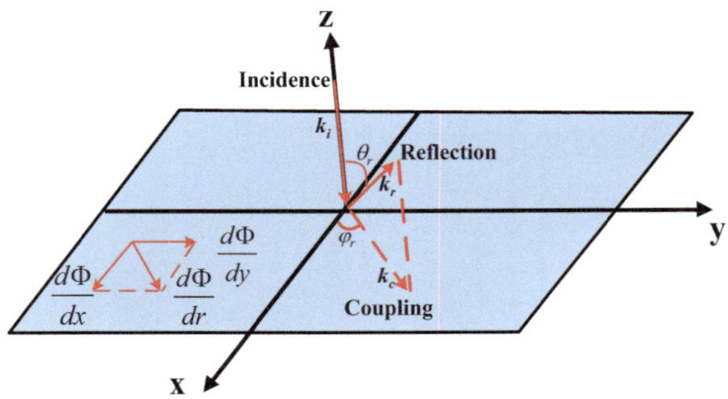

different events may occur—that is, the incident wave may be anomalously reflected into space, or the incident wave may be coupled to the surface wave.

In the first case, the reflected wave vector (k_r) has the same value as the incident wave vector ($\boldsymbol{k_i}$). The relationship between k_i and $d\Phi/dr$ is a key factor in determining which will occur. For $d\Phi/dr < k_i$, the incident wave is reflected into space. The relationship between the incident angle (θ_i) and the reflected angle (θ_r) can be expressed as follows [1]:

$$\sin(\theta_r) - \sin(\theta_i) = \frac{d\boldsymbol{\Phi}/dr}{k_i} \tag{9.1}$$

Only normal incidence is considered here—that is, $\theta_i = 0°$. The angle between the x-axis and the projection of the reflected wave on the xoy-plane can be calculated as follows [7]:

$$\tan(\varphi_r) = \frac{d\boldsymbol{\Phi}/dy}{d\boldsymbol{\Phi}/dx} \tag{9.2}$$

For $d\Phi/dr > k_i$, the incident wave is coupled to the surface wave. The angle between the coupled wave vector (k_c) and the x-axis can be calculated using Eq. (9.2). Owing to the anomalous reflection and the coupling mechanism, specular reflection is reduced when the incident wave normally irradiates the PGM.

Because the PGM is constructed using unit cells with different reflection phases, the phase gradient of the PGM is determined by the phase variations between adjacent unit cells. If the phase differences between them are stable over a wide frequency range, the phase gradient will be constant over a very wide bandwidth. Consequently, the PGM can anomalously reflect an incident wave or couple the incident wave into a surface wave over a wide bandwidth. Based on these principles, wideband RCSR of the PGM can be realized. For convenience of a periodic arrangement along the x-axis or y-axis, 2π was chosen as the phase variation range of the period of the PGM.

Initially, only one square ring was used as the PGM unit structure. The reflection phase was changed by varying the inner side length (l) of the inner ring. As shown in

Fig. 9.2a, the phase variation range has a noticeable distinction between 8–12 GHz. It can be difficult to maintain a phase gradient over a relatively wide bandwidth, so a larger square ring was added to the existing structure. The dimensions of the structure were optimized to obtain a relatively stable phase difference. Accordingly, the width of the larger ring $w_3 = 0.8$ mm, the width of the inner square ring $w_2 = 1.1$ mm, and the gap between the two rings $g = 0.5$ mm.

The reflection phase differences between adjacent unit cells are similar, as shown in Fig. 9.2b. The existence of the cross in the inner square ring improves the reflection phase response at 10 GHz. Finally, the unit cell was determined to be a combination of the cross and two square rings, the unit-cell geometry of the proposed PGM being as shown in Fig. 9.3. The structures were printed on 2.2-permittivity Rogers 5880 substrate with a loss tangent of 0.0009 and a thickness of 3 mm. To obtain a stable phase difference, the unit cells were optimized.

The final parameters were selected as follows: the length of the unit cell $a = 18$ mm, the width of the cross $w_1 = 0.4$ mm, the width of the inner square ring $w_2 = 1.1$ mm, the width of the outer square ring $w_3 = 0.8$ mm, the gap between the two

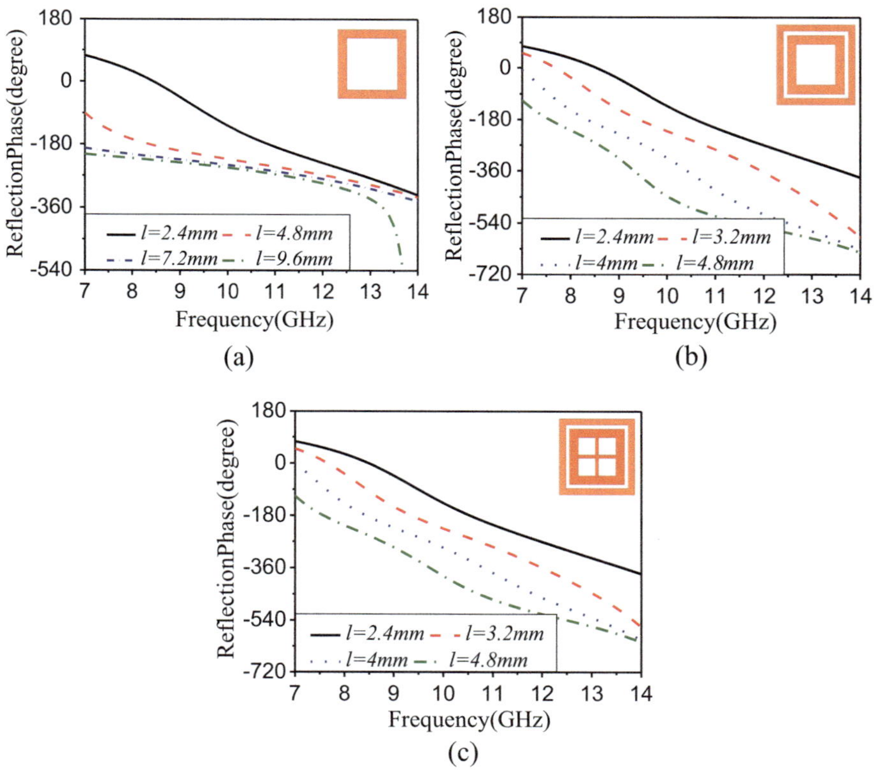

Fig. 9.2 Reflection phases of the unit cells with different lengths (l): **a** only one square ring, **b** two square rings, **c** a cross and two square rings

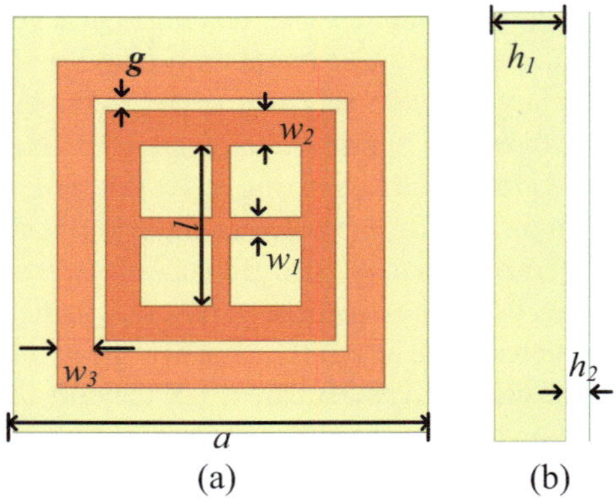

Fig. 9.3 Geometry of the PGM unit cell: **a** top view and **b** side view

rings $g = 0.5$ mm, and the gap between the substrate and the ground of the unit $h_2 =$ 1 mm. Equation (9.1) shows that a larger phase gradient leads to a larger reflection angle. The phase difference between adjacent unit cells was selected to be $\pi/2$. The values of l of the four different unit cells were 2.4, 3.2, 4, and 4.8 mm, respectively. Figure 9.2c shows that a stable phase difference exists between adjacent unit cells from 7–12 GHz under normal incidence.

A subarray of the PGM with phase gradients in two directions—that is, the x-axis and y-axis—was constructed from unit cells, as shown in Fig. 9.4a. Because the PGM is a periodic structure, two cycles were used in the phase gradient directions to form the PGM structure. The monostatic RCS of the proposed PGM was simulated to investigate the characteristics of the PGM. Based on Iriarte Galarregui et al. [9], the monostatic RCSRs of the PGM are shown in Fig. 9.4b. The reduction frequency band from 7–20 GHz is almost the same for the x-polarized and y-polarized incidence.

Figure 9.5 shows the normalized 3D scattering patterns for different frequencies. A diffuse reflection is evident at 15 GHz. The main lobe direction is $\varphi_r = 3\pi/4$, $\theta_r \approx$

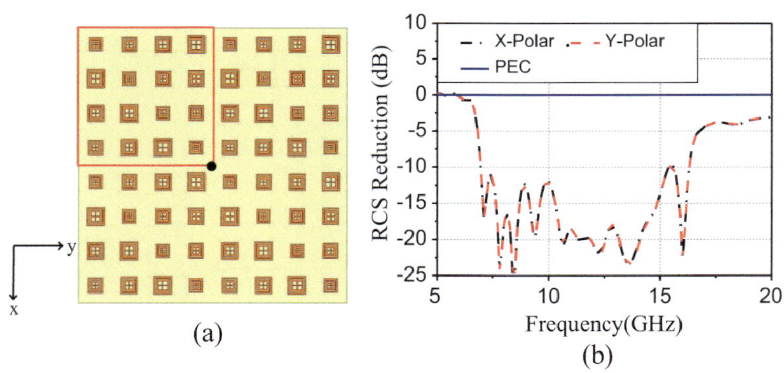

Fig. 9.4 **a** Geometry of the 2D PGM, **b** normalized monostatic RCS with respect to a PEC plate of the same dimensions as the proposed PGM under normal incidence

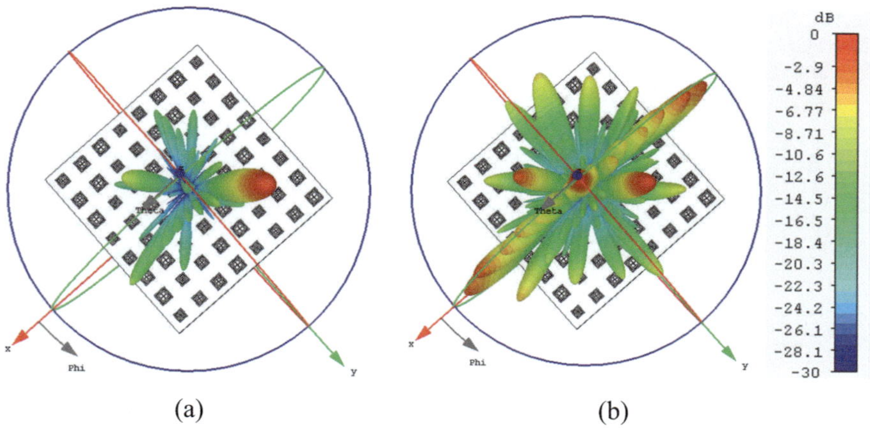

Fig. 9.5 Normalized 3D scattering patterns at different frequencies under x-polarized plane waves normally incident on a 2D PGM, **a** 12 GHz, **b** 15 GHz

$\pi/6$ at 12 GHz. We can then use Eqs. (9.1) and (9.2) to calculate the angles θ_r and φ_r for the proposed PGM at 12 GHz.

For the proposed PGM, $\frac{d\Phi}{dx} + \frac{d\Phi}{dy} = \frac{d\Phi}{dr}$, $\sqrt{2}\left|\frac{d\Phi}{dx}\right| = \sqrt{2}\left|\frac{d\Phi}{dy}\right| = \left|\frac{d\Phi}{dr}\right|$, $\theta = 0°$. Based on Eq. (9.1):

$$\sin\,\theta_r - \sin\,\theta_i = \left|\frac{d\boldsymbol{\Phi}}{dr}\right| / k_i$$

$$\sin\,\theta_r = \sqrt{2}\cdot\left|\frac{d\boldsymbol{\Phi}}{dy}\right| / k_i$$

$$\Delta y = 72\,\text{mm},\ \Delta\Phi = 2\pi,\ k_{i12} = \frac{2\pi}{\lambda_{12GHz}} \Rightarrow \sin\theta_r = \sqrt{2}\cdot\left|\frac{2\pi}{72}\right| / \left(\frac{2\pi}{25}\right) \approx 0.491$$

$$\theta_r \approx \frac{\pi}{6}\left(or\frac{5\pi}{6}\right) + 2n\cdot\pi, n = 0, 1, 2\ldots$$

Based on Eq. (9.2):

$$\tan\varphi_r = \left(\frac{d\boldsymbol{\Phi}}{dy}\right) / \left(\frac{d\boldsymbol{\Phi}}{dx}\right) = -1$$

$$\varphi_r \approx \frac{3\pi}{4} + n\cdot\pi, n = 0, 1, 2\ldots$$

Because the PGM designed above is a type of reflective phase gradient metasurface, the reflection angle (θ_r) is smaller than $\pi/2$. Additionally, the reflected wave is reflected toward the side of the phase lag. The phase gradient along the +ve x-axis is

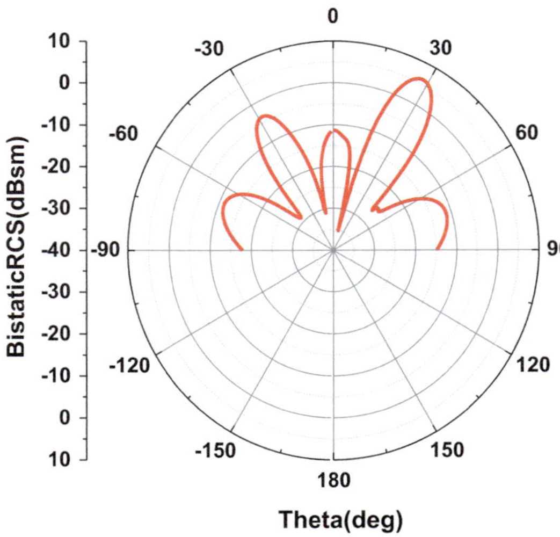

Fig. 9.6 Scattering pattern of the PGM in the $\varphi = \frac{3\pi}{4}$ plane at 12 GHz

positive, and the phase gradient along the +ve *y*-axis is negative. Consequently, we can determine that the main lobe direction of the reflected wave is $\theta_r \approx \frac{\pi}{6}, \varphi_r \approx \frac{3\pi}{4}$. The simulated results are shown in Fig. 9.6. The main lobe direction is approximately $\theta_r = \frac{7\pi}{45}, \varphi_r = \frac{3\pi}{4}$. Because the proposed PGM has only two cycles, there is a small difference between the simulated and calculated results. However, the difference is small enough to demonstrate consistency between them.

9.1.2 Design of the Low-RCS Slot Array Antenna

A traditional 2 × 2 slot array antenna with an operating band from 4.12–4.26 GHz was designed and selected as the reference antenna. Figure 9.7a shows the geometry of the reference antenna. A ground with four slots was printed on the top surface of a 1-mm-thick F4BM dielectric substrate ($\varepsilon_r = 2.65$, $\tan\delta = 0.0023$). Additionally, the bottom surface of the substrate was the feeding network. To make the antenna radiate in the boresight direction, a 4-mm-thick metal-backed substrate was attached to the bottom of the feeding network. A subarray of the 2D PGM was then used as the superstrate of the slot array antenna to reduce its RCS. The PGM used the same ground as the antenna. Owing to the use of the PGM superstrate, the effective permittivity above the radiating slots changed.

Thus, the radiation properties of the antenna were affected, as shown in Fig. 9.8. To reduce the impact of the change in the effective permittivity, the parts of the PGM above the radiating slots were cut, as shown in Fig. 9.7b. Considering the radiation performance and the RCSR effect, the length and width of the PGM slots were optimized to be 3 mm longer than the antenna slot. The antenna with the PGM had the same parameters as the reference antenna except for the PGM parameters.

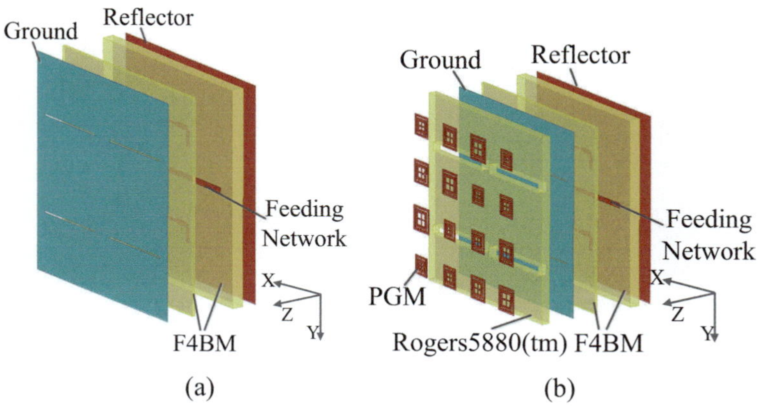

Fig. 9.7 Geometries of **a** the reference antenna (antenna without the PGM) and **b** the proposed antenna (antenna with the PGM)

Fig. 9.8 Comparison of the reflection coefficients

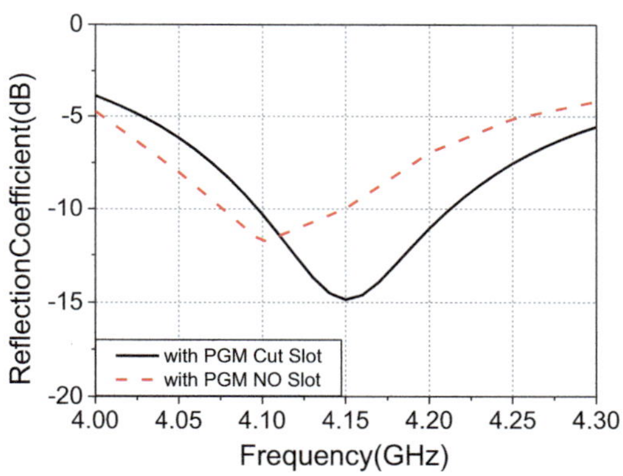

To validate the radiation and scattering characteristics of the proposed antenna, both antennas were simulated and fabricated, as shown in the photographs of Fig. 9.9a and b. The simulated and measured reflection coefficients are shown in Fig. 9.9c. Although four slots were cut on the PGM to reduce the impact caused by PGM loading, the working bandwidth of the proposed antenna still narrows. Owing to fabrication errors, the measured results for both antennas shift toward higher frequencies.

Because of the frequency shifts of the measured reflection coefficients, it is not appropriate to compare the simulated and measured radiation patterns at the same frequency point. The simulated results at 4.15 GHz and the measured results at 4.35 GHz are shown in Figs. 9.10 and 9.11, respectively. Although the measured radiation patterns at 4.35 GHz for these antennas exhibit some differences, the main lobe of the reference antenna is consistent with that of the proposed antenna.

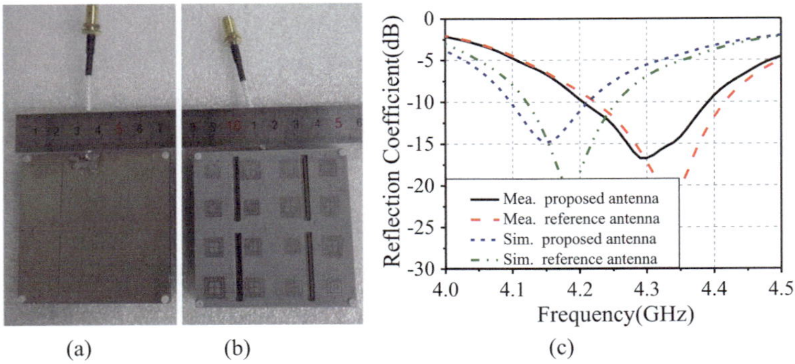

Fig. 9.9 Fabricated prototypes of **a** the reference antenna and **b** the proposed antenna; **c** simulated (Sim.) and measured (Mea.) reflection coefficients of the reference and proposed antennas

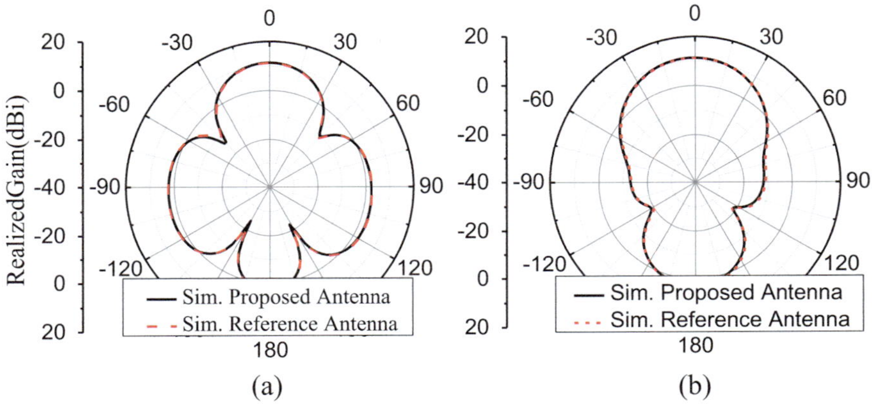

Fig. 9.10 Simulated (Sim.) radiation patterns of the reference antenna and proposed antenna at 4.15 GHz. **a** *E*-plane, **b** *H*-plane

Figure 9.12 shows the simulated monostatic RCSs of the reference antenna and the proposed antenna. The RCS of the reference antenna decreases from 7–20 GHz. The 10-dB RCSR bandwidth ranges from 7–16.8 GHz (82.4%), which is wider than the 10-dB reduction bandwidths in reference [6] (29.7%) and [7] (74%). The monostatic RCS of the proposed antenna under normal incidence was measured to verify the design. Because of the testing limits of the anechoic chamber, the monostatic RCS of the antenna was only measured up to 18 GHz. Good agreement between the simulated and measured results is evident.

The normalized 3D scattering patterns under *x*-polarization incidence at 12 and 15 GHz are shown in Fig. 9.13. Anomalous reflections and diffuse reflections are evident at 12 and 15 GHz, respectively. The main lobe direction is $\varphi_r \approx 3\pi/4$, $\theta_r \approx 3\pi/20$ at 12 GHz. Because the PGM loaded on the antenna only has one cycle, the simulated main lobe direction is slightly different from the theoretical value

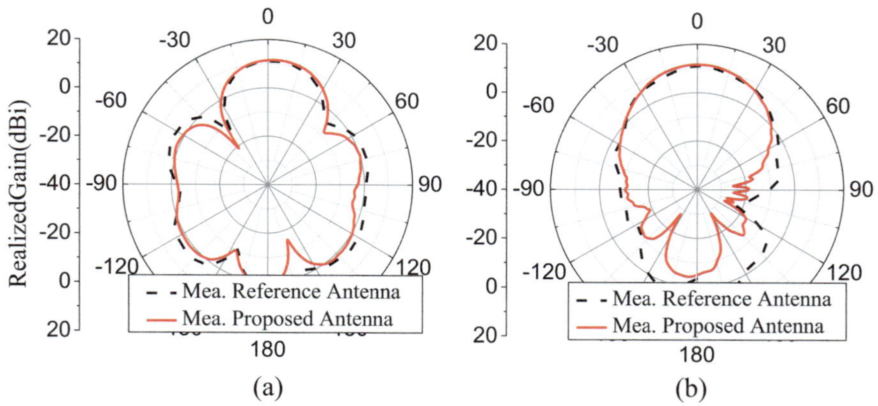

Fig. 9.11 Measured (Mea.) radiation patterns of the reference antenna and proposed antenna at 4.35 GHz. **a** *E*-plane, **b** *H*-plane

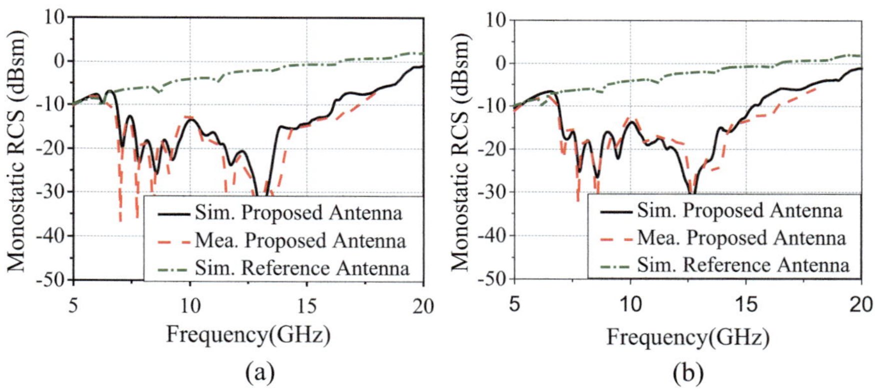

Fig. 9.12 Comparison of the monostatic RCSs of the reference antenna and proposed antenna under normal incidence. **a** *x*-polarization, **b** *y*-polarization

($\varphi_r = 3\pi/4, \theta_r \approx \pi/6$). However, the results still demonstrate the PGM loaded on the antenna acts as a PGM.

Figure 9.14 shows the monostatic RCS at different incident angles. It is evident that the monostatic RCS of the proposed antenna is lower than that of the reference antenna under normal incidence for both polarizations. Owing to the phase superposition effect of the reflection phases from different PGM units, the monostatic RCS of the proposed antenna is higher than that of the reference antenna at several incident angles under oblique incidence conditions.

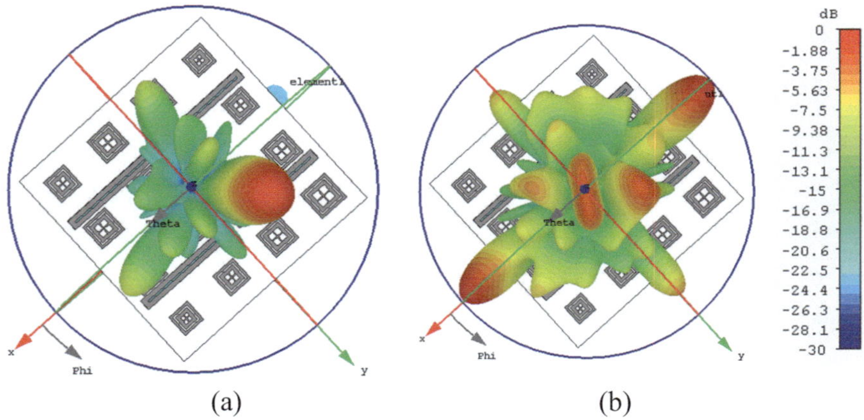

Fig. 9.13 Normalized 3D scattering pattern at different frequencies under an x-polarized plane wave normally incident for the proposed antenna, **a** 12 GHz, **b** 15 GHz

9.2 Phase Gradient Metasurfaces and Their Application to Antenna RCS Reduction [10]

9.2.1 Structure of the FP Antenna Based on a PGM Structure Without an MA

The structure of the proposed FP antenna based on a PGM without an MA is shown in Fig. 9.15a. Two superstrates that work together as a reflective surface were placed above a microstrip antenna. The lower superstrate S_l comprised an 8 × 8 array of square loops of the same size, as shown in Fig. 9.15b. Figure 9.15c shows the superstrate S_u, which comprised an 8 × 8 array of square patches of varying sizes in the x-direction. Both arrays of loops and patches had the same period of 15 mm. The coordinates (x, y) of the feeding points and the center of the microstrip antenna were (X_F, 0) and (X_p, 0), respectively. The values of the parameters are shown in Table 9.1.

9.2.2 Mechanism of the Proposed Low-RCS FP Antenna

For the case of radiation, the two superstrate layers work as a partially reflective surface and form an FP antenna with the microstrip antenna. To investigate the S-parameters of the partially reflective surface (PRS), numerical simulations were conducted using the HFSS_v13 software application. The reflection magnitudes and phases with different values of W_m are shown in Fig. 9.16a and b, respectively. The simulation model for the PRS unit cell is shown in the inset of Fig. 9.16a. The incident

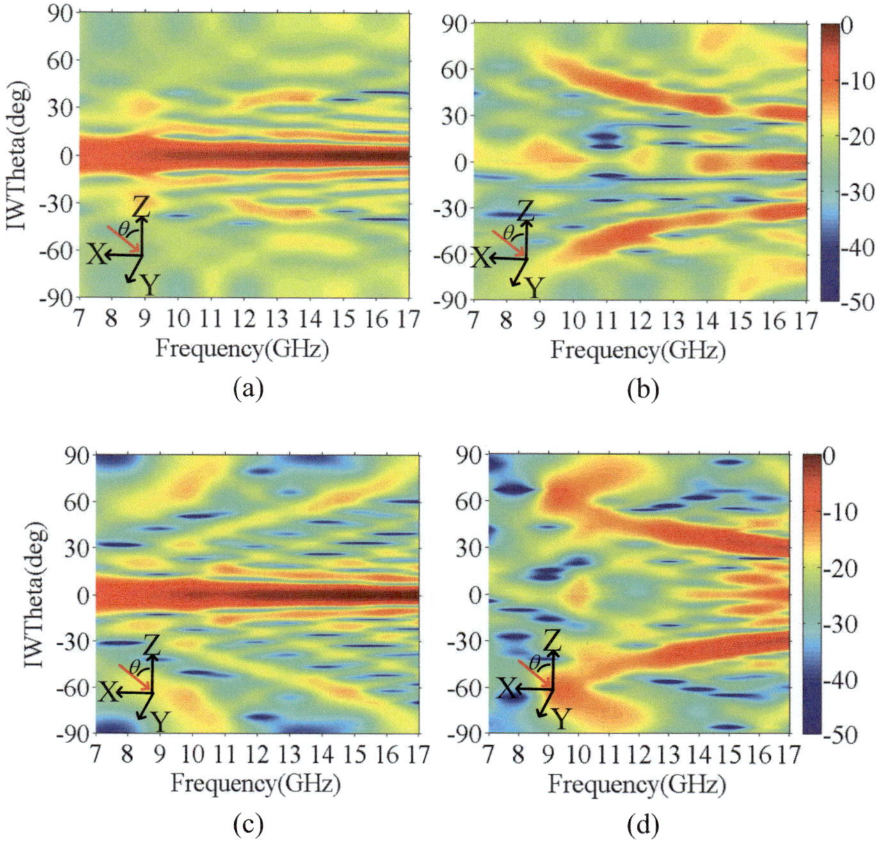

Fig. 9.14 Comparison of the monostatic RCSs of the reference antenna and proposed antenna in the *xoz*-plane under oblique incidence: **a** reference antenna, *TE* polarization; **b** proposed antenna, *TE* polarization, **c** reference antenna, *TM* polarization; **d** proposed antenna, *TM* polarization

wave propagates from Port 1 to Port 2. The unit cell has two layers, and the square loop was printed on the first layer encountered by the incident wave. The reflection magnitudes vary from 0.96–1 with different values of W_m, and the reflection phases vary from $-164°$ to $-167°$ at 8.4 GHz, which shows that the size of the square patches have little influence on the *S*-parameters when the incident wave propagates from Port 1 to Port 2.

The FP antenna can obtain a high gain and aperture efficiency when the spacing between the substrate of the feed antenna and the PRS (d_1) satisfies the following condition [11]:

$$\varphi_{PRS} + \varphi_G - \frac{2\pi}{\lambda} \times 2d_1 = -2N\pi, N = 0, 1, 2 \tag{9.3}$$

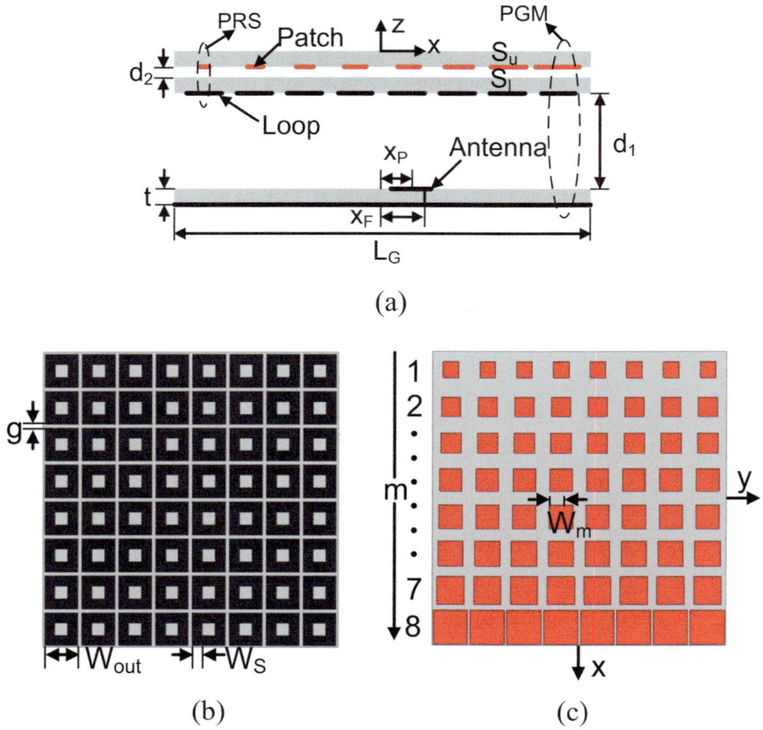

Fig. 9.15 An FP Antenna without an MA. **a** Side view. **b** Square loops printed on the bottom layer of the lower superstrate (S_l). **c** Square patches printed on the bottom layer of the upper superstrate (S_u)

Table 9.1 Parameters of the proposed antenna (unit: mm)

Symbol	Value	Symbol	Value
L_G	120	W_1	6.5
X_F	9.9	W_2	7.9
X_P	7	W_3	8.4
t	1	W_4	9.2
d_1	18	W_5	9.6
d_2	2	W_6	10.3
g	1.5	W_7	11.2
W_{out}	13.5	W_8	14
W_S	4.25		

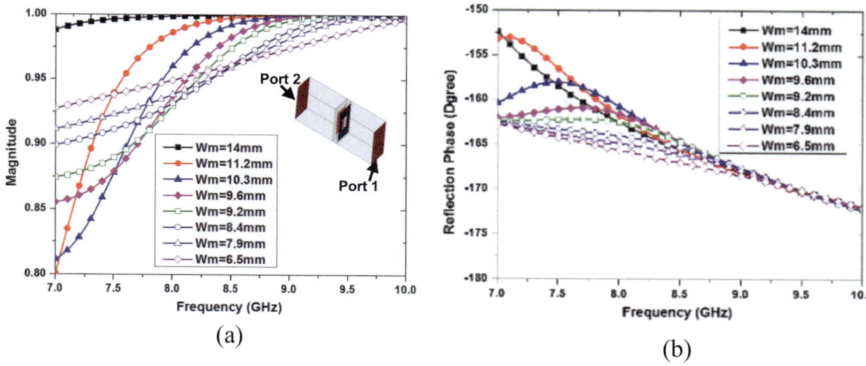

Fig. 9.16 Reflection characteristics of the proposed partially reflective surface with different values of W_m. **a** Reflection magnitudes. **b** Reflection phases

where φ_{PRS} and φ_G denote the reflection phases of the PRS and the metallic ground plane of the microstrip antenna, respectively, ε_A denotes the relative permittivity of the substrate of the microstrip antenna, and λ denotes the operating wavelength. Based on Eq. (9.3), the initial value of d_1 was 18.5 mm, which was then optimized to 18 mm.

The S_{11} and peak gains of the FP antenna with different values of X_P are shown in Fig. 9.17 and it is evident that the proposed FP antenna resonates at 8.4 GHz. The −10-dB impedance bandwidth is approximately 80 MHz (from 8.38–8.46 GHz), and a peak gain of 19.2 dBi is achieved.

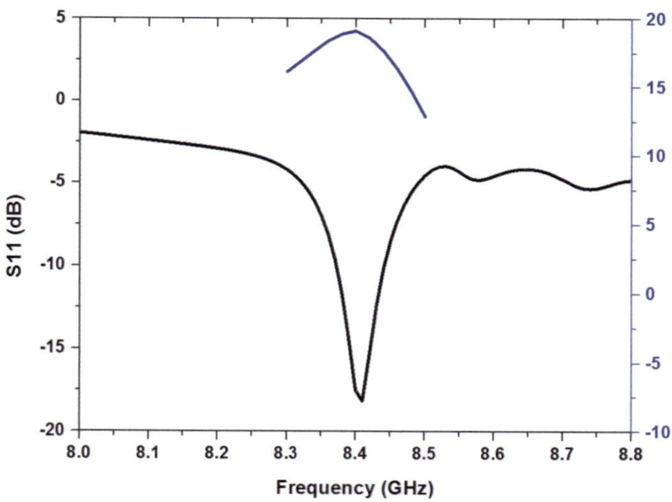

Fig. 9.17 Simulated S_{11} and peak gains of the proposed FP antenna

For the case of scattering, the two superstrate layers and the metallic ground plane of the microstrip antenna work as a PGM. PGMs are two-dimensional structures that manipulate light by imparting local, space-variant phase changes on an incident EM wave. The anomalous reflection of the PGM is governed by the generalized Snell's law, as follows [1]:

$$\sin\theta_r - \sin\theta_i = \frac{\lambda}{2\pi}\frac{\mathrm{d}\emptyset}{\mathrm{d}x} \tag{9.4}$$

where θ_r and θ_i denote the angles of incidence and reflection, respectively (here, θ_r and θ_i denote the angles of incidence and reflection away from the z-axis), $\emptyset$ denotes the phase difference between two adjacent elements of the PGM, and $\mathrm{d}\emptyset/\mathrm{d}x$ denotes the phase gradient in the x-direction.

Because the dimensions of the patches printed on the upper superstrate (S_u) vary in the x-direction, there is a phase gradient in the x-direction. Consequently, the angle of reflection (θ_r) differs from the angle of incidence (θ_i); thus, the monostatic RCS is reduced. For RCSR under a normally incident EM wave ($\theta_i = 0°$), the reflected angle is not larger, but it is better. This improvement is because although the main beam of the reflected wave points at a nonthreatening angle, there are still side lobes. If the side lobe points in the backward direction of the incident wave, the RCS will not be sufficiently reduced. Accordingly, we can use an effective strategy for monostatic RCSR—that is, ensure that the pattern trough between the main lobe and the first side lobe points in the backward direction of the incident wave in the xoz-plane by optimizing the reflection phase of the proposed PGM in the x-direction. Consequently, good RCSR can be obtained without a large phase gradient, which simplifies the design of each unit cell of the PGM.

Because the dimensions of the patches printed on the upper superstrate (S_u) vary only in the x-direction, the proposed PGM can be modeled as a 1×8 reflective array in the x-direction. Thus, the array factor can be calculated as follows [12]:

$$|f_a| = \left|\sum_{i=1}^{8} m_i e^{j(kd_{1i}\cos\theta + \beta_{1i})}\right| \tag{9.5}$$

where m_i denotes the reflection magnitude, and d_{1i} and β_{1i} denote the distance and phase difference between the first element and the ith element, respectively.

Because we are only concerned with the monostatic RCS of the proposed antenna under a normally incident EM wave, the value of θ in Eq. (9.5) can be selected to be 0°. Assuming that the reflection magnitudes m_i are 1 and the phase differences between each adjacent unit are the same, the normalized array factor (f_a) versus the phase difference is as shown in Fig. 9.18. The results show that the first trough is obtained when the phase difference is 45°.

To obtain a phase difference of 45° between each adjacent unit cell, the parameters of the patches were optimized, as listed in Table 9.1. The reflection magnitudes and phases of the PGM unit cells are shown in Fig. 9.19a and b, respectively. The

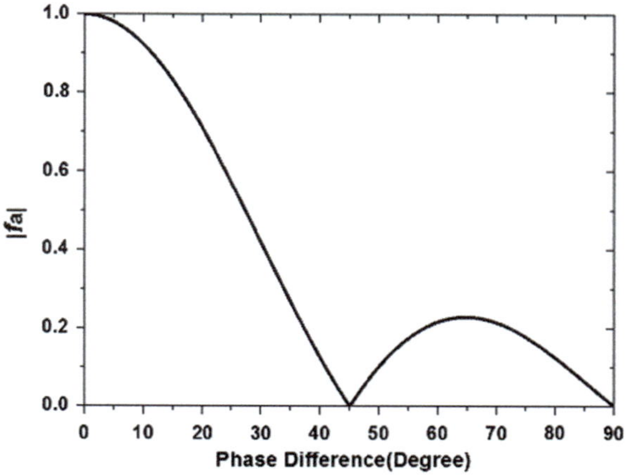

Fig. 9.18 Normalized array factor versus phase difference between adjacent cells at $\theta = 0°$

simulation model for the PGM unit cell is shown in the inset of Fig. 9.19a. The incident wave propagates from Port 1 to Port 2. The unit cell has three layers, and the square patch was printed on the first layer encountered by the incident wave. The length of the metallic ground plane is the same as the period of the unit cell. Figure 9.19a shows that the reflection magnitudes are greater than 0.989 at 8.4 GHz. The reflection phase (φ) increases as W_m decreases: $\varphi = -540°(-180°)$, $-495°$, $-451°$, $-405°$, $-360°$, $-316°$, $-260°$, and $-215°$for $W_m = 14$, 11.2, 10.3, 9.6, 9.2, 8.4, 7.9, and 6.5 mm, respectively. The phase differences between each adjacent unit are all nearly 45°.

After substituting the reflection magnitudes and phases into Eq. (9.5), the normalized array factor (f_a) in the *xoz*-plane is as shown in Fig. 9.20. The results show that the main beam and the first pattern trough point in the directions of $\theta_r = 18°$ and 0°, respectively.

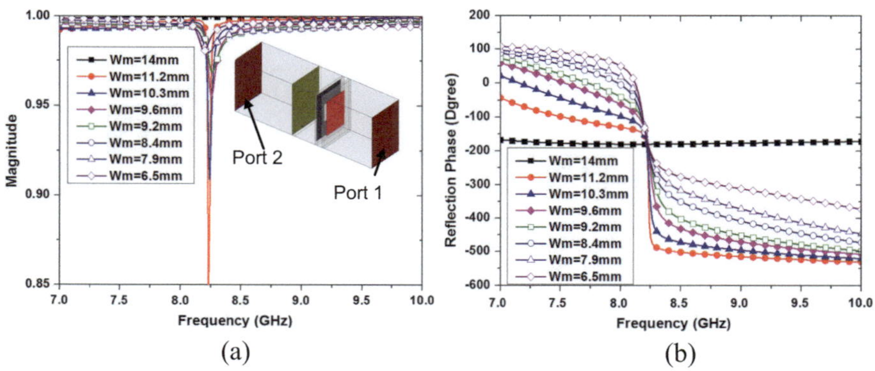

Fig. 9.19 Reflection characteristics of the proposed PGM with different values of W_m. **a** Reflection magnitude. **b** Reflection phase

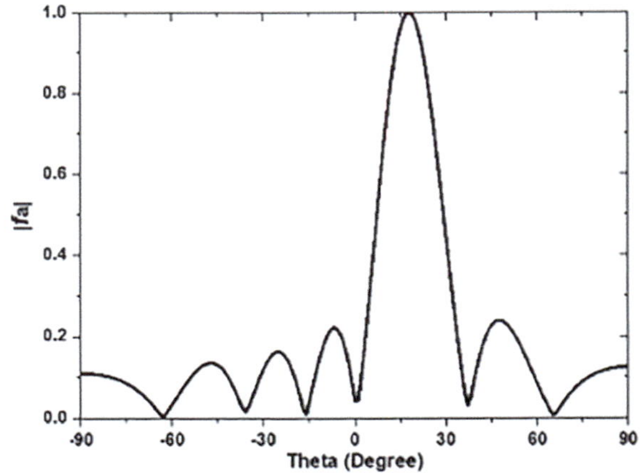

Fig. 9.20 Normalized array factor of the proposed PGM without an MA in the *xoz*-plane

The in-band (at 8.4 GHz) bistatic RCS of the proposed antenna under a normally incident EM wave with *x*-polarization (copolarization RCS; the proposed antenna is *x*-polarized) and *y*-polarization (cross-polarization RCS) in the *xoz*-plane are shown in Figs. 9.21a and b respectively. The microstrip antenna without a PRS was selected as a reference antenna. For the *x*-polarized incident EM wave, the main lobe and pattern trough point in the directions of $\theta_r = 14.5°$ and 2°, respectively. When the incident EM wave is *y*-polarized, the main lobe and pattern trough point in the directions of $\theta_r = 13.5°$ and $-2°$, respectively. Consequently, both the copolarization and cross-polarization RCS of the proposed FP antenna are considerably reduced. According to Fig. 9.20, the main lobe and pattern trough should point in the directions of $\theta_r = 18°$ and 0°. The small differences between the simulated and theoretical results stem from the influence of the microstrip antenna. The reasonable agreement between them indicates the validity of the mechanism of the in-band RCSR.

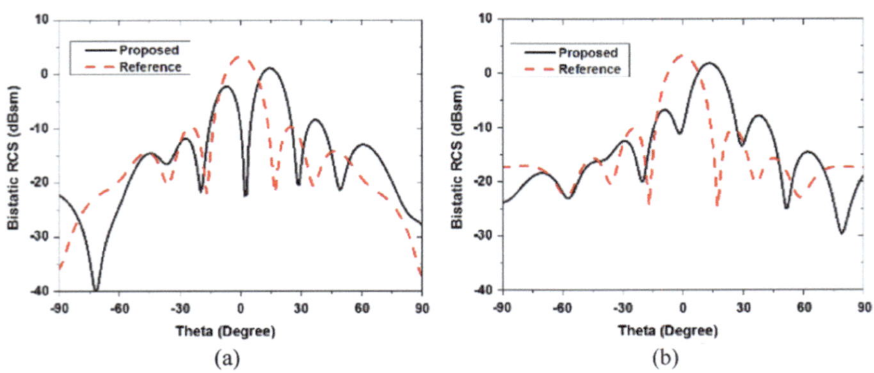

Fig. 9.21 Comparison of in-band bistatic RCSs of the proposed and reference antennas at 8.4 GHz for a normally incident EM wave with **a** *x*-polarization and **b** *y*-polarization

Although the aforementioned discussion focuses on a single frequency point, it can also be applied to wideband RCSR as long as the PGM unit cells have a phase difference. As shown in Fig. 9.19b, the unit cells of the proposed PGM have phase differences over a broad band, which contribute to broadband RCSR. Figure 9.22 compares the monostatic RCSs of the proposed FP antenna and the reference antenna under normally incident EM waves with x- and y-polarization. The RCS of the proposed FP antenna is reduced from 7 to 11 GHz for both x- and y-polarized incident waves. Moreover, the RCSR decreases with decreasing phase gradient.

To investigate the RCS characteristics of the proposed antenna for oblique incident waves, the monostatic RCSs of the reference antenna and the proposed antenna without an MA in the xoz-plane under φ-polarized and θ-polarized incident waves are shown in Fig. 9.23. The monostatic RCS of the proposed antenna is reduced at angles from $-6°$ to $4°$ and from $-7°$ to $6°$ under the φ-polarized and θ-polarized incident waves, respectively.

In this section, a low-RCS FP antenna based on PGM without an MA is proposed. Using the proposed PGM structure, the in-band RCS of the proposed antenna is substantially reduced for both co- and cross-polarized incident EM waves while maintaining a high gain. Additionally, the out-of-band RCS is also reduced from 7 to 11 GHz. To widen the bandwidth of the RCSR, another PGM with an MA structure is proposed and applied to the FP antenna in the next subsection.

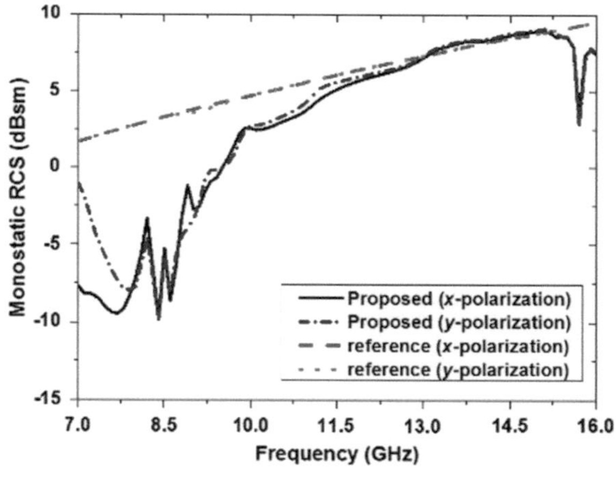

Fig. 9.22 Comparison of the monostatic RCSs of the proposed and reference antennas for normally incident EM waves

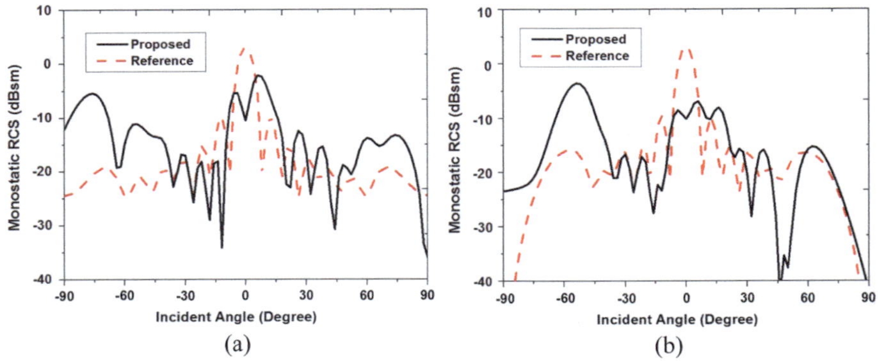

Fig. 9.23 Monostatic RCS of the reference antenna and proposed antenna without an MA in the *xoz*-plane under **a** φ-polarized and **b** θ-polarized incident waves

9.2.3 Design of the FP Antenna Based on a PGM Structure with an MA

The side view of the proposed FP antenna based on a PGM with an MA structure is shown in Fig. 9.24a. The top view of the unit cell on the upper superstrate is shown in Fig. 9.24b. A metallic square loop with four resistors mounted on each side of the loop was printed on the top layer of the upper superstrate S_u-MA. The resistance of all of the resistors was 150 Ω. The parameters of the F–P antenna based on the PGM with the MA that differ from those of the antenna without an MA are listed in Table 9.2.

The two superstrate layers work as a partially reflective surface for the microstrip antenna. The simulation model for the unit cell is shown in the inset of Fig. 9.25a, and the incident wave propagates from Port 1 to Port 2. Figure 9.25 shows that the size of the square patch has little influence on the *S*-parameters of the reflective surface at 8.4 GHz even though the MA structure was added.

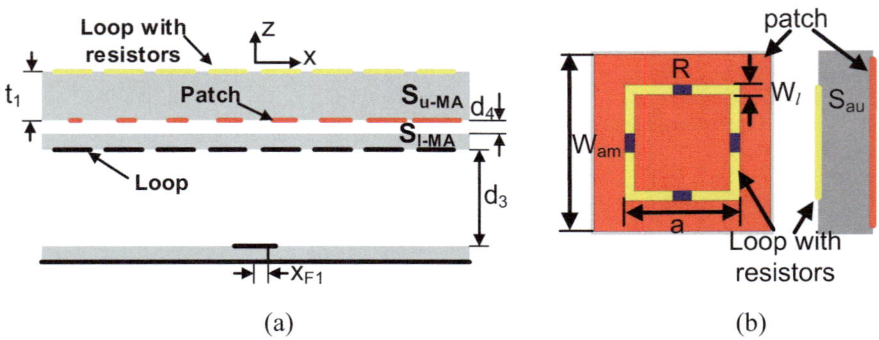

Fig. 9.24 FP Antenna with an MA. (a) Side view. (b) Unit cell printed on the upper superstrate

Table 9.2 Parameters of the proposed antenna with an MA (unit: mm)

Symbol	Value	Symbol	Value
X_{F1}	2.9	Wa_2	6.6
t_1	4	Wa_3	7
d_3	18	Wa_4	7.6
d_4	1	Wa_5	8
a	6	Wa_6	9
W_1	0.5	Wa_7	10
Wa_1	6	Wa_8	11

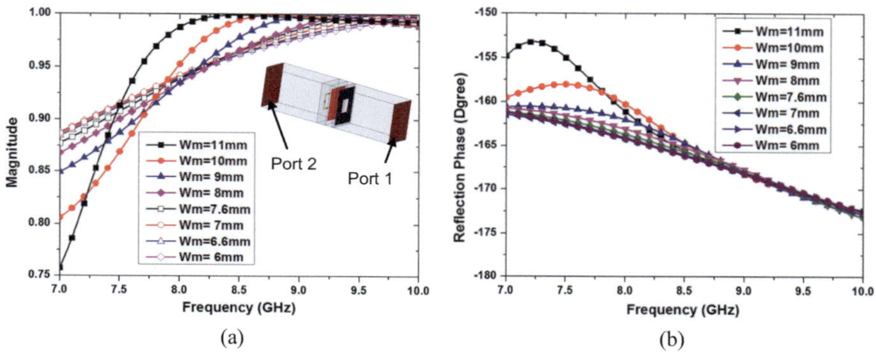

Fig. 9.25 Reflection characteristics of the proposed partially reflective surface consisting of an MA with different values of W_m. **a** Reflection magnitudes. **b** Reflection phases

Figure 9.26 a and b show the reflection magnitudes and phases of the PGM unit cells with the MA, respectively. Unlike the PGM without the MA, the reflection magnitudes of the PGM with the MA vary considerably with variations in W_m, as shown in Fig. 9.26a. Additionally, the reflection magnitudes are small at frequencies from 9–15 GHz, and there is practically no transmitted energy due to the existence of the metallic ground, which indicates good absorbing properties. Consequently, the incident wave in this frequency range can be absorbed by the proposed PGM, and out-of-band RCSR of the proposed antenna can be achieved. To obtain a good RCSR while maintaining the radiation performance, the reflection phases (φ) were optimized as follows: $\varphi = -218.5^{\circ}$, $-534.6°$, $-473°$, $-423.5°$, $-415°$ $-397.7°$, $-396°$, and $-389°$ for $W_m = 11$, 10, 9, 8, 7.6, 7, 6.6, and 6 mm, respectively

To demonstrate the effect of the resistors on the reflection characteristics of the unit cell, Fig. 9.27 shows the reflection magnitudes and phases of the unit cell with different resistor values when $W_m = 9$ mm. As the resistance decreases from 200 to 100 Ω, the absorbing bandwidth increases. However, the linearity of the reflection phase decreases when the resistance is 100 Ω, which is not suitable for wideband RCSR. Consequently, the value of 150 Ω was selected as the resistor value.

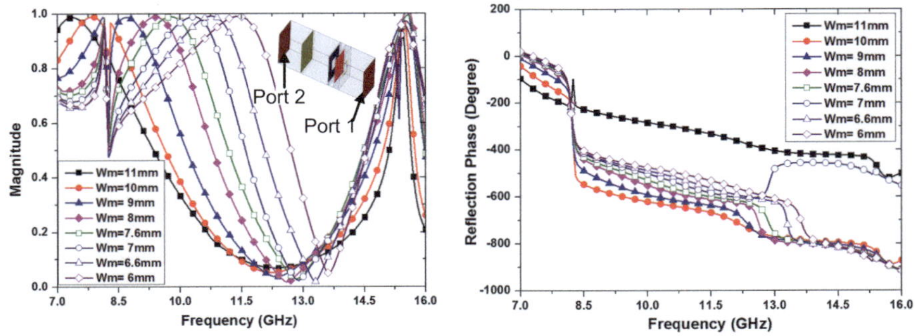

(a) (b)

Fig. 9.26 Reflection characteristics of the proposed PGM consisting of an MA with different values of *Wm*. **a** Reflection magnitude. **b** Reflection phase

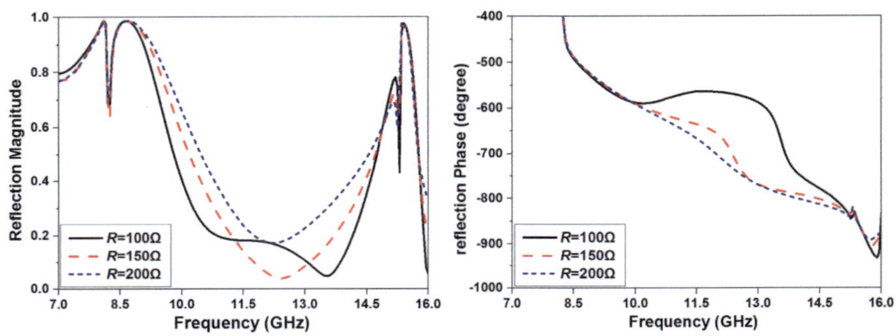

(a) (b)

Fig. 9.27 Reflection characteristics of the proposed PGM consisting of a lossy layer with different resistances. **a** Reflection magnitude. **b** Reflection phase

The S_{11} and radiation patterns of the FP antenna based on the PGM with the MA are shown in Fig. 9.28. The −10 dB impedance bandwidth is approximately 100 MHz (from 8.36–8.46 GHz), and the gain of the antenna is higher than 16.75 dB within this frequency range. The antenna has a peak gain of 17.9 dBi at 8.4 GHz. The gain of this antenna is 1.3 dB lower than that of the antenna proposed in the previous section; this occurs because the two superstrate layers absorb some of the EM waves radiated from the microstrip antenna. Although the gain of the antenna decreases, its aperture efficiency is still 43.3%, which is higher than most other low-RCS FP antennas with an MA presented in previous studies.

After substituting the reflection magnitudes and phases of the PGM with the MA into Eq. (9.5), the normalized array factor (f_a) in the *xoz*-plane is shown in Fig. 9.29. The results show that the main beam and pattern trough point in the directions of $\theta_r = 11°$ and $-4°$, respectively. The bistatic in-band RCSs of the proposed antenna for normally incident EM waves with *x*- and *y*-polarization in the *xoz*-plane are shown

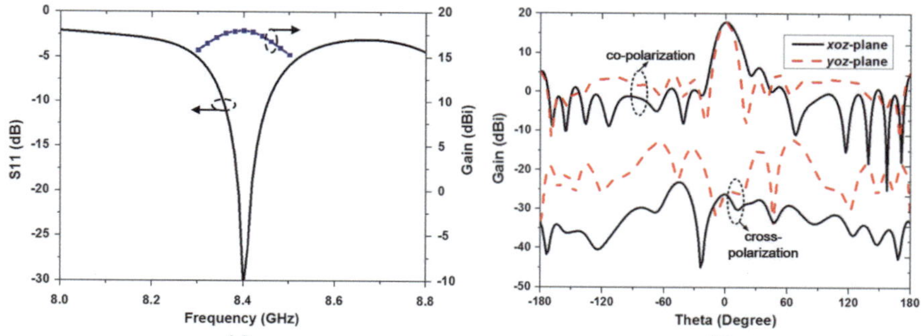

Fig. 9.28 Radiation characteristics of the proposed antenna with the MA. **a** Simulated S_{11} and peak gains versus frequency. **b** Radiation patterns

in Figs. 9.30a and b, respectively. For the x-polarized incident EM wave, the main lobe and pattern trough point in the directions of $\theta_r = 11°$ and $-1°$, respectively. When the incident EM wave is y-polarized, the main lobe and pattern trough point in the directions of $\theta_r = 10°$ and $-3°$, respectively. It is evident that reasonable agreement between the simulated and theoretical results is achieved.

A comparison of the RCSs of the proposed FP antenna with the MA and the reference antenna under normally incident EM waves with x- and y-polarization is shown in Fig. 9.31. The RCS of the proposed FP antenna is reduced from 7 to 15 GHz for both x- and y-polarized incident waves. The out-of-band RCSR of the proposed antenna is realized by using the absorbing property of the MA.

Figure 9.32 shows the monostatic RCSs of the reference antenna and the proposed antenna with the MA in the xoz-plane under oblique incident EM waves with φ- and θ-polarization. The monostatic RCS of the proposed antenna is reduced at angles from

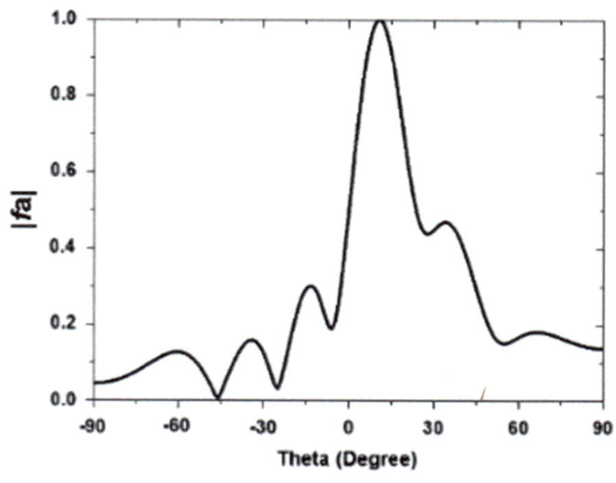

Fig. 9.29 Normalized array factor (f_a) of the proposed PGM with the MA in the xoz-plane

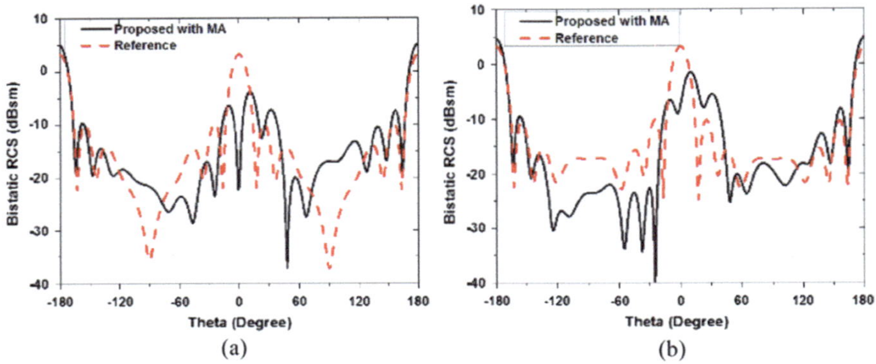

Fig. 9.30 Comparison of the in-band bistatic RCSs of the reference antenna and the proposed antenna with the MA at 8.4 GHz for normally incident EM waves with **a** x-polarization and **b** y-polarization

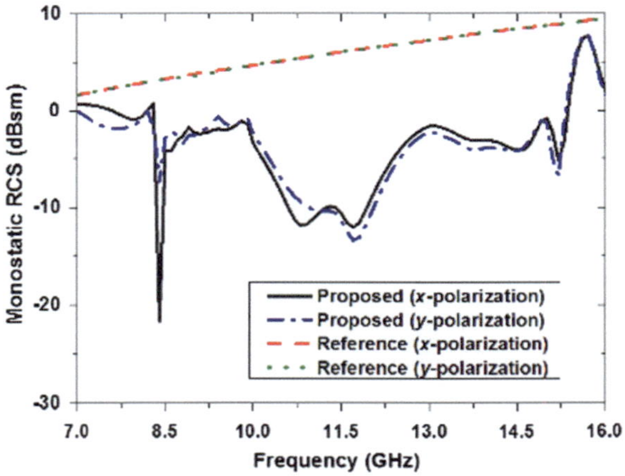

Fig. 9.31 Comparison of the monostatic RCSs of the reference antenna and the proposed antenna with the MA for normally incident EM waves

$-6°$ to $7°$ and from $-7°$ to $6°$ under φ-polarized and θ-polarized incident waves, respectively.

To validate the simulation results, a prototype of the proposed FP antenna based on the PGM with the MA was fabricated and measured. The fabricated FP antenna, as shown in Fig. 9.33, had the same dimensions as those listed in Table 9.2. The S_{11} values of the fabricated antenna were measured using an Agilent N5230C vector network analyzer, and the radiation characteristics were measured in an anechoic chamber.

The simulated and measured S_{11} values are shown in Fig. 9.34. The measured -10 dB impedance bandwidth is 150 MHz (from 8.37–8.52 GHz), and the resonant frequency shifts slightly to a higher frequency, which could be caused by fabrication

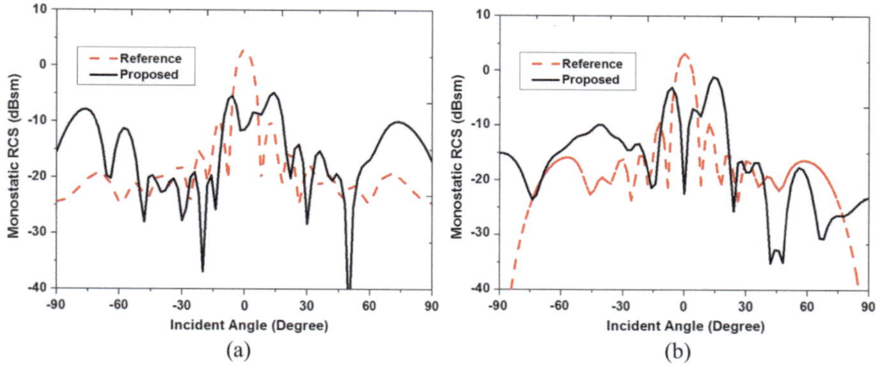

Fig. 9.32 Monostatic RCSs of the reference antenna and proposed antenna with the MA in the *xoz*-plane with **a** φ-polarized and **b** θ-polarized incident waves

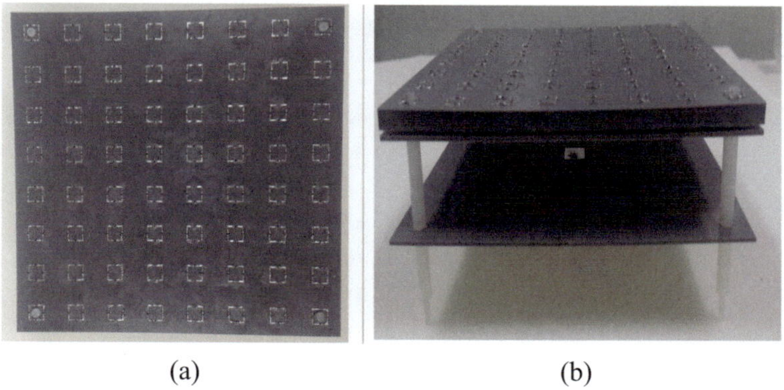

Fig. 9.33 Fabricated FP antenna based on the PGM with the MA. **a** Top view. **b** Side view

and installation errors. Figure 9.35 shows the radiation patterns in the *xoz*- and *yoz*-planes. Reasonable agreement between the measured and simulated results is evident.

The RCS of the proposed antenna was measured in a microwave anechoic chamber. The measured and simulated RCSs of the proposed FP antenna with the MA for normally incident plane waves with *x*-polarization are shown in Fig. 9.36. The RCS is reduced substantially from 8 to 16 GHz, and a reasonable agreement between the simulated and measured results is evident, indicative of good scattering performance of the proposed antenna.

Table 9.3 shows a comparison of the simulated peak gain (aperture efficiency) and the RCSR of previously reported low-RCS FP antennas and the two antennas proposed in this study. The aperture efficiency of the proposed antenna without the MA is 58.6%, which is higher than those of the FP antennas with and without an MA proposed in Pan et al. [13], Huang et al. [14], Jiang et al. [15], Zhang et al. [16] and Mu et al. [17]. The FP antenna in Zhang et al. [16] has a higher efficiency, but

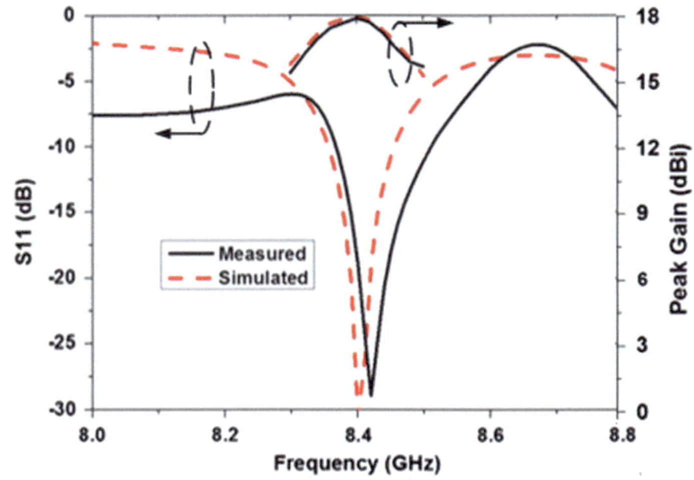

Fig. 9.34 Simulated and measured S_{11} values and peak gain of the proposed FP antenna based on the PGM with the MA

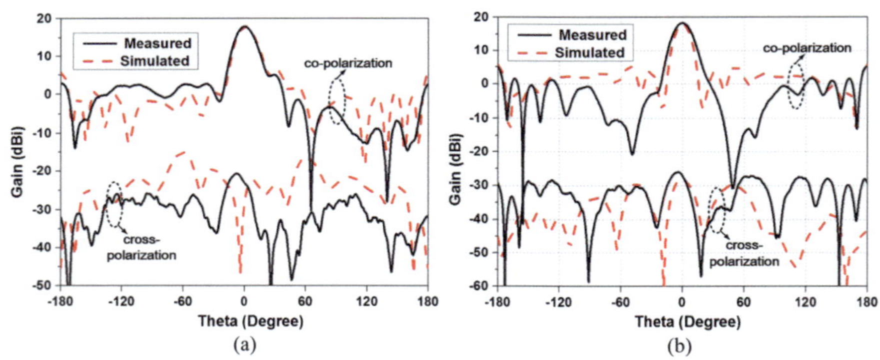

Fig. 9.35 Simulated and measured radiation patterns of the proposed FP antenna based on the PGM with the MA in the **a** *xoz*-plane and **b** *yoz*-plane

it does not achieve good in-band copolarization RCSR. The FP antenna presented in Mu et al. [17] can realize both in-band copolarization and cross-polarization RCSR. However, its aperture efficiency is much lower than that of the antenna proposed in this work. The comparison shows that the proposed FP antenna can simultaneously achieve both in-band copolarization RCSR and high gain.

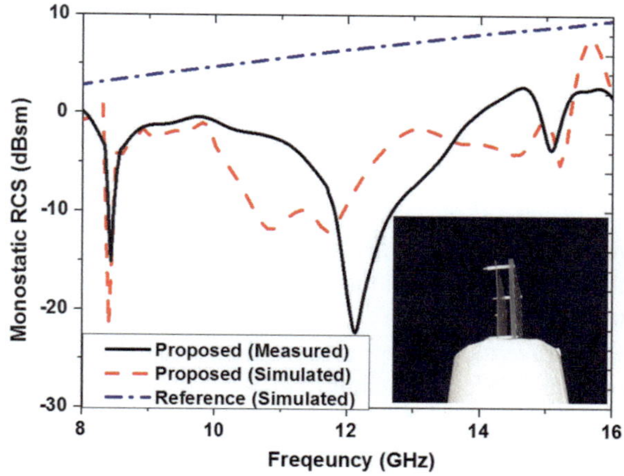

Fig. 9.36 Measured and simulated RCSs of the proposed antenna with the MA for normally incident EM waves with *x*-polarization

Table 9.3 Comparison of the performances between previously reported low-RCS FP antennas and this study

	Electrical size	Gain (aperture efficiency)	In-band copolarization RCSR (dB)	In-band cross-polarization RCSR (dB)	Average RCSR	RCSR bandwidth
Reference [13]	2.3 × 2.3 × 0.58 λ^3	13.2 dB (31.4%)	Approx. 5	Approx.5	10 dB	6–14 GHz
Reference [14]	2.79 × 2.79 × 0.68 λ^3	13.7 dB (23.5%)	Approx.3	Approx.3	/	7–14 GHz
Reference [15]	3.7 × 3.7 × 0.7 $3\lambda^3$	18.4 dB (40.2%)	Approx.2	17	13 dB	8–17 GHz
Reference [16]	3.67 × 3.67 × 0.62 λ^3	19.8 dB (55%)	3.8	16.4	8.76 dB	8–12 GHz
Reference [17]	2.53 × 2.53 × 0.67 λ^3	12.3 dB (20.8%)	25	25	/	5–8.5 GHz/ 11–12.5 GHz
Proposed antenna without MA	3.36 × 3.36 × 0.7 λ^3	19.2 dB (58.6%)	13.1	13	6.4 dB	7–11 GHz
Proposed antenna with MA	3.36 × 3.36 × 0.7 λ^3	17.9 dB (43.3%)	24.8	10.5	9.9 dB	7–15 GHz

9.3 Summary

This chapter introduced methods for reducing the RCS of an antenna using a PGM. By loading the PGM on the surface of a slot array antenna, out-of-band RCS was reduced over a wide bandwidth. Synthetically using the principles of the PGM and FP antenna could achieve in-band and out-of-band RCSR. The in-band copolarization RCS was reduced by 13.1 dB, and the peak gain of the antenna was 17.9 dBi. Compared with the AMC and PRRS, the PGM could reflect the incident wave in a certain direction through modification of the phase gradient of the metasurface. If the incident wave was converted into a surface wave by the PGM, both the monostatic RCS and the bistatic RCS could be reduced.

References

1. N. Yu, *et al.* Light Propagation with Phase Discontinuities: Generalized Laws of Reflection and Refraction. Science, Vol. 334, No. 6054, pp. 333–337, 2011.
2. S. Sun, *et al.* Gradient-index meta-surfaces as a bridge linking propagating waves and surface waves. Nature Materials, Vol. 11, No. 5, pp. 426–431, 2012.
3. M. B. Pu, *et al.* Broadband anomalous reflection based on gradient low-Q meta-surface. AIP Advances, Vol. 3, No. 5, pp. 052136, 2013.
4. C. J. Wu, Y. Z. Cheng, W. Y. Wang, B. He, R. Z. Gong, Ultra-thin and Polarization-independent Phase Gradient Metasurface for High-efficiency Spoof Surface-plasmon-polariton coupling, Applied Physics Express, Vol. 8, No. 12, pp. 122001, 2015.
5. H. Shi *et al.* Gradient metasurface with both polarization-controlled directional surface wave coupling and anomalous reflection. IEEE Antennas and Wireless Propagation Letters, Vol. 14, pp. 104–107, 2015.
6. Kim S H, Kim Y, Yoon Y J. Reflectarray element combined a tripole with double rings and broadband RCS reduction, International Workshop on Antenna Technology, Seoul, pp. 322–323, 2015.
7. Y. LI, J. Zhang, S. Qu, J. Wang, H. Chen, Z. Xu, A. Zhang, Wideband Radar Cross Section Reduction Using Two-dimensional Phase Gradient Metasurfaces, Applied Physics Letters, Vol. 104, pp. 221110, 2014.
8. W. Zhang, Y. Liu, S. Gong, J. Wang and Y. Jiang, Wideband RCS Reduction of a Slot Array Antenna Using Phase Gradient Metasurface, IEEE Antennas and Wireless Propagation Letters, Vol. 17, No. 12, pp. 2193–2197, 2018.
9. J. C. Iriarte Galarregui, A. Tellechea Pereda, J. L. M. de Falcón, I. Ederra, Broadband Radar Cross-Section Reduction Using AMC Technology. IEEE Transactions on Antennas and Propagation, Vol. 61, No. 12, pp. 6136–6143, 2013.
10. Y. Jia, Y. Liu, W. Zhang, J. Wang, S. Gong and G. Liao, High-Gain Fabry–Perot Antennas With Wideband Low Monostatic RCS Using Phase Gradient Metasurface, IEEE Access, Vol. 7, pp. 4816–4824.
11. Trentini G V. Partially reflecting sheet arrays. Ire Transactions on Antennas and Propagation, Vol. 4, No. 4, pp. 666–671, 1956.
12. Y. T. Lo and S. W. Lee, Antenna Handbook: Theory, Applications and Design. New York: van Norstrand Reinhold Company, 1988.
13. W. Pan, C. Huang, P. Chen, X. Ma, C. Hu and X. Luo, A low-RCS and high-gain partially reflecting surface antenna. IEEE Transactions on Antennas and Propagation, Vol. 62, No. 2, pp. 945–949, 2014.

14. C. Huang, W. Pan, X. Ma and X. Luo. A Frequency Reconfigurable Directive Antenna with Wideband Low-RCS Property. IEEE Transactions on Antennas and Propagation, Vol. 64, No. 3, pp. 1173–1178, 2016.
15. H. Jiang, Z. Xue, W. Li, W. Ren, Low-RCS high-gain partially reflecting surface antenna with reconfiguration ground plane, IEEE Transactions on Antennas and Propagation, Vol. 64, No. 9, pp. 4127–4132, 2016.
16. L. Zhang, *et al.* Realization of Low Scattering for a High-Gain Fabry–Perot Antenna Using Coding Metasurface. IEEE Transactions on Antennas and Propagation, Vol. 65, No. 7, pp. 3373–3383, 2017.
17. J. Mu, H. Wang, H. Wang and Y. Huang. Low-RCS and Gain Enhancement Design of a Novel Partially Reflecting and Absorbing Surface Antenna. IEEE Antennas and Wireless Propagation Letters, Vol. 16, pp. 1903–1906, 2017.

MIX
Papier aus verantwortungsvollen Quellen
Paper from responsible sources
FSC® C105338

If you have any concerns about our products,
you can contact us on
ProductSafety@springernature.com

In case Publisher is established outside the EU,
the EU authorized representative is:
Springer Nature Customer Service Center GmbH
Europaplatz 3, 69115 Heidelberg, Germany

Printed by Libri Plureos GmbH
in Hamburg, Germany